Texts in Applied Mathematics 29

Texts in Applied Mathematics

1. *Sirovich:* Introduction to Applied Mathematics.
2. *Wiggins:* Introduction to Applied Nonlinear Dynamical Systems and Chaos.
3. *Hale/Koçak:* Dynamics and Bifurcations.
4. *Chorin/Marsden:* A Mathematical Introduction to Fluid Mechanics, Third Edition.
5. *Hubbard/West:* Differential Equations: A Dynamical Systems Approach: Ordinary Differential Equations.
6. *Sontag:* Mathematical Control Theory: Deterministic Finite Dimensional Systems Second Edition.
7. *Perko:* Differential Equations and Dynamical Systems, Third Edition.
8. *Seaborn:* Hypergeometric Functions and Their Applications.
9. *Pipkin:* A Course on Integral Equations.
10. *Hoppensteadt/Peskin:* Modeling and Simulation in Medicine and the Life Sciences, Second Edition.
11. *Braun:* Differential Equations and Their Applications, Fourth Edition.
12. *Stoer/Bulirsch:* Introduction to Numerical Analysis, Third Edition.
13. *Renardy/Rogers:* An Introduction to Partial Differential Equations.
14. *Banks:* Growth and Diffusion Phenomena: Mathematical Frameworks and Applications.
15. *Brenner/Scott:* The Mathematical Theory of Finite Element Methods, Second Edition.
16. *Van de Velde:* Concurrent Scientific Computing.
17. *Marsden/Ratiu:* Introduction to Mechanics and Symmetry, Second Edition.
18. *Hubbard/West:* Differential Equations: A Dynamical Systems Approach: Higher-Dimensional Systems.
19. *Kaplan/Glass:* Understanding Nonlinear Dynamics.
20. *Holmes:* Introduction to Perturbation Methods.
21. *Curtain/Zwart:* An Introduction to Infinite-Dimensional Linear Systems Theory.
22. *Thomas:* Numerical Partial Differential Equations: Finite Difference Methods.
23. *Taylor:* Partial Differential Equations: Basic Theory.
24. *Merkin:* Introduction to the Theory of Stability of Motion.
25. *Naber:* Topology, Geometry, and Gauge Fields: Foundations.
26. *Polderman/Willems:* Introduction to Mathematical Systems Theory: A Behavioral Approach.
27. *Reddy:* Introductory Functional Analysis: with Applications to Boundary Value Problems and Finite Elements.
28. *Gustafson/Wilcox:* Analytical and Computational Methods of Advanced Engineering Mathematics.

(continued after index)

Aslak Tveito Ragnar Winther

Introduction to Partial Differential Equations

A Computational Approach

With 69 Illustrations

Springer

First South Asian Edition 2019

Aslak Tveito
Simula Research Laboratory
P.O.Box 134
N-1325 Lysaker
Norway
aslak@simula.no

Ragnar Winther
Centre of Mathematics for Applications
and Department of Informatics
University of Oslo
N-0316 Oslo
Norway
rwinther@cma.uio.no

J.E. Marsden
Control and Dynamical Systems, 107-81
California Institute of Technology
Pasadena, CA 91125
USA
marsden@cds.caltech.edu

L. Sirovich
Division of Applied Mathematics
Brown University
Providence, RI 02912
USA
chico@camelot.mssm.edu

M. Golubitsky
Department of Mathematics
University of Houston
Houston, TX 77204-3476
USA
mg@uh.edu

W. Jäger
Department of Applied Mathematics
Universität Heidelberg
Im Neuenheimer Feld 294
69120 Heidelberg
Germany
jaeger@iwr.uni-heidelberg.de

ISBN: 978-3-6625-8467-5

Mathematics Subject Classification (2000): 35-01, 65-01

Libraray of Congress Control Number: 2004111961

ISBN 3-540-22551-X Corr. Second Printing Springer Berlin Heidelberg New York

ISBN 0-387-98327-9 First Edition Springer New York Berlin Heidelberg 1998

Springer is a part of Springer Science+Business Media
springeronline.com

Typesetting: Camera-ready copy produced from the author's output file using a Springer TEX macro package
Production: LE-TeX Jelonek, Schmidt & Vöckler GbR, Leipzig

46/3142YL - 5 4 3 2 1 0

Printed in India by Replika Press Pvt. Ltd.

Series Preface

Mathematics is playing an ever more important role in the physical and biological sciences, provoking a blurring of boundaries between scientific disciplines and a resurgence of interest in the modern as well as the classical techniques of applied mathematics. This renewal of interest, both in research and teaching, has led to the establishment of the series: *Texts in Applied Mathematics* (*TAM*).

The development of new courses is a natural consequence of a high level of excitement on the research frontier as newer techniques, such as numerical and symbolic computer systems, dynamical systems, and chaos mix with and reinforce the traditional methods of applied mathematics. Thus, the purpose of this textbook series is to meet the current and future needs of these advances and encourage the teaching of new courses.

TAM will publish textbooks suitable for use in advanced undergraduate and beginning graduate courses, and will complement the *Applied Mathematical Sciences* (*AMS*) series, which will focus on advanced textbooks and research-level monographs.

Preface

> *"It is impossible to exaggerate the extent to which modern applied mathematics has been shaped and fueled by the general availability of fast computers with large memories. Their impact on mathematics, both applied and pure, is comparable to the role of the telescopes in astronomy and microscopes in biology."*
>
> — *Peter Lax, Siam Rev. Vol. 31 No. 4*

Congratulations! You have chosen to study partial differential equations. That decision is a wise one; the laws of nature are written in the language of partial differential equations. Therefore, these equations arise as models in virtually all branches of science and technology. Our goal in this book is to help you to understand what this vast subject is about. The book is an introduction to the field. We assume only that you are familiar with basic calculus and elementary linear algebra. Some experience with ordinary differential equations would also be an advantage.

Introductory courses in partial differential equations are given all over the world in various forms. The traditional approach to the subject is to introduce a number of analytical techniques, enabling the student to derive exact solutions of some simplified problems. Students who learn about computational techniques on other courses subsequently realize the scope of partial differential equations beyond paper and pencil.

Our approach is different. We introduce analytical and computational techniques in the same book and thus in the same course. The main reason for doing this is that the computer, developed to assist scientists in solving partial differential equations, has become commonly available and is currently used in all practical applications of partial differential equations. Therefore, a modern introduction to this topic must focus on methods suitable for computers. But these methods often rely on deep analytical insight into the equations. We must therefore take great care not to throw away basic analytical methods but seek a sound balance between analytical and computational techniques.

One advantage of introducing computational techniques is that nonlinear problems can be given more attention than is common in a purely analytical introduction. We have included several examples of nonlinear equations in addition to the standard linear models which are present in any introductory text. In particular we have included a discussion of reaction-diffusion equations. The reason for this is their widespread application as important models in various scientific applications.

Our aim is not to discuss the merits of different numerical techniques. There are a huge number of papers in scientific journals comparing different methods to solve various problems. We do not want to include such discussions. Our aim is to demonstrate that computational techniques are simple to use and often give very nice results, not to show that even better results can be obtained if slightly different methods are used. We touch briefly upon some such discussion, but not in any major way, since this really belongs to the field of numerical analysis and should be taught in separate courses. Having said this, we always try to use the simplest possible numerical techniques. This should in no way be interpreted as an attempt to advocate certain methods as opposed to others; they are merely chosen for their simplicity.

Simplicity is also our reason for choosing to present exclusively finite difference techniques. The entire text could just as well be based on finite element techniques, which definitely have greater potential from an application point of view but are slightly harder to understand than their finite difference counterparts.

We have attempted to present the material at an easy pace, explaining carefully both the ideas and details of the derivations. This is particularly the case in the first chapters but subsequently less details are included and some steps are left for the reader to fill in. There are a lot of exercises included, ranging from the straightforward to more challenging ones. Some of them include a bit of implementation and some experiments to be done on the computer. We strongly encourage students not to skip these parts. In addition there are some "projects." These are either included to refresh the student's memory of results needed in this course, or to extend the theories developed in the present text.

Given the fact that we introduce both numerical and analytical tools, we have chosen to put little emphasis on modeling. Certainly, the derivation of models based on partial differential equations is an important topic, but it is also very large and can therefore not be covered in detail here.

The first seven chapters of this book contain an elementary course in partial differential equations. Topics like separation of variables, energy arguments, maximum principles, and finite difference methods are discussed for the three basic linear partial differential equations, i.e. the heat equation, the wave equation, and Poisson's equation. In Chapters 8–10 more theoretical questions related to separation of variables and convergence of Fourier series are discussed. The purpose of Chapter 11 is to introduce nonlinear partial differential equations. In particular, we want to illustrate how easily finite difference methods adopt to such problems, even if these equations may be hard to handle by an analytical approach. In Chapter 12 we give a brief introduction to the Fourier transform and its application to partial differential equations.

Some of the exercises in this text are small computer projects involving a bit of programming. This programming could be done in any language. In order to get started with these projects, you may find it useful to pick up some examples from our web site, http://www.ifi.uio.no/~pde/, where you will find some Matlab code and some simple Java applets.

Acknowledgments

It is a great pleasure for us to thank our friends and colleagues for a lot of help and for numerous discussions throughout this project. In particular, we would like to thank Bent Birkeland and Tom Lyche, who both participated in the development of the basic ideas underpinning this book. Also we would like to thank Are Magnus Bruaset, Helge Holden, Kenneth Hvistendahl Karlsen, Jan Olav Langseth, Hans Petter Langtangen, Glenn Terje Lines, Knut Mørken, Bjørn Fredrik Nielsen, Gunnar Olsen, Klas Samuelsson, Achim Schroll, Wen Shen, Jan Søreng, and Åsmund Ødegård for reading parts of the manuscript. Finally, we would like to thank Hans Birkeland, Truls Flatberg, Roger Hansen, Thomas Skjønhaug, and Fredrik Tyvand for doing an excellent job in typesetting most of this book.

Oslo, Norway, July 1998

Aslak Tveito
Ragnar Winther

Contents

1

Setting the Scene

You are embarking on a journey in a jungle called Partial Differential Equations. Like any other jungle, it is a wonderful place with interesting sights all around, but there are also certain dangerous spots. On your journey, you will need some guidelines and tools, which we will start developing in this introductory chapter.

1.1 What Is a Differential Equation?

The field of differential equations is very rich and contains a large variety of different species. However, there is one basic feature common to all problems defined by a differential equation: the equation relates a function to its derivatives in such a way that the function itself can be determined. This is actually quite different from an algebraic equation, say

$$x^2 - 2x + 1 = 0,$$

whose solution is usually a number. On the other hand, a prototypical differential equation is given by

$$u'(t) = u(t).$$

The solution of this equation is given by the function

$$u(t) = ce^t,$$

where the constant c typically is determined by an extra condition. For instance, if we require

$$u(0) = 1/2,$$

we get $c = 1/2$ and $u(t) = \frac{1}{2}e^t$. So keep this in mind; the solution we seek from a differential equation is a function.

1.1.1 Concepts

We usually subdivide differential equations into partial differential equations (PDEs) and ordinary differential equations (ODEs). PDEs involve partial derivatives, whereas ODEs only involve derivatives with respect to one variable. Typical ordinary differential equations are given by

$$\begin{aligned} (a)\ u'(t) &= u(t), \\ (b)\ u'(t) &= u^2(t), \\ (c)\ u'(t) &= u(t) + \sin(t)\cos(t), \\ (d)\ u''(x) + u'(x) &= x^2, \\ (e)\ u''''(x) &= \sin(x). \end{aligned} \tag{1.1}$$

Here (a), (b) and (c) are "first order" equations, (d) is second order, and (e) is fourth order. So the *order* of an equation refers to the highest order derivative involved in the equation. Typical partial differential equations are given by[1]

$$\begin{aligned} (f)\ u_t(x,t) &= u_{xx}(x,t), \\ (g)\ u_{tt}(x,t) &= u_{xx}(x,t), \\ (h)\ u_{xx}(x,y) + u_{yy}(x,y) &= 0, \\ (i)\ u_t(x,t) &= \big(k(u(x,t))u_x(x,t)\big)_x, \\ (j)\ u_{tt}(x,t) &= u_{xx}(x,t) - u^3(x,t), \\ (k)\ u_t(x,t) + \big(\frac{1}{2}u^2(x,t)\big)_x &= u_{xx}(x,t), \\ (l)\ u_t(x,t) + (x^2+t^2)u_x(x,t) &= 0, \\ (m)\ u_{tt}(x,t) + u_{xxxx}(x,t) &= 0. \end{aligned} \tag{1.2}$$

Again, equations are labeled with orders; (l) is first order, (f), (g), (h), (i), (j) and (k) are second order, and (m) is fourth order.

Equations may have "variable coefficients," i.e. functions not depending on the unknown u but on the independent variables; t, x, or y above. An equation with variable coefficients is given in (l) above.

---

[1] Here $u_t = \frac{\partial u}{\partial t}$, $u_{xx} = \frac{\partial^2 u}{\partial x^2}$, and so on.

Some equations are referred to as nonhomogeneous. They include terms that do not depend on the unknown u. Typically, (c), (d), and (e) are nonhomogeneous equations. Furthermore,

$$u''(x) + u'(x) = 0$$

would be the homogeneous counterpart of **d**). Similarly, the Laplace equation

$$u_{xx}(x,y) + u_{yy}(x,y) = 0$$

is homogeneous, whereas the Poisson equation

$$u_{xx}(x,y) + u_{yy}(x,y) = f(x,y)$$

is nonhomogeneous.

An important distinction is between linear and nonlinear equations. In order to clarify these concepts, it is useful to write the equation in the form

$$L(u) = 0. \tag{1.3}$$

With this notation, (a) takes the form (1.3) with

$$L(u) = u'(t) - u(t).$$

Similarly, (j) can be written in the form (1.3) with

$$L(u) = u_{tt} - u_{xx} + u^3.$$

Using this notation, we refer to an equation of the form (1.3) as *linear* and homogeneous if

$$L(\alpha u + \beta v) = \alpha L(u) + \beta L(v) \tag{1.4}$$

for any constants α and β and any relevant[2] functions u and v. An equation of the form $L(u) = f$ where L is linear and f is independent of u is referred to as linear, but inhomogeneous. Finally, equations which are not linear are called *nonlinear*.

Let us consider (a) above. We have

$$L(u) = u' - u,$$

and thus

---

[2]We have to be a bit careful here in order for the expression $L(u)$ to make sense. For instance, if we choose

$$u = \begin{cases} -1 & x \le 0, \\ 1 & x > 0, \end{cases}$$

then u is not differentiable and it is difficult to interpret $L(u)$. Thus we require a certain amount of differentiability and apply the criterion only to sufficiently smooth functions.

$$\begin{aligned} L(\alpha u + \beta v) &= \alpha u' + \beta v' - \alpha u - \beta v \\ &= \alpha(u' - u) + \beta(v' - v) \\ &= \alpha L(u) + \beta L(v), \end{aligned}$$

for any constants α and β and any differentiable functions u and v. So this equation is linear. But if we consider (j), we have

$$L(u) = u_{tt} - u_{xx} + u^3,$$

and thus

$$L(u+v) = u_{tt} - u_{xx} + v_{tt} - v_{xx} + (u+v)^3,$$

which is not equal to $L(u) + L(v)$ for all functions u and v since, in general,

$$(u+v)^3 \neq u^3 + v^3.$$

So the equation (j) is nonlinear. It is a straightforward exercise to show that also (c), (d), (e), (f), (g), (h), (l) and (m) are linear, whereas (b), (i) and (k), in addition to (j), are nonlinear.

1.2 The Solution and Its Properties

In the previous section we introduced such notions as linear, nonlinear, order, ordinary differential equations, partial differential equations, and homogeneous and nonhomogeneous equations. All these terms can be used to characterize an equation simply by its appearance. In this section we will discuss some properties related to the *solution* of a differential equation.

1.2.1 An Ordinary Differential Equation

Let us consider a prototypical ordinary differential equation,

$$u'(t) = -u(t) \tag{1.5}$$

equipped with an initial condition

$$u(0) = u_0. \tag{1.6}$$

Here u_0 is a given number. Problems of this type are carefully analyzed in introductory courses and we shall therefore not dwell on this subject.[3] The

[3] Boyce and DiPrima [3] and Braun [5] are excellent introductions to ordinary differential equations. If you have not taken an introductory course in this subject, you will find either book a useful reference.

solution of (1.5) and (1.6) is given by

$$u(t) = u_0 e^{-t}.$$

This is easily checked by inspection;

$$u(0) = u_0 e^0 = u_0,$$

and

$$u'(t) = -u_0 e^{-t} = -u(t).$$

Faced with a problem posed by a differential equation and some initial or boundary conditions, we can generally check a solution candidate by determining whether both the differential equation and the extra conditions are satisfied. The tricky part is, of course, finding the candidate.[4]

The motivation for studying differential equations is—to a very large extent—their prominent use as models of various phenomena. Now, if (1.5) is a model of some process, say the density of some population, then u_0 is a measure of the initial density. Since u_0 is a measured quantity, it is only determined to a certain accuracy, and it is therefore important to see if slightly different initial conditions give almost the same solutions. If small perturbations of the initial condition imply small perturbations of the solution, we have a *stable* problem. Otherwise, the problem is referred to as *unstable*.

Let us consider the problem (1.5)–(1.6) with slightly perturbed initial conditions,

$$v'(t) = -v(t), \tag{1.7}$$
$$v(0) = u_0 + \epsilon, \tag{1.8}$$

for some small ϵ. Then

$$v(t) = (u_0 + \epsilon)e^{-t},$$

and

$$|u(t) - v(t)| = |\epsilon| e^{-t}. \tag{1.9}$$

We see that for this problem, a small change in the initial condition leads to small changes in the solution. In fact, the difference between the solutions is reduced at an exponential rate as t increases. This property is illustrated in Fig. 1.1.

---

[4]We will see later that it may also be difficult to check that a certain candidate is in fact a solution. This is the case if, for example, the candidate is defined by an infinite series. Then problems of convergence, existence of derivatives etc. must be considered before a candidate can be accepted as a solution.

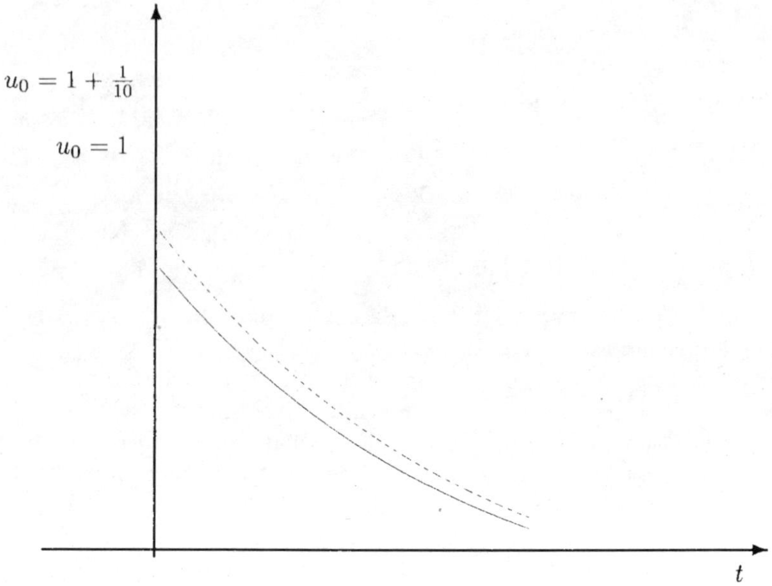

FIGURE 1.1. *The solution of the problem (1.5)–(1.6) with $u_0 = 1$ and $u_0 = 1+1/10$ are plotted. Note that the difference between the solutions decreases as t increases.*

Next we consider a nonlinear problem;

$$\begin{aligned} u'(t) &= tu(t)(u(t)-2), \\ u(0) &= u_0, \end{aligned} \tag{1.10}$$

whose solution is given by

$$u(t) = \frac{2u_0}{u_0 + (2-u_0)e^{t^2}}. \tag{1.11}$$

It follows from (1.11) that if $u_0 = 2$, then $u(t) = 2$ for all $t \geq 0$. Such a state is called an *equilibrium solution.* But this equilibrium is not stable; in Fig. 1.2 we have plotted the solution for $u_0 = 2 - 1/1000$ and $u_0 = 2 + 1/1000$. Although the initial conditions are very close, the difference in the solutions blows up as t approaches a critical time. This critical time is discussed in Exercise 1.3.

1.3 A Numerical Method

Throughout this text, our aim is to teach you both analytical and numerical techniques for studying the solution of differential equations. We

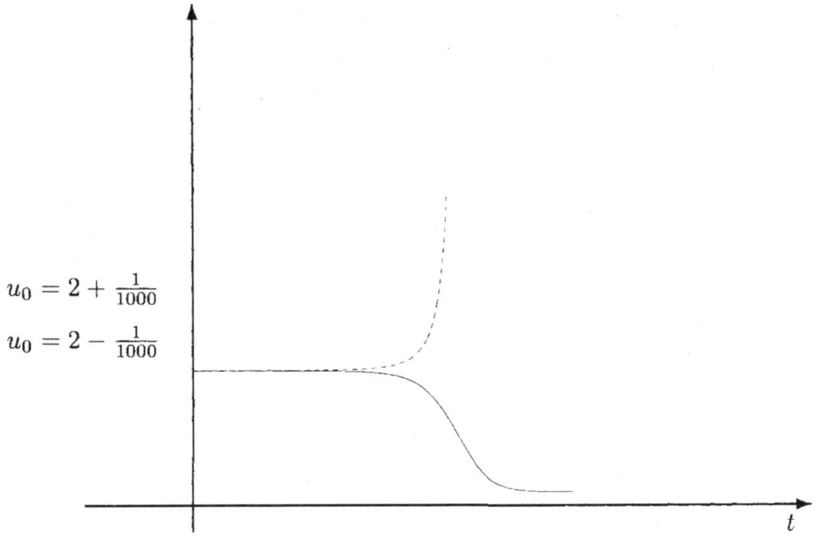

FIGURE 1.2. *Two solutions of (1.11) with almost identical initial conditions are plotted. Note that the difference between the solutions blows up as t increases.*

will emphasize basic principles and ideas, leaving specialties for subsequent courses. Thus we present the simplest methods, not paying much attention to for example computational efficiency.

In order to define a numerical method for a problem of the form

$$\begin{aligned} u'(t) &= f\big(u(t)\big), \\ u(0) &= u_0, \end{aligned} \tag{1.12}$$

for a given function $f = f(u)$, we recall the Taylor series for smooth functions. Suppose that u is a twice continuously differentiable function. Then, for $\Delta t > 0$, we have

$$u(t + \Delta t) = u(t) + \Delta t u'(t) + \frac{1}{2}(\Delta t)^2 u''(t + \xi) \tag{1.13}$$

for some $\xi \in [0, \Delta t]$. Hence, we have[5]

$$u'(t) = \frac{u(t + \Delta t) - u(t)}{\Delta t} + O(\Delta t). \tag{1.14}$$

We will use this relation to put up a scheme for computing approximate solutions of (1.12). In order to define this scheme, we introduce discrete

[5]The O-notation is discussed in Project 1.1.

timelevels

$$t_m = m\Delta t, \qquad m = 0, 1, \ldots,$$

where $\Delta t > 0$ is given. Let v_m, $m = 0, 1, \ldots$ denote approximations of $u(t_m)$. Obviously we put $v_0 = u_0$, which is the given initial condition. Next we assume that v_m is computed for some $m \geq 0$ and we want to compute v_{m+1}. Since, by (1.12) and (1.14),

$$\frac{u(t_{m+1}) - u(t_m)}{\Delta t} \approx u'(t_m) = f(u(t_m)) \tag{1.15}$$

for small Δt, we define v_{m+1} by requiring that

$$\frac{v_{m+1} - v_m}{\Delta t} = f(v_m). \tag{1.16}$$

Hence, we have the scheme

$$v_{m+1} = v_m + \Delta t f(v_m), \qquad m = 0, 1, \ldots. \tag{1.17}$$

This scheme is usually called the forward Euler method. We note that it is a very simple method to implement on a computer for any function f.

Let us consider the accuracy of the numerical approximations computed by this scheme for the following problem:

$$\begin{aligned} u'(t) &= u(t), \\ u(0) &= 1. \end{aligned} \tag{1.18}$$

The exact solution of this problem is $u(t) = e^t$, so we do not really need any approximate solutions. But for the purpose of illustrating properties of the scheme, it is worthwhile addressing simple problems with known solutions. In this problem we have $f(u) = u$, and then (1.17) reads

$$v_{m+1} = (1 + \Delta t)v_m, \qquad m = 0, 1, \ldots. \tag{1.19}$$

By induction we have

$$v_m = (1 + \Delta t)^m.$$

In Fig. 1.3 we have plotted this solution for $0 \leq t_m \leq 1$ using $\Delta t = 1/3$, $1/6$, $1/12$, $1/24$. We see from the plots that v_m approaches $u(t_m)$ as Δt is decreased.

Let us study the error of this scheme in a little more detail. Suppose we are interested in the numerical solution at $t = 1$ computed by a time step Δt given by

$$\Delta t = 1/M,$$

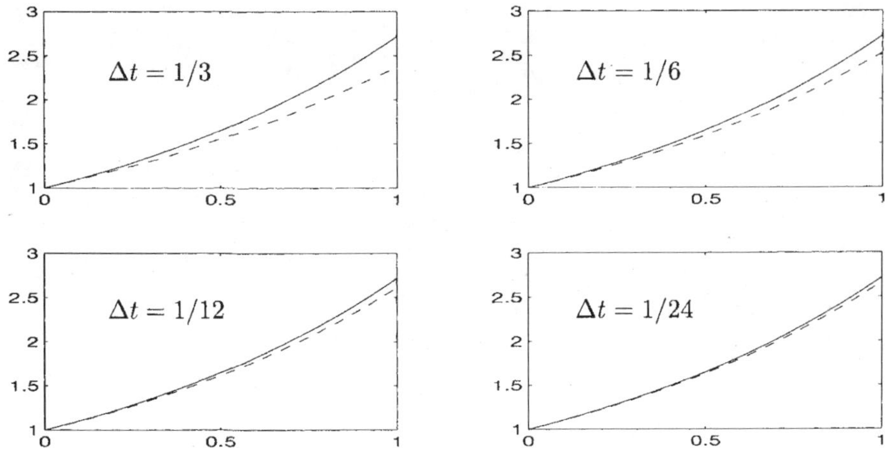

FIGURE 1.3. *The four plots show the convergence of the numerical approximations generated by the forward Euler scheme.*

where $M > 0$ is a given integer. Since the numerical solution at $t = 1$ is given by

$$v_M = (1 + \Delta t)^M = (1 + \Delta t)^{1/\Delta t},$$

the error is given by

$$E(\Delta t) = |e - (1 + \Delta t)^{1/\Delta t}|.$$

From calculus we know that

$$\lim_{\epsilon \to 0}(1 + \epsilon)^{1/\epsilon} = e,$$

so clearly

$$\lim_{\Delta t \to 0} E(\Delta t) = 0,$$

meaning that we get convergence towards the correct solution at $t = 1$. In Table 1.1 we have computed $E(\Delta t)$ and $E(\Delta t)/\Delta t$ for several values of Δt. From the table we can observe that $E(\Delta t) \approx 1.359\Delta t$ and thus conclude that the accuracy of our approximation increases as the number of timesteps M increases.

As mentioned above, the scheme can also be applied to more challenging problems. In Fig. 1.4 we have plotted the exact and numerical solutions of the problem (1.10) on page 6 using $u_0 = 2.1$.

Even though this problem is much harder to solve numerically than the simple problem we considered above, we note that convergence is obtained as Δt is reduced.

Some further discussion concerning numerical methods for ordinary differential equations is given in Project 1.3. A further analysis of the error introduced by the forward Euler method is given in Exercise 1.15.

Δt	$E(\Delta t)$	$E(\Delta t)/\Delta t$
$1/10^1$	$1.245 \cdot 10^{-1}$	1.245
$1/10^2$	$1.347 \cdot 10^{-2}$	1.347
$1/10^3$	$1.358 \cdot 10^{-3}$	1.358
$1/10^4$	$1.359 \cdot 10^{-4}$	1.359
$1/10^5$	$1.359 \cdot 10^{-5}$	1.359
$1/10^6$	$1.359 \cdot 10^{-6}$	1.359

TABLE 1.1. *We observe from this table that the error introduced by the forward Euler scheme (1.17) as applied to (1.18) is about* $1.359\Delta t$ *at* $t = 1$. *Hence the accuracy can be increased by increasing the number of timesteps.*

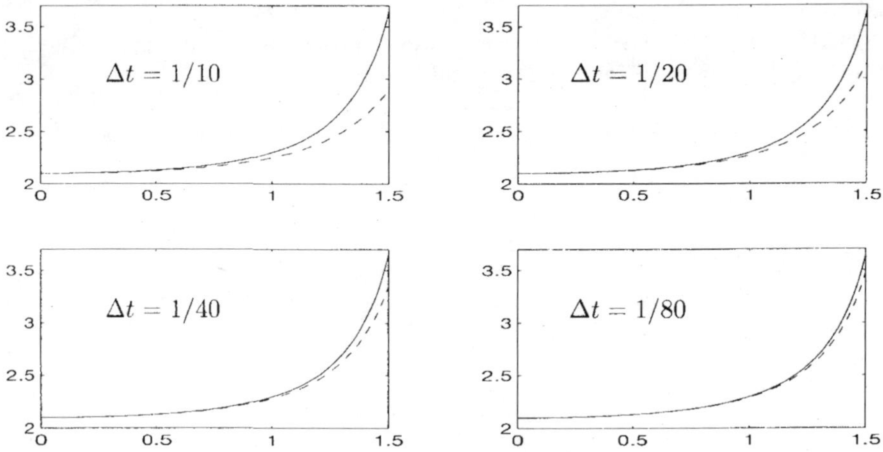

FIGURE 1.4. *Convergence of the forward Euler approximations as applied to problem (1.10) on page 6.*

1.4 Cauchy Problems

In this section we shall derive exact solutions for some partial differential equations. Our purpose is to introduce some basic techniques and show examples of solutions represented by explicit formulas. Most of the problems encountered here will be revisited later in the text.

Since our focus is on ideas and basic principles, we shall consider only the simplest possible equations and extra conditions. In particular, we will focus on pure Cauchy problems. These problems are initial value problems defined on the entire real line. By doing this we are able to derive very simple solutions without having to deal with complications related to boundary values. We also restrict ourselves to one spatial dimension in order to keep things simple. Problems in bounded domains and problems in more than one space dimension are studied in later chapters.

1.4.1 First-Order Homogeneous Equations

Consider the following first-order homogeneous partial differential equation,

$$u_t(x,t) + a(x,t)u_x(x,t) = 0, \qquad x \in \mathbb{R},\ t > 0, \tag{1.20}$$

with the initial condition

$$u(x,0) = \phi(x), \quad x \in \mathbb{R}. \tag{1.21}$$

Here we assume the variable coefficient $a = a(x,t)$ and the initial condition $\phi = \phi(x)$ to be given smooth functions.[6] As mentioned above, a problem of the form (1.20)–(1.21) is referred to as a Cauchy problem. In the problem (1.20)–(1.21), we usually refer to t as the time variable and x as the spatial coordinate. We want to derive a solution of this problem using the method of characteristics. The characteristics of (1.20)–(1.21) are curves in the x–t-plane defined as follows: For a given $x_0 \in \mathbb{R}$, consider the ordinary differential equation

$$\begin{aligned} \frac{dx(t)}{dt} &= a\big(x(t),t\big), \qquad t > 0, \\ x(0) &= x_0. \end{aligned} \tag{1.22}$$

The solution $x = x(t)$ of this problem defines a curve $\big\{\big(x(t),t\big)\,,\ t \geq 0\big\}$ starting in $(x_0, 0)$ at $t = 0$; see Fig. 1.5.

Now we want to consider u along the characteristic; i.e. we want to study the evolution of $u\big(x(t),t\big)$. By differentiating u with respect to t, we get

$$\begin{aligned} \frac{d}{dt}u\big(x(t),t\big) &= u_t + u_x\frac{dx(t)}{dt} \\ &= u_t + a(x,t)u_x = 0, \end{aligned}$$

where we have used the definition of $x(t)$ given by (1.22) and the differential equation (1.20). Since

$$\frac{d}{dt}u\big(x(t),t\big) = 0,$$

[6] A smooth function is continuously differentiable as many times as we find necessary. When we later discuss properties of the various solutions, we shall introduce classes of functions describing exactly how smooth a certain function is. But for the time being it is sufficient to think of smooth functions as functions we can differentiate as much as we like.

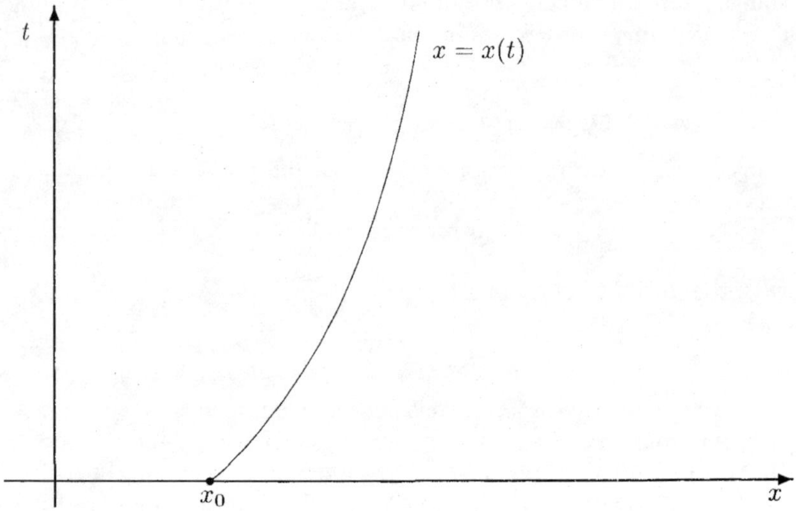

FIGURE 1.5. *The characteristic starting at* $x = x_0$.

the solution u of (1.20)–(1.21) is constant along the characteristic. Hence

$$u\big(x(t),t\big) \;=\; u(x_0,0)$$

or

$$u\big(x(t),t\big) \;=\; \phi(x_0). \tag{1.23}$$

This means that if, for a given $a = a(x,t)$, we are able to solve the ODE given by (1.22), we can compute the solution of the Cauchy problem (1.20)–(1.21). Let us consider two simple examples illustrating the strength of this technique.

EXAMPLE 1.1 Consider the Cauchy problem

$$\begin{aligned} u_t + au_x &= 0, \qquad x \in \mathbb{R},\ t > 0, \\ u(x,0) &= \phi(x),\ \ x \in \mathbb{R}, \end{aligned} \tag{1.24}$$

where a is a constant. For this problem, the ODE (1.22) takes the form

$$x'(t) \;=\; a, \qquad x(0) = x_0,$$

and thus

$$x \;=\; x(t) \;=\; x_0 + at. \tag{1.25}$$

Since, by (1.23), we have

$$u(x,t) \;=\; u\big(x(t),t\big) \;=\; \phi(x_0),$$

and by (1.25) we have

$$x_0 = x - at,$$

consequently

$$u(x,t) = \phi(x - at). \tag{1.26}$$

We conclude that the problem (1.24) is solved by the formula (1.26) for any smooth ϕ and constant a. It is straightforward to check that (1.26) actually solves (1.24);

$$u(x,0) = \phi(x),$$

and

$$\left.\begin{aligned} u_t &= -a\phi'(x - at) \\ u_x &= \phi'(x - at) \end{aligned}\right\} \implies u_t + au_x = 0.$$

Hence both the initial condition and the differential equation are fulfilled. ■

EXAMPLE 1.2 Consider the Cauchy problem

$$\begin{aligned} u_t + xu_x &= 0, \quad x \in \mathbb{R},\ t > 0, \\ u(x,0) &= \phi(x), \quad x \in \mathbb{R}. \end{aligned} \tag{1.27}$$

Now the characteristics are defined by

$$x'(t) = x(t), \qquad x(0) = x_0$$

so

$$x(t) = x_0 e^t \quad \text{and} \quad x_0 = xe^{-t}.$$

Since

$$u\big(x(t),t\big) = \phi(x_0)$$

(see (1.23)), we get

$$u(x,t) = \phi\left(xe^{-t}\right). \tag{1.28}$$

As above, it is a straightforward task to check that (1.28) solves (1.27). ■

1.4.2 First-Order Nonhomogeneous Equations

The method of characteristics can also be utilized for nonhomogeneous problems. Consider the Cauchy problem

$$\begin{aligned} u_t + a(x,t)u_x &= b(x,t), \quad && x \in \mathbb{R},\ t > 0, \\ u(x,0) &= \phi(x), && x \in \mathbb{R}. \end{aligned} \tag{1.29}$$

Here a, b, and ϕ are given smooth functions. Again we define the characteristic by

$$\begin{aligned} x'(t) &= a\big(x(t),t\big), \\ x(0) &= x_0, \end{aligned} \tag{1.30}$$

and study the evolution of u along $x = x(t)$,

$$\begin{aligned} \frac{d}{dt}u\big(x(t),t\big) &= u_t + u_x\frac{dx}{dt} \\ &= u_t + a(x,t)u_x \\ &= b\big(x(t),t\big). \end{aligned}$$

Hence, the solution is given by

$$u\big(x(t),t\big) \;=\; \phi(x_0) + \int_0^t b\big(x(\tau),\tau\big)d\tau \tag{1.31}$$

along the characteristic given by $x = x(t)$. So the procedure for solving (1.29) by the method of characteristics is to first find the characteristics defined by (1.30) and then use (1.31) to compute the solutions along the characteristics.

EXAMPLE 1.3 Consider the following nonhomogeneous Cauchy problem:

$$\begin{aligned} u_t + u_x &= x, \quad && x \in \mathbb{R},\ t > 0 \\ u(x,0) &= \phi(x), && x \in \mathbb{R}. \end{aligned} \tag{1.32}$$

Here, the characteristics defined by (1.30) are given by

$$x(t) \;=\; x_0 + t,$$

and along a characteristic we have

$$\begin{aligned} u\big(x(t),t\big) &= \phi(x_0) + \int_0^t x(\tau)d\tau \\ &= \phi(x_0) + x_0 t + \frac{1}{2}t^2; \end{aligned}$$

cf. (1.31). Since $x_0 = x - t$, we get

$$u(x,t) \;=\; \phi(x-t) + \left(x - \frac{t}{2}\right)t.$$

■

1.4.3 The Wave Equation

The wave equation

$$u_{tt}(x,t) = u_{xx}(x,t) \tag{1.33}$$

arises in for example modeling the motion of a uniform string; see Weinberger [28]. Here, we want to solve the Cauchy problem[7] for the wave equation, i.e. (1.33) with initial data

$$u(x,0) = \phi(x) \tag{1.34}$$

and

$$u_t(x,0) = \psi(x). \tag{1.35}$$

But let us first concentrate on the equation (1.33) and derive possible solutions of this equation. To this end, we introduce the new variables

$$\xi = x + t \quad \text{and} \quad \eta = x - t,$$

and define the function

$$v(\xi,\eta) = u(x,t). \tag{1.36}$$

By the chain rule, we get

$$u_x = v_\xi \frac{\partial \xi}{\partial x} + v_\eta \frac{\partial \eta}{\partial x} = v_\xi + v_\eta$$

and

$$u_{xx} = v_{\xi\xi} + 2v_{\xi\eta} + v_{\eta\eta}.$$

Similarly, we have

$$u_{tt} = v_{\xi\xi} - 2v_{\xi\eta} + v_{\eta\eta},$$

and thus (1.33) implies that

$$0 = u_{tt} - u_{xx} = -4v_{\xi\eta}.$$

Since

$$v_{\xi\eta} = 0 \tag{1.37}$$

we easily see that

$$v(\xi,\eta) = f(\xi) + g(\eta). \tag{1.38}$$

---

[7] Initial-boundary value problems for the wave equation are studied in Chapter 5.

solves (1.37) for any smooth functions f and g. In fact, all solutions of (1.37) can be written in the form (1.38); see Exercise 1.12. Now it follows from (1.36) that

$$u(x,t) = f(x+t) + g(x-t) \tag{1.39}$$

solves (1.33) for any smooth f and g. This can be verified by direct derivation:

$$\left.\begin{aligned} u_{tt} &= f'' + g'' \\ u_{xx} &= f'' + g'' \end{aligned}\right\} \Longrightarrow u_{tt} = u_{xx}.$$

Next we turn our attention to the initial data (1.33) and (1.34). We want to determine the functions f and g in (1.39) such that (1.33) and (1.34) are satisfied. Of course, ϕ and ψ are supposed to be given functions.

By (1.39) we have

$$u(x,t) = f(x+t) + g(x-t)$$

and

$$u_t(x,t) = f'(x+t) - g'(x-t).$$

Inserting $t = 0$, (1.34) and (1.35) imply that

$$\phi(x) = f(x) + g(x) \tag{1.40}$$

and

$$\psi(x) = f'(x) - g'(x). \tag{1.41}$$

By differentiating (1.40) with respect to x, we get

$$\phi'(x) = f'(x) + g'(x). \tag{1.42}$$

Combining (1.41) and (1.42) yields

$$f' = \frac{1}{2}(\phi' + \psi)$$

and

$$g' = \frac{1}{2}(\phi' - \psi),$$

and thus, by integration, we have

$$f(s) = c_1 + \frac{1}{2}\phi(s) + \frac{1}{2}\int_0^s \psi(\theta)d\theta \tag{1.43}$$

and

$$g(s) = c_2 + \frac{1}{2}\phi(s) - \frac{1}{2}\int_0^s \psi(\theta)d\theta, \tag{1.44}$$

where c_1 and c_2 are constants of integration. From (1.40) we note that

$$\phi(x) = f(x) + g(x),$$

and thus by adding (1.43) and (1.44), we observe that

$$c_1 + c_2 = 0.$$

Putting $s = x + t$ in (1.43) and $s = x - t$ in (1.44), it follows from (1.39) that

$$u(x,t) = \frac{1}{2}\big(\phi(x+t) + \phi(x-t)\big) + \frac{1}{2}\int_0^{x+t} \psi(\theta)d\theta - \frac{1}{2}\int_0^{x-t} \psi(\theta)d\theta,$$

or

$$u(x,t) = \frac{1}{2}\big(\phi(x+t) + \phi(x-t)\big) + \frac{1}{2}\int_{x-t}^{x+t} \psi(\theta)d\theta. \tag{1.45}$$

This formula is referred to as the d'Alembert solution. Let us use it to compute the solution of one Cauchy problem.

EXAMPLE 1.4 Consider the Cauchy problem

$$\begin{aligned} u_{tt} &= u_{xx}\,, && x \in \mathbb{R}\,,\ t > 0\,,\\ u(x,0) &= 0\,, && x \in \mathbb{R}\,,\\ u_t(x,0) &= \cos(x)\,, && x \in \mathbb{R}\,. \end{aligned} \tag{1.46}$$

Since $\phi(x) = 0$ and $\psi(x) = \cos(x)$, it follows by (1.45) that

$$\begin{aligned} u(x,t) &= \frac{1}{2}\int_{x-t}^{x+t} \cos(\theta)d\theta\\ &= \frac{1}{2}\big[\sin(\theta)\big]_{x-t}^{x+t}\\ &= \frac{1}{2}\big(\sin(x+t) - \sin(x-t)\big), \end{aligned}$$

so

$$u(x,t) = \cos(x)\sin(t). \tag{1.47}$$

It is straightforward to check by direct computation that (1.47) in fact solves (1.46). ■

1.4.4 The Heat Equation

The heat equation,

$$u_t(x,t) = u_{xx}(x,t), \qquad x \in \mathbb{R},\ t > 0, \tag{1.48}$$

arises in models of temperature evolution in uniform materials; see e.g. Weinberger [28]. The same equation also models diffusion processes — say the evolution of a piece of ink in a glass of water. It is therefore often referred to as the diffusion equation.

Since our purpose in this introductory chapter is to explain basic features of PDEs, we shall study (1.48) equipped with the simplest possible initial data,

$$u(x,0) = H(x) = \begin{cases} 0 & x \leq 0, \\ 1 & x > 0. \end{cases} \tag{1.49}$$

Here $H = H(x)$ is usually referred to as the Heavyside function. The Cauchy problem (1.48)–(1.49) can be interpreted as a model of the temperature in a uniform rod of infinite length. At $t = 0$, the rod is cold to the left and hot to the right. How will the temperature evolve as t increases?

Intuitively you know approximately how this will develop, but let us compute it.

First we observe that the solution of the Cauchy problem (1.48)–(1.49) is actually only a function of one variable. To see this, define the function

$$v(x,t) = u(cx, c^2 t) \tag{1.50}$$

for any $c > 0$. Then

$$v(x,0) = u(cx,0) = \begin{cases} 0 & x \leq 0, \\ 1 & x > 0, \end{cases}$$

and

$$\left.\begin{aligned} v_t &= c^2 u_t \\ v_{xx} &= c^2 u_{xx} \end{aligned}\right\} \overset{(1.48)}{\Longrightarrow} v_t = v_{xx},$$

so we conclude that also v solves the Cauchy problem for any $c > 0$. However, the solution of the problem (1.48)–(1.49) is unique. Uniqueness of the solution of the heat equation will be discussed later in the text. But then, since v given by (1.50) solves (1.48)–(1.49) for any $c > 0$, the solution $u = u(x,t)$ has to be constant along the line parameterized by $(cx, c^2 t)$ for c running from zero to plus infinity. Thus, u is constant along lines where

$$x/\sqrt{t} = \text{constant}.$$

We therefore define $y = x/\sqrt{t}$, introduce

$$w(y) = w(x/\sqrt{t}) = u(x,t), \tag{1.51}$$

and observe that the initial condition (1.49) implies

$$w(-\infty) = 0 \quad \text{and} \quad w(\infty) = 1.$$

Using the chain rule, we get

$$\begin{aligned} u_t &= w'(y)\frac{\partial y}{\partial t} = -\frac{1}{2}yt^{-1}w'(y), \\ u_{xx} &= \frac{\partial}{\partial x}\left(t^{-1/2}w'(y)\right) = t^{-1}w''(y), \end{aligned}$$

and since $u_t = u_{xx}$, we get the ordinary differential equation

$$w''(y) + (y/2)w'(y) = 0 \tag{1.52}$$

with boundary conditions

$$w(-\infty) = 0 \quad \text{and} \quad w(\infty) = 1. \tag{1.53}$$

This problem can be solved analytically. Upon multiplying (1.52) by $e^{y^2/4}$, we get

$$\begin{aligned} 0 &= e^{y^2/4}w''(y) + \frac{y}{2}e^{y^2/4}w'(y) \\ &= \left(e^{y^2/4}w'(y)\right)'. \end{aligned}$$

We integrate this relation and get

$$e^{y^2/4}w'(y) = \alpha,$$

where α is a constant of integration. If we now integrate

$$w'(z) = \alpha e^{-z^2/4}$$

from $-\infty$ to y, we obtain

$$\begin{aligned} \left[w(z)\right]_{-\infty}^{y} &= \alpha \int_{-\infty}^{y} e^{-z^2/4}dz \\ &= 2\alpha \int_{-\infty}^{y/2} e^{-\theta^2}\,d\theta. \end{aligned}$$

Since $w(-\infty) = 0$, we have

$$w(y) = 2\alpha \int_{-\infty}^{y/2} e^{-\theta^2}\,d\theta. \tag{1.54}$$

Using the boundary condition $w(\infty) = 1$, it follows from (1.54) that

$$1 = 2\alpha \int_{-\infty}^{\infty} e^{-\theta^2}\,d\theta = 2\alpha\sqrt{\pi}$$

or

$$2\alpha = 1/\sqrt{\pi};$$

see Exercise 1.11. Hence

$$w(y) = \frac{1}{\sqrt{\pi}} \int_{-\infty}^{y/2} e^{-\theta^2} d\theta$$

and

$$u(x,t) = \frac{1}{\sqrt{\pi}} \int_{-\infty}^{x/2\sqrt{t}} e^{-\theta^2} d\theta. \tag{1.55}$$

We show in Exercise 1.13 that u tends to the Heavyside function as $t \to 0$, $t > 0$.

In Fig. 1.6 we have plotted this solution for $x \in [-2, 2]$ and $t = 0, 1/4, 1$. Note the smoothing property of this solution. Even when the initial function $u(x, 0)$ is discontinuous as a function of x, $u(x,t)$ is continuous as function of x for any $t > 0$; see Exercise 1.13. This feature is very characteristic for the heat equation and other equations of the same form.

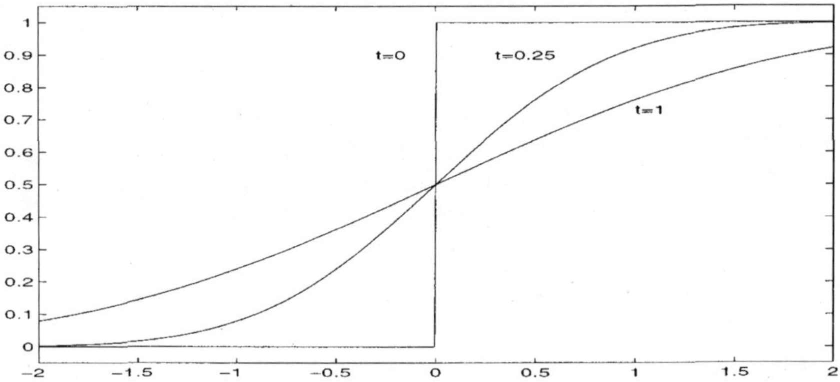

FIGURE 1.6. *The solution of the heat equation for* $t = 0, 1/4, 1$.

1.5 Exercises

EXERCISE 1.1 Consider the following differential equations:

$$\begin{aligned}
(i)\quad & u'(t) = e^t u(t), \\
(ii)\quad & u''(x) = u(x)\sqrt{x}, \\
(iii)\quad & u_{xx}(x,y) + u_{yy}(x,y)e^{\sin(x)} = 1, \\
(iv)\quad & u_t(x,t) + u_x(x,t) = u_{xx}(x,t) + u^2(x,t), \\
(v)\quad & (u'(t))^2 + u(t) = e^t.
\end{aligned}$$

Characterize these equations as:

(a) PDEs or ODEs,

(b) linear or nonlinear,

(c) homogeneous or nonhomogeneous.

EXERCISE 1.2 Consider

$$u'(t) = -\alpha u(t),$$
$$u(0) = u_0,$$

for a given $\alpha > 0$. Show that this problem is stable with respect to perturbation in u_0.

EXERCISE 1.3 Consider the ordinary differential equation

$$\begin{aligned} u'(t) &= tu(t)\big(u(t) - 2\big), \\ u(0) &= u_0. \end{aligned} \tag{1.56}$$

(a) Verify that

$$u(t) = \frac{2u_0}{u_0 + (2 - u_0)e^{t^2}}$$

solves (1.56).

(b) Show that if $0 \le u_0 \le 2$, then $0 \le u(t) \le 2$ for all $t \ge 0$.

(c) Show that if $u_0 > 2$, then $u(t) \to \infty$ as

$$t \to \left(\ln\left(\frac{u_0}{u_0 - 2}\right)\right)^{1/2}.$$

(d) Suppose we are interested in (1.56) for u_0 close to 1, say $u_0 \in [0.9, 1.1]$. Would you say that the problem (1.56) is stable for such data?

EXERCISE 1.4 We have discussed the question of stability with respect to perturbations in the initial conditions. A model which is expressed as a differential equation may also involve coefficients based on measurements. Hence, it is also relevant to ask whether a solution is stable with respect to changes in coefficients. One example can be based on the problem of Exercise 1.2,

$$\begin{aligned} u'(t) &= -\alpha u(t), \\ u(0) &= u_0. \end{aligned} \tag{1.57}$$

We assume that $\alpha > 0$ is a measured number, and we consider a slightly perturbed problem

$$\begin{aligned} v'(t) &= -(\alpha + \epsilon)v(t), \\ v(0) &= u_0. \end{aligned}$$

(a) We are interested in the solution at $t = 1$. Do small changes in α imply small changes in the solution?

(b) Next we assume that both u_0 and α are measured. Discuss the stability of the problem (1.57) in this context.

EXERCISE 1.5 Find the exact solution of the following Cauchy problems:

(a)

$$\begin{aligned} u_t + 2xu_x &= 0 \qquad x \in \mathbb{R},\ t > 0, \\ u(x,0) &= e^{-x^2}. \end{aligned}$$

(b)

$$\begin{aligned} u_t - xu_x &= 0 \qquad x \in \mathbb{R},\ t > 0, \\ u(x,0) &= \sin(87x). \end{aligned}$$

(c)

$$\begin{aligned} u_t + xu_x &= x \qquad x \in \mathbb{R},\ t > 0, \\ u(x,0) &= \cos(90x). \end{aligned}$$

(d)

$$\begin{aligned} u_t + xu_x &= x^2 \qquad x \in \mathbb{R},\ t > 0, \\ u(x,0) &= \sin(87x)\cos(90x). \end{aligned}$$

EXERCISE 1.6 Compute the exact solution of the following Cauchy problem:

$$\begin{aligned} u_t + u_x &= u, && x \in \mathbb{R},\ t > 0, \\ u(x,0) &= \phi(x), && x \in \mathbb{R}, \end{aligned}$$

where ϕ is a given smooth function.

EXERCISE 1.7 We want to consider the stability of first-order nonhomogeneous Cauchy problems

$$\begin{aligned} u_t + a u_x &= b(x,t), & x \in \mathbb{R},\ t > 0, \\ u(x,0) &= \phi(x), & x \in \mathbb{R}. \end{aligned} \tag{1.58}$$

We assume that a is a constant and that b and ϕ are given smooth functions. Consider also the Cauchy problem

$$\begin{aligned} v_t + a v_x &= b(x,t), & x \in \mathbb{R},\ t > 0, \\ v(x,0) &= \phi(x) + \epsilon(x), \end{aligned}$$

where $\epsilon = \epsilon(x)$ is a smooth function. Show that

$$\sup_{x\in\mathbb{R},t\geq 0} |u(x,t) - v(x,t)| = \sup_{x\in\mathbb{R}} |\epsilon(x)|,$$

and conclude that the Cauchy problem (1.58) is stable with respect to perturbations in the initial data.

EXERCISE 1.8 Consider the wave equation

$$\begin{aligned} u_{tt} &= c^2 u_{xx}, & x \in \mathbb{R},\ t > 0, \\ u(x,0) &= \phi(x), \\ u_t(x,0) &= \psi(x), \end{aligned} \tag{1.59}$$

for a given $c > 0$. Follow the steps used to derive the solution in the case of $c = 1$ and show that

$$u(x,t) = \frac{1}{2}\left(\phi(x+ct) \pm \phi(x-ct)\right) + \frac{1}{2c}\int_{x-ct}^{x+ct} \psi(\theta)d\theta \tag{1.59a}$$

solves (1.59).

EXERCISE 1.9 Use the solution derived above to solve the Cauchy problem

$$\begin{aligned} u_{tt} &= 16 u_{xx}, & x \in \mathbb{R},\ t > 0, \\ u(x,0) &= 6\sin^2(x), & x \in \mathbb{R}, \\ u_t(x,0) &= \cos(6x), & x \in \mathbb{R}. \end{aligned}$$

EXERCISE 1.10 Find the solution of the Cauchy problem

$$\begin{aligned} u_t &= \epsilon u_{xx}, & x \in \mathbb{R},\ t > 0 \\ u(x,0) &= \begin{cases} 0 & x \leq 0 \\ 1 & x > 0 \end{cases} \end{aligned}$$

for any given constant $\epsilon > 0$. Use the solution formula to plot the solution at $t = 1$ for $x \in [-1,1]$ using $\epsilon = 1/10, 1/2, 1, 10$. In order to use the solution formula you will have to apply numerical integration. Those not familiar with this subject may consult Project 2.1.

EXERCISE 1.11 Let I denote the integral

$$I = \int_{-\infty}^{\infty} e^{-x^2}\, dx.$$

(a) Explain why

$$I^2 = \int_{-\infty}^{\infty}\int_{-\infty}^{\infty} e^{-(x^2+y^2)}\, dxdy.$$

(b) Use polar coordinates to show that $I = \sqrt{\pi}$.

EXERCISE 1.12 Show that any solution of (1.37) can be written in the form (1.38).

EXERCISE 1.13 Consider the function $u(x,t)$ given by (1.55).

(a) Verify directly that u satisfies the heat equation (1.48) for any $x \in \mathbb{R}$ and $t > 0$.

(b) Let $t > 0$ be fixed. Show that $u(\cdot, t) \in C^\infty(\mathbb{R})$, i.e. u is a C^∞-function with respect to x for any fixed $t > 0$.

(c) Show that

$$u(0,t) = \frac{1}{2} \qquad \text{for all } t > 0.$$

(d) Let $x \neq 0$ be fixed. Show that

$$\lim_{t \to 0+} u(x,t) = H(x).$$

EXERCISE 1.14 Consider the initial value problem (1.59) for the wave equation, i.e.

$$\begin{aligned} u_{tt} &= c^2 u_{xx}, \qquad x \in \mathbb{R},\ t > 0, \\ u(x,0) &= \phi(x), \\ u_t(x,0) &= \psi(x). \end{aligned} \tag{1.60}$$

The purpose of this exercise is to give an alternative derivation of the d'Alembert solution (1.59a), based on the method of characteristics for first order equations.

(a) Assume that $u = u(x,t)$ solves (1.60) and let $v = u_t + cu_x$. Show that

$$v_t - cv_x = 0.$$

(b) Find $v(x,t)$ expressed by ϕ and ψ.

(c) Explain why

$$u(x,t) = \phi(x - ct) + \int_0^t v[x - c(t - \tau), \tau]\, d\tau.$$

(d) Derive the expression (1.59a) for $u(x,t)$.

EXERCISE 1.15 The purpose of this exercise is to perform a theoretical analysis of the numerical experiments reported in Table 1.1. There we studied the forward Euler method applied to the initial value problem (1.18), and the experiments indicated that the error $E(\Delta t)$ at $t = 1$ satisfies

$$E(\Delta t) \approx 1.359\Delta t.$$

(a) Let $0 \leq (m+1)\Delta t \leq T$ and let $u(t)$ be the solution of (1.18). Show that if $t_m = m\Delta t$, then

$$\frac{u(t_{m+1}) - u(t_m)}{\Delta t} = u(t_m) + \tau_m,$$

where the truncation error τ_m satisfies

$$|\tau_m| \leq \frac{\Delta t}{2} e^T \quad \text{for } 0 \leq (m+1)\Delta t \leq T.$$

(b) Assume that $\{v_m\}$ is the corresponding forward Euler solution given by

$$v_{m+1} = (1 + \Delta t)v_m, \qquad v_0 = 1,$$

and let $w_m = u_m - v_m$ be the error at time $t_m = m\Delta t$. Explain why $\{w_m\}$ satisfies the difference equation

$$w_{m+1} = (1 + \Delta t)w_m + \Delta t\, \tau_m, \qquad w_0 = 0.$$

(c) Use induction on m to prove that

$$|w_m| \leq \frac{\Delta t}{2} e^T (e^{t_m} - 1) \qquad \text{for } 0 \leq t_m \leq T.$$

How does this result compare to what was obtained in Table 1.1?

EXERCISE 1.16 Let $u(x,t)$ be a solution of the heat equation (1.48) with initial data

$$u(x,0) = f(x).$$

(a) Let $a \in \mathbb{R}$ and define a function

$$v(x,t) = u(x-a,t).$$

Show that v solves the heat equation with initial data $v(x,0) = f(x-a)$.

(b) Let $k > 0$ be given and define

$$w(x,t) = u(k^{1/2}x, kt).$$

Show that w solves the heat equation with initial data $w(x,0) = f(k^{1/2}x)$.

(c) Assume that $u^1(x,t), u^2(x,t), \ldots, u^n(x,t)$ are solutions of the heat equation (1.48) with initial functions

$$u^k(x,0) = f^k(x) \quad \text{for } k = 1,2,\ldots,n.$$

Furthermore, let $c_1, c_2, \ldots, c_n \in \mathbb{R}$ and define a new function $u(x,t)$ by

$$u(x,t) = \sum_{k=1}^{n} c_k u^k(x,t).$$

Show that u solves (1.48) with initial data

$$u(x,0) = \sum_{k=1}^{n} c_k f^k(x).$$

Exercise 1.17 Consider the function $S(x,t)$ given by

$$S(x,t) = \frac{1}{\sqrt{4\pi t}} e^{-\frac{x^2}{4t}} \qquad \text{for } x \in \mathbb{R}, \quad t > 0.$$

This function is well known in probability theory. It corresponds to the density function for the normal distribution with variance $2t$. As we shall see below, this function also appears naturally in the analysis of the Cauchy problem for the heat equation. In the context of differential equations the function S is therefore frequently referred to as the Gaussian kernel function or the fundamental solution of the heat equation.

(a) Use the result of Exercise 1.11 to show that

$$\int_{\mathbb{R}} S(x,t)\, dx = 1 \qquad \text{for any } t > 0.$$

(b) Consider the solution (1.55) of the heat equation (1.48) with the Heavyside function H as a initial function. Show that $u(x,t)$ can be expressed as

$$u(x,t) = \int_{\mathbb{R}} S(x-y,t)H(y)\,dy.$$

(c) Let $a \in \mathbb{R}$ be given and define

$$v(x,t) = \int_{\mathbb{R}} S(x-y,t)H(y-a)\,dy.$$

Use the result of Exercise 1.16 (a) to show that v solves (1.48) with initial condition

$$u(x,0) = H(x-a).$$

(d) Let $a, b \in \mathbb{R}$, $a < b$, be given and define

$$\chi_{a,b}(x) = \begin{cases} 1 & \text{for } x \in [a,b], \\ 0 & \text{otherwise.} \end{cases}$$

Show that the function

$$u(x,t) = \int_{\mathbb{R}} S(x-y,t)\chi_{a,b}(y)\,dy$$

solves (1.48) with initial condition

$$u(x,0) = \chi_{a,b}(x).$$

Hint: Observe that $\chi_{a,b}(x) = H(x-a) - H(x-b)$ and use Exercise 1.16 (c).

(e) Let $f(x)$ be a step function of the form

$$f(x) = \begin{cases} 0 & \text{for } x \leq a_0, \\ c_1 & \text{for } x \in [a_0, a_1], \\ \vdots & \\ c_n & \text{for } x \in [a_{n-1}, a_n], \\ 0 & \text{for } x > a_n, \end{cases}$$

where $c_1, c_2, \ldots, c_n$ and $a_0 < a_1 < \cdots < a_n$ are real numbers. Show that the function $u(x,t)$ given by

$$u(x,t) = \int_{\mathbb{R}} S(x-y,t)f(y)\,dy \tag{1.61}$$

solves the heat equation (1.48) with initial condition

$$u(x, 0) = f(x).$$

In fact, the solution formula (1.61) is not restricted to piecewise constant initial functions f. This formula is true for general initial functions f, as long as f satisfies some weak smoothness requirements. We will return to a further discussion of the formula (1.61) in Chapter 12.

1.6 Projects

Project 1.1 *Convergence of Sequences*

In dealing with numerical approximations of various kinds, we are often interested in assessing the quality of the numerical estimates. Proving error bounds in order to obtain such estimates might be a very difficult task,[8] but in many cases empirical estimates can be obtained using simple computer experiments. The purpose of this project is thus to develop a "quick and dirty" way of investigating the convergence of schemes under some fortunate circumstances. More precisely, the exact solution has to be available in addition to the numerical approximation. Of course, one might ask why a numerical approximation is needed in such cases, but the general idea is that if we know how one method converges for one particular problem, this will guide us in learning how the scheme handles more delicate problems.

Let us start by defining some basic concepts concerning convergence of an infinite sequence of real numbers $\{z_n\}_{n\geq 1}$.

Convergence of Sequences. If, for any $\epsilon > 0$, there is an integer N such that

$$|z_n - z| < \epsilon \quad \text{for all} \quad n \geq N,$$

we say that the sequence $\{z_n\}$ converges towards z, and we write

$$\lim_{n\to\infty} z_n = z.$$

Rate of Convergence. We say that the sequence $\{z_n\}$ converges towards a real number z with the rate α if there is a finite constant c, not depending on n, such that

$$|z_n - z| \leq c\left(\frac{1}{n}\right)^{\alpha}.$$

---

[8]Some argue strongly that this is the very core of numerical analysis.

If $\alpha = 1$, we have first-order, or linear convergence, $\alpha = 2$ is referred to as second-order, or quadratic convergence, and so on.

Superlinear Convergence. We say that the sequence $\{z_n\}$ converges superlinearly towards a real number z if there is a sequence of positive real numbers $\{c_n\}$ such that

$$\lim_{n\to\infty} c_n = 0$$

and

$$|z_n - z| \leq c_n/n.$$

The O-Notation. Let $\{y_n\}_{n\geq 1}$ and $\{z_n\}_{n\geq 1}$ be two sequences of positive real numbers. If there is a finite constant c, not depending on n, such that

$$y_n \leq c z_n \quad \text{for all} \quad n \geq 1,$$

we say that the sequence $\{y_n\}$ is of order $\{z_n\}$, and we write,

$$y_n = O(z_n)$$

(a) Estimate the rate of convergence, as n tends to infinity, for the following sequences:

1. $z_n = \sqrt{1/n}$
2. $z_n = \sin(1/n)$
3. $z_n = \sqrt{1/n}\sin^2(1/n)$
4. $z_n = n(e^{(1/n)} - 1 - \frac{1}{n})$

(b) Determine whether the following sequences converge linearly or superlinearly toward zero as n tends to infinity:

1. $z_n = 1/n$
2. $z_n = \frac{1}{n\log(n)}$
3. $z_n = \frac{e^{1/n}}{n}$

(c) In some cases, we consider a parameter h tending to zero, rather than n tending to infinity. Typically, $h \approx 1/n$ in many of our applications. Restate the definitions above for sequences $\{z_h\}$ where $h > 0$, and estimate the rate of convergence, as $h \to 0$, for the following sequences:

1. $z_h = \sqrt{h}\sin(h)$
2. $z_h = \sqrt{h}\cos(h)$

3. $z_h = \sqrt{h}e^h$

(d) Let $f = f(x)$ be a smooth function, and show that for small h we have[9]:

1. $\frac{f(x+h)-f(x)}{h} = f'(x) + O(h)$
2. $\frac{f(x)-f(x-h)}{h} = f'(x) + O(h)$
3. $\frac{f(x+h)-f(x-h)}{2h} = f'(x) + O(h^2)$
4. $\frac{f(x+h)-2f(x)+f(x-h)}{h^2} = f''(x) + O(h^2)$

(e) In many cases, the sequence $\{z_n\}$ is not known by a formula. It might for example be given as the result of numerical experiments. In such cases, we want a procedure for estimating the rate of convergence numerically. To do this, we define the error by

$$e_h = |z_h - z|,$$

and assume that there exist real numbers c and α, which are independent of h, such that

$$e_h = ch^\alpha. \tag{1.62}$$

Let $h_1 \neq h_2$, and use (1.62) to derive that the rate α can be estimated by

$$\alpha = \frac{\log(e_{h_1}/e_{h_2})}{\log(h_1/h_2)} \tag{1.63}$$

provided that the model (1.62) holds.

Consider the sequences given in (c) above, and compute e_h for

$$h = 1/100, 1/200, 1/400, \text{ and } 1/800,$$

and estimate the rate α given by (1.63) by comparing subsequent values of e_h. How do your results compare with those obtained in (c) above?

(f) Use the procedure described above to estimate the rate of convergence for the sequence given by

$$z_h = |h \log(h)|.$$

Try to explain the difficulties you encounter, and note the dangers of blindly applying the procedure.

[9]The Taylor series is useful here.

(g) In this course, Fourier series for approximating functions will be used in the analysis of partial differential equations. One peculiar property of these series is that they provide series expansions for some irrational numbers. Try to estimate the order of convergence for the following series by applying the technique developed above.

$$(i)\ z = \pi/4 \quad \text{and} \quad z_n = \sum_{j=0}^{n} \frac{(-1)^j}{2j+1}$$

$$(ii)\ z = \pi^2/8 \quad \text{and} \quad z_n = \sum_{j=0}^{n} \frac{1}{(2j+1)^2}$$

$$(iii)\ z = \pi^2/6 \quad \text{and} \quad z_n = \sum_{j=1}^{n} \frac{1}{j^2}$$

Project 1.2 *Linear Algebra*

Throughout this course we will need some basic concepts of linear algebra; matrices, vectors, norms and so on. Familarity with elementary linear algebra is assumed; this project is intended to refresh your memory. We simply state a series of facts about matrices and vectors, followed by some fairly simple problems showing possible applications of the results. Proofs of the properties can be found in any introductory book on linear algebra; see e.g. H. Anton [1].

Linear Independent Vectors. Let $V = \{v_1, v_2, \dots, v_k\}$ be a collection of vectors in $\mathbb{R}^n$. If there exist scalars $c_1, c_2, \dots, c_k$ such that at least one of the c_js is nonzero and

$$c_1 v_1 + c_2 v_2 + \cdots + c_k v_k = 0, \tag{1.64}$$

we say that the collection of vectors V is a *linearly dependent* set. If the requirement (1.64) implies that all the scalars $c_1, c_2, \dots, c_k$ have to be zero, the vectors are referred to as a *linearly independent* set.

PROBLEMS

(a) Define the vectors

$$u_1 = \begin{pmatrix} 2 \\ -1 \\ 0 \end{pmatrix}, \quad u_2 = \begin{pmatrix} -1 \\ 2 \\ -1 \end{pmatrix}, \quad \text{and} \quad u_3 = \begin{pmatrix} 0 \\ -1 \\ 2 \end{pmatrix}.$$

Show that $\{u_1, u_2, u_3\}$ is a linearly independent set of vectors.

(b) Show that the vectors

$$v_1 = \begin{pmatrix} 1 \\ 2 \\ 3 \end{pmatrix}, \quad v_2 = \begin{pmatrix} 4 \\ 5 \\ 6 \end{pmatrix}, \quad \text{and} \quad v_3 = \begin{pmatrix} 7 \\ 8 \\ 9 \end{pmatrix}$$

form a linear dependent set.

(c) Show that any collection of $n+1$ vectors in $\mathbb{R}^n$ form linearly dependent sets.

Singular and Nonsingular Matrices. Let A be an $n \times n$ matrix, i.e. $A \in \mathbb{R}^{n,n}$. Then A is said to be *nonsingular* if there is another $n \times n$ matrix A^{-1} such that

$$A^{-1}A = I.$$

where $I \in \mathbb{R}^{n,n}$ is the identity matrix. If no such matrix exists, A is called *singular*.

There are several ways of characterizing a nonsingular matrix; the following statements are equivalent:

- The matrix A is nonsingular.
- The determinant of A is nonzero.
- The vectors defined by the rows of A form a linearly independent set.
- The vectors defined by the columns of A form a linearly independent set.
- The linear system $Ax = 0$ has only one solution; $x = 0$.
- The linear system $Ax = b$ has a unique solution $x = A^{-1}b$ for any $b \in \mathbb{R}^n$.

Similarly, a singular matrix can be characterized by the following equivalent statements:

- The matrix A is singular.
- The determinant of A is zero.
- The vectors defined by the rows of A form a linearly dependent set.
- The vectors defined by the columns of A form a linearly dependent set.
- There exists at least one nonzero vector $x \in \mathbb{R}^n$ such that $Ax = 0$.
- There exists a vector $b \in \mathbb{R}^n$ such that the linear system $Ax = b$ has no solution.

The *rank* of a matrix is the number of linearly independent columns (or rows) in the matrix. Obviously, the rank of a nonsingular $n \times n$ matrix is n.

PROBLEMS (CONTINUED)

(d) Let

$$A_1 = \begin{pmatrix} 2 & -1 & 0 \\ -1 & 2 & -1 \\ 0 & -1 & 2 \end{pmatrix}, \qquad A_2 = \begin{pmatrix} 1 & 1/2 & 1/3 \\ 1/2 & 1/3 & 1/4 \\ 1/3 & 1/4 & 1/5 \end{pmatrix},$$

and

$$A_3 = \begin{pmatrix} 1 & 4 & 7 \\ 2 & 5 & 8 \\ 3 & 6 & 9 \end{pmatrix}.$$

Show that

$$A_1^{-1} = \begin{pmatrix} 3/4 & 1/2 & 1/4 \\ 1/2 & 1 & 1/2 \\ 1/4 & 1/2 & 3/4 \end{pmatrix}, \qquad A_2^{-1} = \begin{pmatrix} 9 & -36 & 30 \\ -36 & 192 & -180 \\ 30 & -180 & 180 \end{pmatrix}$$

and that A_3 is singular.

(e) Solve the linear systems

$$A_1 x_1 = b_1 \quad \text{and} \quad A_2 x_2 = b_2$$

where $b_1 = (1, 2, 1)^T$ and $b_2 = (-1, 2, -4)^T$.

(f) Show that the rank of A_1, A_2, and A_3 is 3, 3, and 2 respectively.

(g) Show that if $ad \neq bc$, then

$$\begin{pmatrix} a & b \\ c & d \end{pmatrix}^{-1} = \frac{1}{ad - bc} \begin{pmatrix} d & -b \\ -c & a \end{pmatrix}.$$

The Euclidean Inner Product and the Associated Norm. For two vectors x and y in $\mathbb{R}^n$, the *Euclidean inner product* is defined by

$$(x, y) = \sum_{j=1}^{n} x_j y_j,$$

and the associated norm is defined by

$$||x|| = (x, x)^{1/2}.$$

Two vectors x and y are said to be *orthogonal* if $(x, y) = 0$. A collection of vectors $\{v_1, v_2 \ldots, v_k\}$ is said to be an *orthogonal set* if $(v_i, v_j) = 0$ for all $i \neq j$. If, in addition, $||v_i|| = 1$ for all $i = 1, 2, \ldots, k$, the set is called *orthonormal.*

The norm $|| \cdot ||$ satisfies the following properties:

1. $||x|| \geq 0$ for all vectors $x \in \mathbb{R}^n$.
2. $||x|| = 0$ if and only if $x = 0$.
3. $||\alpha x|| = |\alpha| \, ||x||$ for all scalars α and vectors $x \in \mathbb{R}^n$.
4. $||x + y|| \leq ||x|| + ||y||$ for any vectors $x, y \in \mathbb{R}^n$.
5. $(x, y) \leq ||x|| \, ||y||$ for any vectors $x, y \in \mathbb{R}^n$.

Here, (4) is referred to as the triangle inequality and (5) is the Cauchy-Schwarz inequality.

PROBLEMS (CONTINUED)

(h) Consider the vectors defined in (*a*) and (*b*) above, and compute the inner products (u_1, u_2), (u_1, u_3), (u_2, u_3), and (v_1, v_3). Compute the norms $||u_1||$, $||u_2||$, $||v_1||$, and $||v_3||$.

(i) Suppose the vectors $x \in \mathbb{R}^n$ and $y \in \mathbb{R}^n$ are orthogonal. Show that

$$||x + y||^2 = ||x||^2 + ||y||^2.$$

This is referred to as the theorem of Pythagoras.

(j) Show that a set of orthonormal vectors forms a linearly independent set.

(k) Show that the usual basis of $\mathbb{R}^n$ forms a orthonormal set.

(l) Suppose $Y = \{y_1, y_2, \ldots, y_n\}$ is an orthonormal set in $\mathbb{R}^n$. Show that any vector $z \in \mathbb{R}^n$ can be written as a linear combination of the vectors in Y. More precisely, determine the coefficients $\{c_1, c_2, \ldots, c_n\}$ such that

$$z = \sum_{j=1}^{n} c_j y_j.$$

Is this representation of the vector z in terms of the vectors in y unique?

Eigenvalues and Eigenvectors. Let $A \in \mathbb{R}^{n,n}$, and suppose that there exists a scalar value λ and a nonzero vector x such that

$$Ax = \lambda x.$$

Then λ is an *eigenvalue* and x is a corresponding *eigenvector* of the matrix A. Basic facts about eigenvalues and eigenvectors:

- Any matrix $A \in \mathbb{R}^{n,n}$ has at most n eigenvalues.
- If the matrix is symmetric, i.e. $A^T = A$, all the eigenvalues are real and the corresponding eigenvectors form an orthogonal set.

- A matrix is nonsingular if and only if all eigenvalues are nonzero.

PROBLEMS (CONTINUED)

(m) Find the eigenvalues and the eigenvectors of the matrices

$$A_4 = \begin{pmatrix} 2 & -1 \\ -1 & 2 \end{pmatrix}, \qquad A_5 = \begin{pmatrix} 1 & -1 \\ -1 & 1 \end{pmatrix},$$

and A_1 above.

(n) Verify for the matrices A_1 and A_4 that the eigenvectors are orthogonal.

(o) Suppose λ is an eigenvalue and x is the corresponding eigenvector for a nonsingular matrix $A \in \mathbb{R}^{n,n}$. Define the matrix $B_1 = I + \alpha_1 A$ where I is the identity matrix and α_1 is a scalar. Show that $\mu_1 = 1 + \alpha_1 \lambda$ is an eigenvalue of the matrix B_1, and that x is the corresponding eigenvector.

(p) Let $B_2 = \alpha_0 I + \alpha_1 A + \alpha_2 A^2$, and show that $\mu_2 = \alpha_0 + \alpha_1 \lambda + \alpha_2 \lambda^2$ is an eigenvalue of B_2 and that x is the corresponding eigenvector.

(q) Try to generalize these observations to find a formula for the eigenvalues for a general matrix polynomial of the form

$$P(A) = \sum_{j=0}^{m} \alpha_j A^j.$$

(r) Show that $1/\lambda$ is an eigenvalue and x is an eigenvector for the inverse of A, i.e. for A^{-1}.

Positive Definite Matrices. A symmetric matrix $A \in \mathbb{R}^{n,n}$ is called *positive definite* if

$$x^T A x > 0 \quad \text{for all nonzero } x \in \mathbb{R}^n.$$

Similarly, it is called *positive semidefinite* if

$$x^T A x \geq 0 \quad \text{for all } \ x \in \mathbb{R}^n.$$

Basic facts about positive definite matrices:

- A symmetric and positive definite matrix is nonsingular.
- A symmetric matrix is positive definite if and only if all the eigenvalues are real and strictly positive.
- A symmetric matrix is positive semidefinite if and only if all the eigenvalues are real and nonnegative.

PROBLEMS (CONTINUED)

(s) Show that the matrices A_1 and A_4 are symmetric and positive definite, and that the matrix A_5 is symmetric and positive semidefinite.

(t) Show that a sum of symmetric and positive definite matrices is also symmetric and positive definite.

(u) Let $A \in \mathbb{R}^{n,n}$ be a nonsingular matrix and define $B = A^T A$. Show that B is symmetric positive definite.

(v) A matrix $A \in \mathbb{R}^{n,n}$, not necessarily symmetric, is called *positive real* if

$$x^T Ax > 0 \quad \text{for all nonzero } x \in \mathbb{R}^n.$$

Show that if A is positive real, then the matrix $B = A + A^T$ is symmetric and positive definite.

Project 1.3 *Numerical Methods for ODEs*

The purpose of this project is to illustrate that there is more to life than forward Euler. Numerical methods for ordinary differential equations is a vast subject reaching far beyond our scope. However, some ideas applied in that field will appear later in the text, so we use this project to present them in a simple framework.

We start by considering the problem

$$\begin{aligned} u'(t) &= -u(t), \\ u(0) &= 1, \end{aligned} \tag{1.65}$$

which we know has the analytical solution $u(t) = e^{-t}$.

(a) Show that the numerical solution computed by the forward Euler method (see (1.17) page 8) is given by

$$v_m = (1 - \Delta t)^m, \qquad m = 0, 1, \ldots \tag{1.66}$$

(b) Show that v_m converges toward the correct solution at $t = 1$ as Δt tends to zero.

(c) In the derivation of the forward Euler method on page 8, we argued that

$$\frac{u(t_{m+1}) - u(t_m)}{\Delta t} \approx u'(t_m) = f\big(u(t_m)\big); \tag{1.67}$$

see (1.15). Show, in a similar manner, that we have

$$\frac{u(t_{m+1}) - u(t_m)}{\Delta t} \approx u'(t_{m+1}) = f\big(u(t_{m+1})\big). \tag{1.68}$$

(d) Use (1.68) to derive the backward Euler method,

$$v_{m+1} - \Delta t f(v_{m+1}) = v_m, \qquad m = 0, 1, \ldots . \tag{1.69}$$

(e) Apply the backward Euler method to the problem (1.65) and show that

$$v_m = \frac{1}{(1+\Delta t)^m}, \qquad m = 0, 1, \ldots \tag{1.70}$$

(f) Explain why

$$\frac{u(t_{m+1}) - u(t_m)}{\Delta t} \approx \frac{1}{2}\big(f(u(t_{m+1})) + f(u(t_m))\big)$$

and use this to derive the scheme

$$v_{m+1} - \frac{1}{2}\Delta t f(v_{m+1}) = v_m + \frac{1}{2}\Delta t f(v_m), \qquad m = 0, 1, \ldots \tag{1.71}$$

(g) Apply (1.71) to (1.65) and show that

$$v_m = \left(\frac{2 - \Delta t}{2 + \Delta t}\right)^m . \tag{1.72}$$

(h) Compare the accuracy of the three methods by computing approximations to the solutions of (1.65) at $t = 1$. Use the technique displayed in Table 1.1 and Project 1.1 to argue that the errors when using the schemes (1.66), (1.70), and (1.72) are $O(\Delta t)$, $O(\Delta t)$, and $O((\Delta t)^2)$ respectively.

(i) Implement the schemes discussed above for $f(v) = -v$. Check the correctness of your implementation by using your code to generate approximations of (1.65).

(j) Generalize your codes to the problem

$$\begin{aligned} u'(t) &= -u^2(t), \\ u(0) &= 1. \end{aligned} \tag{1.73}$$

(k) Derive the exact solution of (1.73) and use this to study the error of three schemes at $t = 1$. Do the conclusions of (h) above also apply to this nonlinear problem?

2

Two-Point Boundary Value Problems

In Chapter 1 above we encountered the wave equation in Section 1.4.3 and the heat equation in Section 1.4.4. These equations occur rather frequently in applications, and are therefore often referred to as fundamental equations. We will return to these equations in later chapters. Another fundamental equation is Poisson's equation, given by

$$-\sum_{j=1}^{n} \frac{\partial^2 u}{\partial x_j^2} = f,$$

where the unknown function u is a function of n spatial variables $x_1, \ldots, x_n$.

The main purpose of this chapter is to study Poisson's equation in one space dimension with Dirichlet boundary conditions, i.e. we consider the two-point boundary value problem given by

$$-u''(x) = f(x), \quad x \in (0,1), \quad u(0) = u(1) = 0. \tag{2.1}$$

Although the emphasis of this text is on partial differential equations, we must first pay attention to a simple ordinary differential equation of second order, since the properties of such equations are important building blocks in the analysis of certain partial differential equations. Moreover, the techniques introduced for this problem also apply, to some extent, to the case of partial differential equations.

We will start the analysis of (2.1) by investigating the analytical properties of this problem. Existence and uniqueness of a solution will be demonstrated, and some qualitative properties will be derived. Then we will turn our attention to numerical methods for solving this simple problem, and we will carefully study how well the numerical solutions mimic the properties of the exact solutions. Finally, we will study eigenvalue problems associated with the boundary value problem (2.1). The results of this analysis will be a fundamental tool in later chapters.

Although the equations investigated in this chapter are very simple and allow analytical solution formulas, we find it appropriate to start our study of numerical methods by considering these problems. Clearly, numerical values of the solutions of these problems could have been generated without the brute force of finite difference schemes. However, as we will encounter more complicated equations later on, it will be useful to have a feeling for how finite difference methods handle the very simplest equations.

2.1 Poisson's Equation in One Dimension

In this section we will show that the problem (2.1) has a unique solution. Moreover, we will find a representation formula for this solution.

We start by recalling a fundamental theorem of calculus: There is a constant c_1 such that

$$u(x) = c_1 + \int_0^x u'(y)\,dy, \tag{2.2}$$

and similarly, there is a constant c_2 such that

$$u'(y) = c_2 + \int_0^y u''(z)\,dz. \tag{2.3}$$

This is true for any twice continuously differentiable function u. Suppose now that u satisfies the differential equation (2.1). Then (2.3) implies that

$$u'(y) = c_2 - \int_0^y f(z)\,dz. \tag{2.4}$$

Then, inserting this equation into (2.2), we obtain

$$u(x) = c_1 + c_2 x - \int_0^x \left(\int_0^y f(z)\,dz \right) dy. \tag{2.5}$$

In order to rewrite this expression in a more convenient form, we define

$$F(y) = \int_0^y f(z)\,dz,$$

and observe that

$$\begin{aligned}\int_0^x \left(\int_0^y f(z)\,dz \right) dy &= \int_0^x F(y)\,dy \\ &= [yF(y)]_0^x - \int_0^x yF'(y)\,dy \\ &= xF(x) - \int_0^x yf(y)\,dy \\ &= \int_0^x (x-y)f(y)\,dy,\end{aligned}$$

where we have used integration by parts. Now (2.5) can be rewritten in the following form:

$$u(x) = c_1 + c_2 x - \int_0^x (x-y)f(y)\,dy. \tag{2.6}$$

Note that c_1 and c_2 are arbitrary constants, and that so far we have only used the differential equation of (2.1) and not the boundary conditions given by

$$u(0) = u(1) = 0.$$

These conditions are taken into account by choosing c_1 and c_2 properly. The condition $u(0) = 0$ implies that $c_1 = 0$, and then $u(1) = 0$ implies that

$$c_2 = \int_0^1 (1-y)f(y)\,dy.$$

Hence, the constants c_1 and c_2 are uniquely determined from the boundary conditions. This observation is an important one; since any solution of the differential equation

$$-u''(x) = f(x)$$

can be written on the form (2.6) and the constants involved in (2.6) are uniquely determined by the boundary conditions of (2.1), it follows that the problem (2.1) has a unique solution.

We observe that if we use the derived expressions for c_1 and c_2 in (2.6), we are allowed to write the solution u in the following form:

$$u(x) = x\int_0^1 (1-y)f(y)\,dy - \int_0^x (x-y)f(y)\,dy. \tag{2.7}$$

EXAMPLE 2.1 Consider the problem (2.1) with $f(x) = 1$. From (2.7) we easily obtain

$$u(x) = x\int_0^1 (1-y)\,dy - \int_0^x (x-y)\,dy = \frac{1}{2}x(1-x).$$

■

EXAMPLE 2.2 Consider the problem (2.1) with $f(x) = x$. Again, from (2.7) we get

$$u(x) = x\int_0^1 (1-y)y\,dy - \int_0^x (x-y)y\,dy = \frac{1}{6}x(1-x^2).$$

■

Further examples of how to compute the exact solution formulas for two-point boundary value problems are given in the exercises. In Project 2.1 we will also see how the exact representation of the solution can be used to derive numerical approximations when the integrals involved cannot be evaluated analytically.

2.1.1 Green's Function

The unique solution of (2.1) can be represented in a very compact way by introducing an auxiliary function: the Green's function.

Introduce the function

$$G(x,y) = \begin{cases} y(1-x) & \text{if } 0 \le y \le x, \\ x(1-y) & \text{if } x \le y \le 1. \end{cases} \tag{2.8}$$

It follows that the representation (2.7) can be written simply as

$$u(x) = \int_0^1 G(x,y)f(y)\,dy. \tag{2.9}$$

The function G is called the Green's function for the boundary value problem (2.1), and it has the following properties:

- G is continuous,
- G is symmetric in the sense that $G(x,y) = G(y,x)$,
- $G(0,y) = G(1,y) = G(x,0) = G(x,1) = 0$,
- G is a piecewise linear function of x for fixed y, and vice versa,
- $G(x,y) \ge 0$ for all $x, y \in [0,1]$.

These properties follow directly from (2.8). The function is plotted in Fig. 2.1.

Of course, the representation (2.9) is only a reformulation of (2.7). However, the representation (2.9) is very convenient when we want to derive various properties of the solution u.

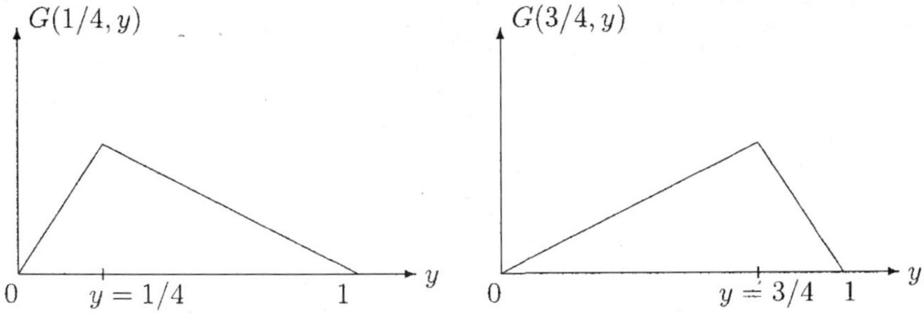

FIGURE 2.1. *Green's function $G(x,y)$ for two values of x. To the left we have used $x = 1/4$, and to the right we have used $x = 3/4$.*

2.1.2 Smoothness of the Solution

Having an exact representation of the solution, we are in a position to analyze the properties of the solution of the boundary value problem. In particular, we shall see that the solution is smoother than the "data," i.e. the solution $u = u(x)$ is smoother than the right-hand side f.

Assume that the right-hand side f of (2.1) is a continuous function, and let u be the corresponding solution given by (2.9). Since u can be represented as an integral of a continuous function, u is differentiable and hence continuous. Let $C\big((0,1)\big)$ denote the set of continuous functions on the open unit interval $(0,1)$ and $C[0,1]$ the space of continuous functions on the corresponding closed interval $[0,1]$.[1] Then the mapping

$$f \mapsto u, \tag{2.10}$$

where u is given by (2.9), maps from $C\big([0,1]\big)$ into $C\big([0,1]\big)$. From (2.7) we obtain that

$$u'(x) = \int_0^1 (1-y)f(y)\,dy - \int_0^x f(y)\,dy$$

and (not surprisingly!)

$$u''(x) = -f(x).$$

Therefore, if $f \in C\big([0,1]\big)$, then $u \in C^2\big((0,1)\big)$, where for an integer $m \geq 0$, $C^m\big((0,1)\big)$ denotes the set of m-times continuously differentiable functions on $(0,1)$. The solution u is therefore smoother than the right-hand side f.

In order to save space we will introduce a symbol for those functions that have a certain smoothness, and in addition vanish at the boundaries. For this purpose, we let

$$C_0^2\big((0,1)\big) = \big\{g \in C^2\big((0,1)\big) \cap C\big([0,1]\big) \,|\, g(0) = g(1) = 0\big\}.$$

[1] A continuous function g on $(0,1)$ is continuous on the closed interval $[0,1]$, i.e. in $C\big([0,1]\big)$, if the limits $\lim_{x\to 0^+} g(x)$ and $\lim_{x\to 1^-} g(x)$ both exist.

With this notation at hand, we notice that the formula for the exact solution given by (2.9) defines a mapping from $C([0,1])$ into $C_0^2((0,1))$.

The following result is a summary of the discussion so far.

Theorem 2.1 *For every $f \in C([0,1])$ there is a unique solution $u \in C_0^2((0,1))$ of the boundary value problem (2.1). Furthermore, the solution u admits the representation (2.9) above.*

Having established this result, further smoothness of the solution can be derived by using the differential equation. More precisely, if $f \in C^m((0,1))$, for $m \geq 1$, then $u \in C^{m+2}((0,1))$ and

$$u^{(m+2)} = -f^{(m)},$$

Hence, the solution is always smoother than the "data," and for $f \in C^\infty$, we have $u \in C^\infty$.

EXAMPLE 2.3 Consider the problem (2.1) with $f(x) = 1/x$. Note that $f \in C((0,1))$, but $f \notin C([0,1])$ since $f(0)$ does not exist. It is easy to verify directly that the solution u is given by

$$u(x) = -x \ln(x),$$

and

$$u'(x) = -1 - \ln(x).$$

Hence, $u \in C_0^2((0,1))$. However, note that u' and u'' are not continuous at zero. This example shows that the map $f \mapsto u$, given by (2.9), is also well defined for some functions f which are not in $C([0,1])$. On the other hand, if $f(x) = 1/x^2$ then $G(x,y) = (1-x)/y$ for $y \leq x$, and hence the integral (2.9) does not exist. So the map $f \mapsto u$ can be extended from $C([0,1])$ to functions in $C((0,1))$ with a modest singularity at the endpoints. ■

2.1.3 A Maximum Principle

The solution of (2.1) has several interesting properties. First we shall consider what is often referred to as a monotonicity property. It states that nonnegative data, represented by the right-hand side f, is mapped into a nonnegative solution. Secondly, we will derive a maximum principle for the solution of the two-point boundary value problem. This principle states how large the solution of the problem, measured by its absolute value, can be for a given right-hand side f.

The following monotonicity property is derived using the representation of the solution given by (2.9).

Proposition 2.1 *Assume that $f \in C([0,1])$ is a nonnegative function. Then the corresponding solution u of (2.1) is also nonnegative.*

Proof: Since $G(x,y) \geq 0$ for all $x, y \in [0,1]$, this follows directly from (2.9). ■

In order to state the next property, we introduce a norm on the set $C([0,1])$. For any function $f \in C([0,1])$, let

$$||f||_\infty = \sup_{x\in[0,1]} |f(x)|.$$

The scalar value $||f||_\infty$, which we will refer to as the sup-norm of f, measures, in some sense, the size of the function f. Let us look at some examples clarifying this concept.

EXAMPLE 2.4 Let $f(x) = x$, $g(x) = x(1-x)$, and $h(x) = e^{\sqrt{x}}$. The sup-norm of these functions, considered on the unit interval $[0,1]$, are given by $||f||_\infty = 1$, $||g||_\infty = 1/4$, and finally $||h||_\infty = e$. ■

The following result relates the size of the solution u of the problem (2.1) to the size of the corresponding data given by the right-hand side f.

Proposition 2.2 *Assume that $f \in C([0,1])$ and let u be the unique solution of (2.1). Then*

$$||u||_\infty \leq (1/8)||f||_\infty.$$

Proof: Since G is nonnegative, it follows from (2.9) that for any $x \in [0,1]$,

$$|u(x)| \leq \int_0^1 G(x,y)|f(y)|\,dy.$$

From the definition of $||f||_\infty$ above, it therefore follows that

$$|u(x)| \leq ||f||_\infty \int_0^1 G(x,y)\,dy = ||f||_\infty \frac{1}{2}x(1-x),$$

and hence

$$||u||_\infty = \sup_{x\in[0,1]} |u(x)| \leq (1/8)||f||_\infty.$$

■

2.2 A Finite Difference Approximation

The basic idea of almost any numerical method for solving equations of the form (2.1) is to approximate the differential equation by a system of algebraic equations. The system of algebraic equations is set up in a clever way such that the corresponding solution provides a good approximation of the solution of the differential equation. The simplest way of generating such a system is to replace the derivatives in the equation by finite differences.

In fact, the basic idea of any finite difference scheme stems from a very familiar definition; the definition of the derivative of a smooth function:

$$u'(x) = \lim_{h \to 0} \frac{u(x+h) - u(x)}{h}.$$

This indicates that in order to get good approximations, h must be sufficiently small. Typically, the number of unknowns in the algebraic system is of order[2] $O(1/h)$. Thus, in order to compute good approximations, we have to solve very large systems of algebraic equations.[3] From this point of view, the differential equation may be regarded as a linear system of infinitely many unknowns; the solution is known at the endpoints and determined by the differential equation in the interior solution domain.

In this section we will introduce a finite difference scheme approximating a two-point boundary value problem. We shall observe that such schemes can provide quite accurate approximations, and that they are, in fact, very simple to deal with on a computer. A more elaborate analysis of the approximation properties will be the topic of subsequent sections.

2.2.1 *Taylor Series*

In order to define the finite difference approximation of problem (2.1), we first recall how Taylor's theorem can be used to provide approximations of derivatives. Assume that $g = g(x)$ is a four-times continuously differentiable function. For any $h > 0$ we have

$$g(x+h) = g(x) + hg'(x) + \frac{h^2}{2}g''(x) + \frac{h^3}{6}g^{(3)}(x) + \frac{h^4}{24}g^{(4)}(x+h_1),$$

where h_1 is some number between 0 and h. Similarly,

$$g(x-h) = g(x) - hg'(x) + \frac{h^2}{2}g''(x) - \frac{h^3}{6}g^{(3)}(x) + \frac{h^4}{24}g^{(4)}(x-h_2),$$

for $0 \leq h_2 \leq h$. In particular, this implies that

$$\frac{g(x+h) - 2g(x) + g(x-h)}{h^2} = g''(x) + E_h(x), \tag{2.11}$$

where the error term E_h satisfies

$$|E_h(x)| \leq \frac{M_g h^2}{12}. \tag{2.12}$$

---

[2]The O notation is discussed in Project 1.1.

[3]This is currently a very active field of research, and the advent of high-speed computing facilities has dramatically increased the applicability of numerical methods. In fact, the numerical solution of partial differential equations has been a major motivation for developing high-speed computers ever since World War II. A thorough discussion of this issue can be found in Aspray [2].

Here the constant M_g is given by

$$M_g = \sup_x |g^{(4)}(x)|.$$

We observe that for a fixed function g, the error term E_h tends to zero as h tends to zero. In particular, if g is a polynomial of degree ≤ 3, such that $g^{(4)} \equiv 0$, the error term satisfies $E_h(x) = 0$ for all x. This property will be discussed in Exercise 2.16. Further discussions on Taylor series can be found in Project 1.1.

2.2.2 A System of Algebraic Equations

The first step in deriving a finite difference approximation of (2.1) is to partition the unit interval $[0, 1]$ into a finite number of subintervals. We introduce the grid points $\{x_j\}_{j=0}^{n+1}$ given by $x_j = jh$, where $n \geq 1$ is an integer and the spacing h is given by $h = 1/(n+1)$. Typically n will be large, and hence the spacing h is small. The solution v of the discrete problem is defined only at the grid points x_j where the values of the approximation are given by v_j. Between these points, an approximation can be defined by, for example, piecewise linear interpolation.

As usual, we let u denote the solution of the two-point boundary value problem

$$-u''(x) = f(x), \quad x \in (0, 1), \quad u(0) = u(1) = 0,$$

and we define the approximation $\{v_j\}_{j=0}^{n+1}$ by requiring

$$-\frac{v_{j-1} - 2v_j + v_{j+1}}{h^2} = f(x_j) \quad \text{for} \quad j = 1, \ldots, n, \quad \text{and} \quad v_0 = v_{n+1} = 0. \tag{2.13}$$

Obviously, the second-order derivative in the differential equation is approximated by the finite difference derived above; see (2.11). The system of n equations and n unknowns $\{v_j\}_{j=1}^{n}$ defined by (2.13) can be written in a more compact form by introducing the $n \times n$ matrix

$$A = \begin{pmatrix} 2 & -1 & 0 & \ldots & 0 \\ -1 & 2 & -1 & \ddots & \vdots \\ 0 & \ddots & \ddots & \ddots & 0 \\ \vdots & \ddots & -1 & 2 & -1 \\ 0 & \ldots & 0 & -1 & 2 \end{pmatrix}. \tag{2.14}$$

Furthermore, let $b = (b_1, b_2, \ldots, b_n)^T$ be an n-vector with components given by

$$b_j = h^2 f(x_j) \quad \text{for} \quad j = 1, 2, \ldots, n.$$

Grouping the unknowns in the vector $v = (v_1, v_2, \ldots, v_n)^T$, the system (2.13) can be rewritten as a system of equations in the standard form

$$Av = b. \tag{2.15}$$

Below we will show that the matrix A is nonsingular,[4] implying that the system (2.15) has a unique solution. We will also discuss how systems of this form can be solved numerically. However, for the time being, we find it more interesting to turn to an example showing how the approximation (2.13) actually works for a particular case.

EXAMPLE 2.5 Let us consider the following two-point boundary value problem:

$$-u''(x) = (3x + x^2)e^x, \quad x \in (0,1), \quad u(0) = u(1) = 0,$$

where the exact solution is given by

$$u(x) = x(1-x)e^x.$$

For this problem, we let

$$b_j = h^2(3x_j + x_j{}^2)e^{x_j} \quad \text{for} \quad j = 1, \ldots, n,$$

and solve the system of equations defined by (2.15) for different grid sizes, i.e. for some values of n. In Fig. 2.2 we have plotted the exact solution (solid line) and numerical solution (dashed line). For the numerical solution we used $n = 5$. We notice that, even for this very coarse grid, the finite difference approximation captures the form of the exact solution remarkably well. In the next figure, the grid is refined using $n = 15$, and we notice that, within the current scaling, the numerical and analytical solutions are almost identical.

How good is the approximation actually? What is the rate of convergence? Since the exact solution is available for this problem, the rate of convergence can be estimated simply by running some experiments. We define the error to be

$$E_h = \max_{j=0,\ldots,n+1} |u(x_j) - v_j| \tag{2.16}$$

and compute this value for some grid sizes. The results are given in Table 2.1. We have also estimated the rate of convergence by comparing the results of subsequent grid sizes. Exactly how this computation is done is discussed in Project 1.1. From the table, we observe that the error seems to satisfy a bound of the form

$$E_h = O(h^2).$$

[4]The basic concepts of linear algebra are reviewed in Project 1.2.

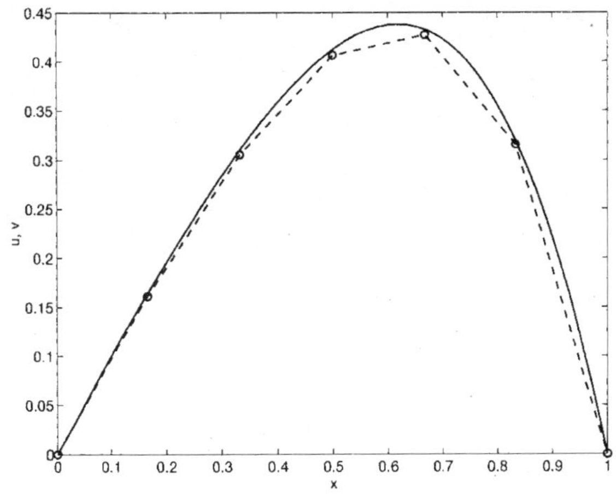

FIGURE 2.2. *The figure shows the numerical solution (dashed line) and the exact solution (solid line) of the boundary value problem. For the numerical scheme, we have used* $n = 5$ *interior grid points, and drawn a linear interpolation between the values on the grid. The solution at the grid points are marked by 'o'.*

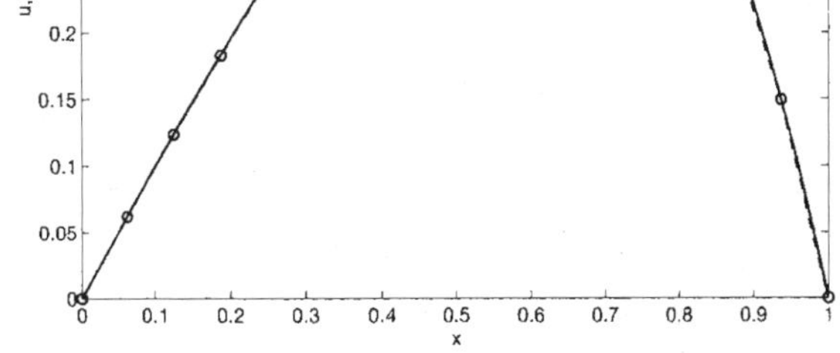

FIGURE 2.3. *The figure shows the numerical solution (dashed line) and the exact solution (solid line) of the boundary value problem. For the numerical scheme, we have used* $n = 15$ *interior grid points.*

n	h	E_h	Rate of convergence
5	1/6	0.0058853	
10	1/11	0.0017847	1.969
20	1/21	0.0004910	1.996
40	1/41	0.0001288	2.000
80	1/81	0.0000330	2.000

TABLE 2.1. *The table shows the maximum error measured at the grid points for several values of h.*

Later, we will return to the problem of determining the rate of convergence for this numerical method and prove that the observed rate of convergence in the present example holds for a wide class of functions f.

■

2.2.3 Gaussian Elimination for Tridiagonal Linear Systems

The purpose of this section is to derive a numerical algorithm which can be used to compute the solution of tridiagonal systems of the form (2.14), (2.15). Furthermore, we shall derive conditions which can be used to verify that a given system has a unique solution. These criteria and the algorithm developed in this section will be useful throughout this course. We warn the reader that this section may be a bit technical — in fact Gaussian elimination is rather technical — and we urge you to keep track of the basic steps and not get lost in the forest of indices.

We consider a system of the form

$$Av = b, \tag{2.17}$$

where the coefficient matrix A has the form

$$A = \begin{pmatrix} \alpha_1 & \gamma_1 & 0 & \dots & 0 \\ \beta_2 & \alpha_2 & \gamma_2 & \ddots & \vdots \\ 0 & \ddots & \ddots & \ddots & 0 \\ \vdots & \ddots & \beta_{n-1} & \alpha_{n-1} & \gamma_{n-1} \\ 0 & \dots & 0 & \beta_n & \alpha_n \end{pmatrix}. \tag{2.18}$$

This system can be equivalently written in component form, i.e.

$$\begin{array}{llll} \alpha_1 v_1 + \gamma_1 v_2 & & & = b_1, \\ \beta_2 v_1 + \alpha_2 v_2 + \gamma_2 v_3 & & & = b_2, \\ \quad \beta_3 v_2 + \alpha_3 v_3 + & \gamma_3 v_4 & & = b_3, \\ & \ddots & & \vdots \\ & & \beta_{n-1} v_{n-2} + \alpha_{n-1} v_{n-1} + \gamma_{n-1} v_n & = b_{n-1}, \\ & & \beta_n v_{n-1} + \alpha_n v_n & = b_n. \end{array} \tag{2.19}$$

Here the coefficients $\beta_2, \dots, \beta_n$, $\alpha_1, \dots, \alpha_n$, $\gamma_1, \dots, \gamma_{n-1}$, and the right-hand side $b_1, \dots, b_n$ are given real numbers and $v_1, v_2, \dots, v_n$ are the unknowns. Note that by choosing $\alpha_j = 2$, $\beta_j = \gamma_j = -1$, we get the "second-order difference" matrix defined in (2.14).

The basic idea in Gaussian elimination for this system is to use the first equation to eliminate the first variable, i.e. v_1, from the second equation. Then, the new version of the second equation is used to eliminate v_2 from the third equation, and so on. After $n-1$ steps, we are left with one equation containing only the last unknown v_n. This first part of the method is often referred to as the "forward sweep."

Then, starting at the bottom with the last equation, we compute the value of v_n, which is used to find v_{n-1} from the second from last equation, and so on. This latter part of the method is referred to as the "backward sweep."

With an eye to this overview, we dive into the details. Observe first that if we subtract $m_2 = \beta_2/\alpha_1$ times the first equation in (2.19) from the second equation, the second equation is replaced by

$$\delta_2 v_2 + \gamma_2 v_3 = c_2,$$

where

$$\delta_2 = \alpha_2 - m_2 \gamma_1$$

and

$$c_2 = b_2 - m_2 b_1 .$$

Hence, the variable v_1 has been eliminated from the second equation.

By a similar process the variable v_{j-1} can be eliminated from equation j. Assume for example that equation $j-1$ has been replaced by

$$\delta_{j-1} v_{j-1} + \gamma_{j-1} v_j = c_{j-1} . \tag{2.20}$$

Equation j of the original system (2.19) has the form

$$\beta_j v_{j-1} + \alpha_j v_j + \gamma_j v_{j+1} = b_j .$$

Then, if $m_j = \beta_j/\delta_{j-1}$ times (2.20) is subtracted from this equation, we get

$$\delta_j v_j + \gamma_j v_{j+1} = c_j,$$

where

$$\begin{aligned} \delta_j &= \alpha_j - m_j \gamma_{j-1} \ , \\ c_j &= b_j - m_j c_{j-1} \ . \end{aligned}$$

After $k-1$ iterations of this procedure we obtain a system of the form

$$\begin{array}{lllll} \delta_1 v_1 + \gamma_1 v_2 & & & & = c_1, \\ \ddots \quad \ddots & & & & \vdots \\ & \delta_k v_k + \gamma_k v_{k+1} & & & = c_k, \\ & \beta_{k+1} v_k + \alpha_{k+1} v_{k+1} & + \gamma_{k+1} v_{k+2} & & = b_{k+1}, \\ & & \ddots & & \vdots \\ & & \beta_{n-1} v_{n-2} + \alpha_{n-1} v_{n-1} & + \gamma_{n-1} v_n & = b_{n-1}, \\ & & & \beta_n v_{n-1} + \alpha_n v_n & = b_n. \end{array} \tag{2.21}$$

Here the variables δ_j and c_j are defined from the given coefficients $\alpha_j, \beta_j, \gamma_j$, and b_j of (2.19) by the recurrence relations

$$\begin{aligned} \delta_1 &= \alpha_1 \ , \quad c_1 = b_1 \ , \\ m_j &= \frac{\beta_j}{\delta_{j-1}} \ , \\ \delta_j &= \alpha_j - m_j \gamma_{j-1} \ , \qquad j = 2, 3, \ldots, k \ , \\ c_j &= b_j - m_j c_{j-1} \ . \end{aligned} \tag{2.22}$$

Note that in the derivation of the system (2.21) from the original system (2.19) we have implicitly assumed that the variables $\delta_1, \delta_2, \ldots, \delta_{k-1}$ will be nonzero. Furthermore, if $\delta_1, \delta_2, \ldots, \delta_{k-1}$ are nonzero, the two systems (2.19) and (2.21) are equivalent in the sense that they have the same solutions.

If the computed values of the δ_ks are always nonzero, we can continue to derive a system of the form (2.21) until $k = n$. Hence, in this case we obtain a system of the form

$$\begin{array}{llll} \delta_1 v_1 + \gamma_1 v_2 & & & = c_1 \ , \\ & \delta_2 v_2 + \gamma_2 v_3 & & = c_2 \ , \\ & \ddots & & \vdots \\ & & \delta_{n-1} v_{n-1} + \gamma_{n-1} v_n & = c_{n-1} \ , \\ & & \delta_n v_n & = c_n \ . \end{array} \tag{2.23}$$

However, from this bidiagonal system we can easily compute the solution v. From the last equation, we have

$$v_n = \frac{c_n}{\delta_n}, \tag{2.24}$$

and by tracking the system backwards we find

$$v_k = \frac{c_k - \gamma_k v_{k+1}}{\delta_k}, \qquad k = n-1, n-2, \dots, 1. \tag{2.25}$$

Hence, we have derived an algorithm for computing the solution v of the original tridiagonal system (2.19). First we compute the variables δ_j and c_j from the relations (2.22) with $k = n$, and then we compute the solution v from (2.24) and (2.25).

Algorithm 2.1

$$\begin{array}{l}
\delta_1 = \alpha_1 \\
c_1 = b_1 \\
\quad \text{for } k = 2, 3, \dots, n \\
\qquad m_k = \beta_k / \delta_{k-1} \\
\qquad \delta_k = \alpha_k - m_k \gamma_{k-1} \\
\qquad c_k = b_k - m_k c_{k-1} \\
v_n = c_n / \delta_n \\
\quad \text{for } k = n-1, n-2, \dots, 1 \\
\qquad v_k = (c_k - \gamma_k v_{k+1}) / \delta_k
\end{array}$$

However, as we have observed above, this procedure breaks down if one of the δ_ks becomes zero. Hence, we have to give conditions which guarantee that this does not happen.

2.2.4 Diagonal Dominant Matrices

One way to check whether a matrix is nonsingular is to see if the entries on the main diagonal of the matrix dominate the off-diagonal elements in the following sense:

Definition 2.1 *A tridiagonal matrix A of the form (2.18) is said to be diagonal dominant*[5] *if*

$$|\alpha_1| > |\gamma_1|, \quad |\alpha_k| \geq |\beta_k| + |\gamma_k| \quad \text{for } k = 2, 3, \dots, n,$$

where γ_n is taken to be zero.

[5] In numerical analysis, there are several different definitions of diagonal dominant matrices. This definition is useful in the present course.

Diagonal dominant matrices occur frequently in numerical analysis.

EXAMPLE 2.6 The matrix given by (2.14), derived in the previous section, is diagonal dominant. This follows since the desired inequality holds with equality for all rows, except for the first and the last, while we have strict inequality in these two rows. ■

Lemma 2.1 *Assume that the coefficient matrix A of the triangular system (2.19) is diagonal dominant and that $\beta_k \neq 0$ for $k = 2, 3, \ldots, n$. Then the variables $\delta_k, k = 1, 2, \ldots, n$ determined by Algorithm 2.1 are well defined and nonzero.*

Proof: We prove by induction that

$$|\delta_k| > |\gamma_k| \quad \text{for } k = 1, 2, \ldots, n.$$

By assumption this holds for $k = 1$. Assume now

$$|\delta_{k-1}| > |\gamma_{k-1}| \qquad \text{for some } k \text{ such that } 2 \leq k \leq n\ .$$

Since $\delta_{k-1} \neq 0$, m_k, and hence δ_k, is well defined and

$$\delta_k = \alpha_k - \frac{\beta_k}{\delta_{k-1}}\gamma_{k-1}\ .$$

By the induction hypothesis $|\gamma_{k-1}/\delta_{k-1}| < 1$, and hence, since $\beta_k \neq 0$,

$$|\beta_k||\frac{\gamma_{k-1}}{\delta_{k-1}}| < |\beta_k|.$$

Therefore, by the triangle inequality and since the system is diagonal dominant we obtain

$$|\delta_k| \geq |\alpha_k| - |\beta_k||\frac{\gamma_{k-1}}{\delta_{k-1}}| > |\alpha_k| - |\beta_k| \geq |\gamma_k|\ .$$

■

Assume that the system (2.19) satisfies the assumptions given in Lemma 2.1 above. Then, if the vector $b = 0$, also the vector $c = 0$, and hence, by tracking the system (2.23) backwards, the unique solution of (2.23) is $v = 0$. However, since the systems (2.19) and (2.23) are equivalent, this means that $v = 0$ is the only solution of (2.19) when $b = 0$. Hence, A is nonsingular. We have therefore obtained the following result:

Proposition 2.3 *Assume that the coefficient matrix A of (2.19) satisfies the properties specified in Lemma 2.1 above. Then, the system has a unique solution which can be computed by Algorithm 2.1.*

As a direct consequence of this proposition, and the result of Example 2.6, we reach the following result:

Corollary 2.1 *The system of equations defined by (2.14)–(2.15), has a unique solution that can be computed using Algorithm 2.1.*

At this point it should be noted that this result is valid only in the presence of exact arithmetics. On computers with a fixed number of digits representing each real number, round-off errors may accumulate and destroy the results of the algorithm. Precise results are available stating sufficient conditions on the matrix in order for Gaussian elimination to provide a good approximation to the solution of the linear system. Techniques also exist to reduce the effect of round-off errors. These issues are discussed in books on numerical linear algebra. If you are interested, you should consult e.g. Golub and van Loan [11]. In the present course we regard these difficulties, or more precisely, potential difficulties, as beyond our scope.

2.2.5 Positive Definite Matrices

Above, we showed that if the system is diagonal dominant, then Algorithm 2.1 is applicable. Now we will show that a similar result holds for positive definite matrices.

Let us first briefly recall some basic facts concerning positive definite matrices.

- A symmetric matrix $A \in \mathbb{R}^{n,n}$ is referred to as positive definite if
$$v^T A v \geq 0 \quad \text{for all } v \in \mathbb{R}^n,$$
with equality only if $v = 0$.
- A symmetric and positive definite matrix is nonsingular.
- A symmetric matrix is positive definite if and only if all the eigenvalues are real and strictly positive.

These and other properties of matrices are discussed in Project 1.2, and can, of course, be found in any textbook on linear algebra.[6]

The properties of symmetric and positive definite matrices are closely connected to the similar properties for differential operators. These connections will be studied below. In the present section we will prove that if the matrix is symmetric and positive definite, the linear system of equations can be solved by Algorithm 2.1.

Let us start by observing that a symmetric and positive definite matrix is not necessarily diagonal dominant. Consider the 2×2 matrix
$$A = \begin{pmatrix} 5 & 2 \\ 2 & 1 \end{pmatrix}.$$

---

[6] The basic concepts of linear algebra are introduced e.g. in the book of H. Anton [1] In numerical linear algebra, the book of Golub and van Loan [11] is a standard reference.

This matrix is obviously symmetric, and one easily computes the eigenvalues

$$\lambda = 3 \pm 2\sqrt{2},$$

which are both positive. Hence A is positive definite. However, we observe that A is not diagonal dominant. Therefore, Proposition 2.1 is not sufficient to guarantee that Algorithm 2.1 will work for all positive definite systems. But, as mentioned above, we shall prove that all symmetric tridiagonal matrices that are positive definite can be handled by the method.

Proposition 2.4 *Consider a tridiagonal system of the form (2.19) and assume that the corresponding coefficient matrix (2.18) is symmetric and positive definite. Then the system has a unique solution that can be computed by Algorithm 2.1.*

Proof: We claim that $\delta_k > 0$ for $k = 1, 2, \ldots, n$. Assume on the contrary that $\delta_1, \delta_2, \ldots, \delta_{k-1} > 0$ and that $\delta_k \leq 0$ for some index k, $1 \leq k \leq n$. We will show that this assumption leads to a contradiction.

Define the vector $v \in \mathbb{R}^n$ by

$$v_k = 1 \quad \text{and} \quad v_{k+1} = v_{k+2} = \cdots = v_n = 0$$

and

$$v_j = -\frac{\gamma_j}{\delta_j} v_{j+1} \quad \text{for} \quad j = k-1, k-2, \ldots, 1.$$

This vector v satisfies the system (2.21) with

$$c_1 = c_2 = \cdots = c_{k-1} = 0, \quad c_k = \delta_k \leq 0, \quad b_{k+1} = \beta_{k+1}, \quad b_{k+2} = \cdots = b_n = 0.$$

However, since (2.19) and (2.21) are equivalent, we obtain by (2.22) that $Av = b$, where

$$b_1 = b_2 = \cdots = b_{k-1} = 0 \quad \text{and} \quad b_k = c_k = \delta_k \leq 0.$$

Since A is positive definite and $v_k = 1$, we know that

$$v^T Av > 0.$$

On the other hand, from the properties of the vectors v and b above, we have

$$v^T Av = v^T b = \sum_{j=1}^{n} v_j b_j = b_k \leq 0.$$

This is the desired contradiction. ■

2.3 Continuous and Discrete Solutions

In the previous section, we saw that a finite difference scheme can produce numerical solutions quite close to the exact solutions of two-point boundary value problems. In this section, we shall go deeper into these matters and show that almost all essential properties of the exact, or continuous, solution are somehow present in the approximate solution. For this purpose, we will need a bit more notation for the discrete solutions; in fact, we find it useful to introduce a rather suggestive notation that can help us in realizing the close relations. When this more convenient notation is introduced, we will see that it is actually quite easy to derive properties such as symmetry and positive definiteness in the discrete case simply by following the steps of the proof for the continuous case. At the end of this section, we will also prove that the finite difference solutions converge towards the continuous solution as the mesh size h tends to zero.

2.3.1 Difference and Differential Equations

Let us start by recalling our standard two-point boundary value problem. We let L denote the differential operator

$$(Lu)(x) = -u''(x),$$

and let $f \in C\big([0,1]\big)$. Then, (2.1) can be written in the following form: Find $u \in C_0^2\big((0,1)\big)$ such that

$$(Lu)(x) = f(x) \quad \text{for all } x \in (0,1). \tag{2.26}$$

Recall here that $u \in C_0^2\big((0,1)\big)$ means that we want the solution to be twice continuously differentiable, and to be zero at the boundaries. Thus, we capture the boundary conditions in the definition of the class where we seek solutions.

Now, let us introduce a similar formalism for the discrete case. First, we let D_h be a collection of discrete functions defined at the grid points x_j for $j = 0, \dots, n+1$. Thus, if $v \in D_h$, it means that $v(x_j)$ is defined for all $j = 0, \dots, n+1$. Sometimes we will write v_j as an abbreviation for $v(x_j)$. This should cause no confusion. Next, we let $D_{h,0}$ be the subset of D_h containing discrete functions that are defined in each grid point, but with the special property that they are zero at the boundary.

Note that a discrete function $y \in D_h$ has $n+2$ degrees of freedom $y_0, y_1, \dots, y_{n+1}$. This means that we have to specify $n+2$ real numbers in order to define such a function. A discrete function $z \in D_{h,0}$ has only n degrees of freedom $z_1, \dots, z_n$, since the boundary values are known.

For a function w we define the operator L_h by

$$(L_h w)(x_j) = -\frac{w(x_{j+1}) - 2w(x_j) + w(x_{j-1})}{h^2},$$

which we recognize as the finite difference approximation of the second derivative. Notice that this definition is valid both for discrete and continuous functions.

Now we can formulate the discrete problem (2.13) as follows: Find a discrete function $v \in D_{h,0}$ such that

$$(L_h v)(x_j) = f(x_j) \quad \text{for all } j = 1, \dots, n. \tag{2.27}$$

In this formulation, we take care of the boundary conditions in the requirement that $v \in D_{h,0}$. This is exactly how we did it in the continuous case.

Some of the properties of the two operators L and L_h that we shall derive are connected to the inner product of functions. These inner products are defined by integration for continuous functions and by summation for discrete functions. For two continuous functions u and v, we define the inner product of the functions by

$$\langle u, v \rangle = \int_0^1 u(x)v(x)\,dx. \tag{2.28}$$

Similarly, for two discrete functions, i.e. for u and v in D_h, we define the inner product to be

$$\langle u, v \rangle_h = h\Big(\frac{u_0 v_0 + u_{n+1} v_{n+1}}{2} + \sum_{j=1}^{n} u_j v_j\Big), \tag{2.29}$$

where we have used the shorthand notation v_j for $v(x_j)$. Clearly, (2.29) is an approximation of (2.28). In the language of numerical integration, this is referred to as the *trapezoidal rule*; you will find more about this in Exercise 2.20.

Having established a suitable notation for the continuous and the discrete problem, we are in position to start deriving some properties.

2.3.2 *Symmetry*

The first property we will show is that both the operators L and L_h are symmetric. For matrices we are used to saying that a matrix $A \in \mathbb{R}^{n,n}$ is symmetric if the transpose of the matrix equals the matrix itself, i.e. if

$$A^T = A.$$

It turns out that this is equivalent to the requirement[7] that

$$(Ax, y) = (x, Ay)$$

[7]Note that $(\cdot,\cdot)$ denotes the usual Euclidean inner product of vectors in $\mathbb{R}^n$; see Exercise 2.21 on page 79 or Project 1.2 on page 31.

for all vectors x and y in $\mathbb{R}^n$. The problem of proving this equivalence is left to the reader in Exercise 2.21.

As we turn our attention to operators not representable by matrices, the latter requirement suggests a generalized notion of symmetry.

Lemma 2.2 *The operator L given in (2.26) is symmetric in the sense that*

$$\langle Lu, v\rangle = \langle u, Lv\rangle \quad \text{for all } u, v \in C_0^2\big((0,1)\big).$$

Proof: The property follows from integration by parts. For $u, v \in C_0^2\big((0,1)\big)$ we have

$$\langle Lu, v\rangle = -\int_0^1 u''(x)v(x)\,dx = -u'(x)v(x)|_0^1 + \int_0^1 u'(x)v'(x)\,dx$$

Since $v(0) = v(1) = 0$, this implies that

$$\langle Lu, v\rangle = \int_0^1 u'(x)v'(x)\,dx. \tag{2.30}$$

However, by performing one more integration by parts, we obtain as above that

$$\int_0^1 u'(x)v'(x)\,dx = -\int_0^1 u(x)v''(x)\,dx = \langle u, Lv\rangle,$$

which is the desired result. ■

Before we derive a similar property for the discrete operator L_h, let us look more closely at the main step of the proof above; no doubt the trick is integration by parts. In order to derive a similar "summation by parts" for discrete functions, we start by reminding ourselves how integration by parts is derived. To this end, let u and v be continuously differentiable functions and recall how we differentiate a product of two functions;

$$(u(x)v(x))' = u'(x)v(x) + u(x)v'(x).$$

Now, by integrating this identity on the unit interval, we get

$$\int_0^1 u'(x)v(x)\,dx = [uv]_0^1 - \int_0^1 u(x)v'(x)\,dx.$$

Then we turn our attention to discrete functions and start by deriving a product rule for differences. Let y and z be two members of D_h, i.e. discrete functions, and observe that

$$y_{j+1}z_{j+1} - y_j z_j = (y_{j+1} - y_j)z_j + (z_{j+1} - z_j)y_{j+1}.$$

By summing this identity from $j = 0$ to $j = n$, we get

$$\sum_{j=0}^{n}(y_{j+1} - y_j)z_j = y_{n+1}z_{n+1} - y_0 z_0 - \sum_{j=0}^{n}(z_{j+1} - z_j)y_{j+1}. \tag{2.31}$$

This identity is referred to as *summation by parts,* and it is exactly the tool we need to prove that L_h is symmetric.

Lemma 2.3 *The operator L_h is symmetric in the sense that*

$$\langle L_h u, v\rangle_h = \langle u, L_h v\rangle_h \quad \text{for all } u, v \in D_{h,0}.$$

Proof: Note that $u_0 = v_0 = u_{n+1} = v_{n+1} = 0$, and define also $u_{-1} = v_{-1} = 0$. Then, using summation by parts twice, we get

$$\begin{aligned}
\langle L_h u, v\rangle_h &= -h^{-1}\sum_{j=0}^{n}((u_{j+1} - u_j) - (u_j - u_{j-1}))v_j \\
&= h^{-1}\sum_{j=0}^{n}(u_{j+1} - u_j)(v_{j+1} - v_j) \\
&= -h^{-1}\sum_{j=0}^{n}((v_{j+1} - v_j) - (v_j - v_{j-1}))u_j \\
&= \langle u, L_h v\rangle_h.
\end{aligned}$$

■

The next property we would like to establish is the fact that the two operators L and L_h are positive definite.

Lemma 2.4 *The operators L and L_h are positive definite in the following sense:*

(i) For any $u \in C_0^2\big((0,1)\big)$ we have

$$\langle Lu, u\rangle \geq 0,$$

with equality only if $u \equiv 0$.

(ii) For any $v \in D_{h,0}$ we have

$$\langle L_h v, v\rangle_h \geq 0,$$

with equality only if $v \equiv 0$.

Proof: Assume that $u \in C_0^2\big((0,1)\big)$. From (2.30) we have that

$$\langle Lu, u\rangle = \int_0^1 (u'(x))^2\, dx,$$

which is clearly nonnegative. Furthermore, if $\langle Lu, u\rangle = 0$, then $u' \equiv 0$. Hence, u is a constant, and since $u(0) = 0$, we have $u \equiv 0$. This establishes the desired property for the operator L.

The result for the operator L_h follows by similar discrete arguments. From the proof of the symmetry property of L_h above, we note that

$$\langle L_h v, v\rangle_h = h^{-1}\sum_{j=0}^{n}(v_{j+1} - v_j)^2 \geq 0,$$

for any $v \in D_{h,0}$. Furthermore, if $\langle L_h v, v\rangle_h = 0$, we have

$$v_{j+1} = v_j \quad \text{for } j = 0, 1, \ldots, n.$$

and then, since $v_0 = 0$, this implies that $v \equiv 0$. ■

2.3.3 Uniqueness

We have already seen that the continuous problem (2.26) and the discrete problem (2.27) have unique solutions. This is stated in Theorem 2.1, page 44, and Corollary 2.1, page 55, respectively. In this section, we shall use the results on positive definiteness derived above to give an alternative proof of these facts.

Lemma 2.5 *The solution u of (2.26) and the solution v of (2.27) are unique solutions of the continuous and the discrete problems, respectively.*

Proof: Let $f \in C([0,1])$ be given and assume that $u^1, u^2 \in C_0^2((0,1))$ are two solutions of (2.26), thus

$$Lu^1 = f \quad \text{and} \quad Lu^2 = f.$$

In order to show that $u^1 \equiv u^2$, we let $e = u^1 - u^2$. Then

$$Le = L(u^1 - u^2) = Lu^1 - Lu^2 = 0.$$

Hence, by multiplying this identity by the error e and integrating over the unit interval, we get

$$\langle Le, e\rangle = 0.$$

By Lemma 2.4 we therefore derive that $e(x) \equiv 0$, and thus $u^1 \equiv u^2$.

A similar argument can be given in the discrete case. ■

2.3.4 A Maximum Principle for the Discrete Problem

Let us recall the representation (2.8)–(2.9) of the solution u of problem (2.26), i.e.

$$u(x) = \int_0^1 G(x, y) f(y)dy,$$

where the Green's function is given by

$$G(x,y) = \begin{cases} y(1-x) & 0 \le y \le x, \\ x(1-y) & x \le y \le 1. \end{cases}$$

In this section we shall derive a similar representation for the solution of the discrete problem (2.27), and then use this to prove a discrete analog of the maximum principle (see Proposition 2.1, page 44).

For a given grid point $x_k = kh$ define a grid function $G^k \in D_{h,0}$ by $G^k(x_j) = G(x_j, x_k)$. Since $G(x,y)$ is linear in x for $x \neq y$, it follows by a straightforward calculation that

$$(L_h G^k)(x_j) = 0 \quad \text{for} \quad j \neq k,$$

while

$$(L_h G^k)(x_k) = -\frac{1}{h^2}\big((x_k - h)(1-x_k) - 2x_k(1-x_k) + x_k(1-x_k-h)\big) = \frac{1}{h}.$$

Hence,

$$L_h G^k = \frac{1}{h} e^k, \tag{2.32}$$

where $e^k \in D_{h,0}$ satisfies

$$e^k(x_j) = \begin{cases} 1 & \text{if } k = j, \\ 0 & \text{otherwise.} \end{cases}$$

For any arbitrary $f \in D_{h,0}$, define $w \in D_{h,0}$ by

$$w = h \sum_{k=1}^{n} f(x_k) G^k.$$

By linearity of the operator L_h, we obtain from (2.32) that

$$L_h w = h \sum_{k=1}^{n} f(x_k)(L_h G^k) = \sum_{k=1}^{n} f(x_k) e^k = f.$$

Hence, w is exactly the unique solution v of problem (2.27). We have therefore established the representation

$$v(x_j) = h \sum_{k=1}^{n} G(x_j, x_k) f(x_k) \tag{2.33}$$

for the solution v of problem (2.27). This representation is the desired discrete analog of (2.9). The following result is a discrete analog of Proposition 2.1 on page 44.

Proposition 2.5 *Assume that* $f(x) \geq 0$ *for all* $x \in [0,1]$, *and let* $v \in D_{h,0}$ *be the solution of (2.27). Then* $v(x_j) \geq 0$ *for all* $j = 1, \ldots, n$.

Proof: Since $G(x, y) \geq 0$ this follows directly from (2.33). ∎

Let us recall from Exercise 2.16 that the solution of problem (2.27) with $f \equiv 1$ is $v(x_j) = \frac{1}{2}x_j(1 - x_j)$. By combining this with (2.33), we obtain

$$h \sum_{k=1}^{n} G(x_j, x_k) = \frac{1}{2} x_j(1 - x_j). \tag{2.34}$$

Before we present the maximum principle, we introduce a norm on discrete functions similar to the sup-norm of continuous function. For any discrete function $v \in D_h$ we define this norm by

$$||v||_{h,\infty} = \max_{j=0,\ldots,n+1} |v(x_j)|. \tag{2.35}$$

In fact, we have met this norm[8] before under the pseudonym E_h; cf. Example 2.5 on page 48.

The following property corresponds to the property stated in Proposition 2.2 for the continuous problem.

Proposition 2.6 *The solution* $v \in D_{h,0}$ *of (2.27) satisfies*

$$||v||_{h,\infty} \leq (1/8)||f||_{h,\infty}.$$

Proof: Since $G(x, y) \geq 0$, it follows from (2.33) and (2.34) that

$$\begin{aligned} |v(x_j)| &\leq h \sum_{k=1}^{n} G(x_j, x_k)|f(x_k)| \\ &\leq ||f||_{h,\infty} \left(h \sum_{k=1}^{n} G(x_j, x_k) \right) \\ &= ||f||_{h,\infty} \frac{x_j(1 - x_j)}{2} \;\leq\; \frac{1}{8}||f||_{h,\infty}. \end{aligned}$$

∎

2.3.5 Convergence of the Discrete Solutions

So far we have established several similarities between the continuous problem (2.26) and the corresponding numerical approximation (2.27). Our final

[8] For continuous functions this is a seminorm. This is so because we can have $||g||_{h,\infty} = 0$ for a continuous function g not identically equal to zero.

goal in this section is to show that the discrete solution v will indeed converge to the continuous solution u when the spacing h approaches zero. This problem was discussed in Example 2.5 on page 48, where we observed that the error of the approximation was of order $O(h^2)$. Another example indicating the same rate is given in Project 2.2. But in this section we shall prove that this property holds for a large class of functions f.

Before we start proving convergence, we want to introduce the concepts of truncation error and consistency. These terms are quite essential in the general analysis of finite difference schemes.

Definition 2.2 *Let $f \in ([0,1])$, and let $u \in C_0^2((0,1))$ be the solution of (2.26). Then we define the discrete vector τ_h, called the* truncation error, *by*

$$\tau_h(x_j) = (L_h u)(x_j) - f(x_j) \quad \text{for all} \quad j = 1, \ldots, n.$$

We say that the finite difference scheme (2.27) is consistent *with the differential equation (2.26) if*

$$\lim_{h \to 0} ||\tau_h||_{h,\infty} = 0.$$

■

You should note here that the truncation error is defined by applying the difference operator L_h to the *exact solution* u. Thus, a scheme is consistent if the exact solution almost solves the discrete problem.

For sufficiently smooth functions f, the scheme (2.27) is consistent.

Lemma 2.6 *Suppose $f \in C^2([0,1])$. Then the truncation error defined above satisfies*

$$||\tau_h||_{h,\infty} \le \frac{||f''||_\infty}{12} h^2.$$

Proof: By using the fact that $-u'' = f$ and $-u'''' = f''$, we derive from the Taylor series expansion (2.11) and the error estimate (2.12) that

$$\begin{aligned} |\tau_h(x_j)| &= |\frac{u(x_{j-1}) - 2u(x_j) + u(x_{j+1})}{h^2} + f(x_j)| \\ &\le |u''(x_j) + f(x_j)| + \frac{||u''''||_\infty}{12} h^2 = \frac{||f''||_\infty}{12} h^2. \end{aligned}$$

■

By using this bound on the truncation error, we can prove that the numerical solution converges towards the exact solution as the grid size tends to zero.

Theorem 2.2 *Assume that $f \in C^2([0,1])$ is given. Let u and v be the corresponding solutions of (2.26) and (2.27), respectively. Then*

$$||u - v||_{h,\infty} \leq \frac{||f''||_\infty}{96} h^2.$$

Proof: Define the discrete error function $e \in D_{h,0}$ by $e(x_j) = u(x_j) - v(x_j)$ for $j = 1, \ldots, n$. Observe that

$$L_h e = L_h u - L_h v = L_h u - f_h = \tau_h,$$

where f_h denotes the discrete function with elements $(f(x_1), \ldots, f(x_n))$. Then it follows from Proposition 2.6 that

$$||e||_{h,\infty} \leq (1/8)||\tau_h||_{h,\infty} \leq \frac{||f''||_\infty}{96} h^2.$$

■

This theorem guarantees that the error measured in each grid point tends to zero as the mesh parameter h tends to zero. Moreover, the rate of convergence is 2. In Exercise 2.23, we study how to define an approximation of the solution for values between the grid points.

2.4 Eigenvalue Problems

In this final section of this chapter we shall study eigenvalue problems associated with the operators L and L_h. The results of this discussion will be used frequently in later chapters.

2.4.1 The Continuous Eigenvalue Problem

A real number[9] λ is said to be an *eigenvalue* associated with the boundary value problem (2.1) if

$$Lu = \lambda u \tag{2.36}$$

for a suitable nonzero[10] function $u \in C_0^2((0,1))$. Here, as above, $Lu = -u''$. The function u is referred to as an *eigenfunction*.

[9] In general, eigenvalues are allowed to be complex. However, due to the symmetry property of L given in Lemma 2.2, all eigenvalues will be real in the present case; cf. Exercise 2.28.

[10] The term "a nonzero function" refers to a function that is not identically equal to zero. Thus it is allowed to vanish at certain points, and even on a subinterval, but not for all $x \in [0,1]$. Sometimes we also use the term "nontrivial" for such functions.

At this point you should notice that this is quite similar to the eigenvalue/eigenvector relations for matrices. Suppose that $A \in \mathbb{R}^{n,n}$ and $v \in \mathbb{R}^n$. Then, if

$$Av = \lambda v$$

for some scalar value λ and nonzero vector v, we refer to λ and v as an eigenvalue/eigenvector pair for the matrix A. We recall that if v is an eigenvector for A, then, for any scalar $c \neq 0$, the vector cv is also an eigenvector with the same eigenvalue. The same property holds for the eigenvalue problem (2.36). If u is an eigenfunction for (2.36) and $c \in \mathbb{R}$, $c \neq 0$, then, by the linearity of L, the function cu is also an eigenfunction with the same eigenvalue. Hence, eigenfunctions are only determined modulo multiplication by a constant.

Before finding the actual eigenvalues and eigenfunctions for the problem (2.36), let us restrict the possible values that λ can attain by using the properties of L derived in Lemma 2.4. Here we proved that the operator L is positive definite, thus

$$\langle Lu, u \rangle > 0,$$

for all nonzero functions $u \in C_0^2\big((0,1)\big)$. Suppose now that λ and u solve (2.36). Then, upon multiplying both sides of the equation by u and integrating, we obtain

$$\langle Lu, u \rangle = \lambda \langle u, u \rangle.$$

Since the operator L is positive definite and the eigenfunction u is nonzero, it follows that

$$\lambda > 0. \tag{2.37}$$

Given the sign of the eigenvalue, we proceed by finding explicit formulas for both the eigenvalues as well as the eigenfunctions.

Since we know that the eigenvalues are positive, we can define

$$\beta = \sqrt{\lambda},$$

and study the equation

$$u''(x) + \beta^2 u(x) = 0,$$

which has general solutions of the form

$$u(x) = c_1 \cos(\beta x) + c_2 \sin(\beta x). \tag{2.38}$$

Here c_1 and c_2 are constants. Using the boundary condition $u(0) = 0$, we get $c_1 = 0$. The other boundary condition, $u(1) = 0$, implies that

$$c_2 \sin(\beta) = 0.$$

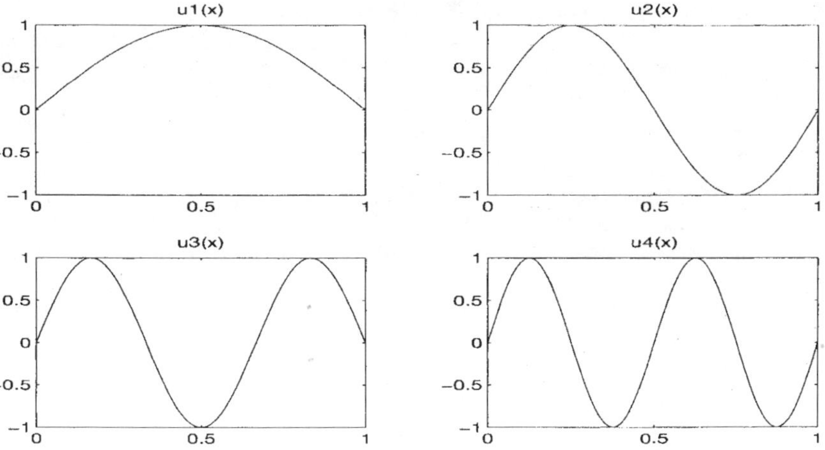

FIGURE 2.4. *The first four eigenfunctions $u_k(x)$.*

Since we are only interested in nontrivial solutions, we have to choose β such that $\sin(\beta) = 0$, hence

$$\beta = \beta_k = k\pi \qquad \text{for} \quad k = 1, 2, \ldots \ . \tag{2.39}$$

We can summarize these results as follows:

Lemma 2.7 *The eigenvalues and eigenfunctions of the problem (2.36) are given by*

$$\lambda_k = (k\pi)^2 \qquad \textit{for} \quad k = 1, 2, \ldots, \tag{2.40}$$

and

$$u_k(x) = \sin(k\pi x) \qquad \textit{for} \quad k = 1, 2, \ldots \ . \tag{2.41}$$

We observe, in particular, that the eigenvalue problem (2.36) has infinitely many eigenvalues. The first four eigenfunctions are plotted in Fig. 2.4.

Let us make a remark concerning (2.39). Why do we only use positive values of k? Of course, $k = 0$ is ruled out by requiring nontrivial solutions; but what about negative values? Note that for any $\alpha \in \mathbb{R}$ we have

$$\sin(-\alpha) = -\sin(\alpha).$$

Hence, negative values of k do not introduce new eigenfunctions. This simply corresponds to multiplying one of the eigenfunctions given above by -1.

A fundamental property of the eigenfunctions $\{u_k\}_{k=1}^{\infty}$ is that these functions are orthogonal with respect to the inner product $\langle \cdot, \cdot \rangle$. This property will be very useful in later chapters, where we use these eigenfunctions to derive analytical solutions of some linear partial differential equations.

Lemma 2.8 *The functions* $\{\sin(k\pi x)\}_{k\geq 1}$ *satisfy*

$$\langle \sin(k\pi x), \sin(m\pi x)\rangle = \begin{cases} 0 & k \neq m, \\ 1/2 & k = m. \end{cases} \tag{2.42}$$

Proof: Recall the following trigonometric identity:

$$\sin(\alpha)\sin(\beta) = \frac{1}{2}(\cos(\alpha-\beta) - \cos(\alpha+\beta)), \tag{2.43}$$

which holds for any real numbers α and β. By using this identity, (2.42) is proved by direct integration. Suppose $k \neq m$; then

$$\begin{aligned}
&\int_0^1 \sin(k\pi x)\sin(m\pi x)\,dx \\
&= \frac{1}{2}\int_0^1 (\cos((k-m)\pi x) - \cos((k+m)\pi x))\,dx \\
&= \frac{1}{2}\left[\frac{1}{(k-m)\pi}\sin((k-m)\pi x) - \frac{1}{(k+m)\pi}\sin((k+m)\pi x)\right]_0^1 \\
&= 0,
\end{aligned}$$

since $\sin(l\pi) = 0$ for any integer l.

For $k = m$, the identity (2.43) gives

$$\int_0^1 \sin^2(k\pi x)\,dx = \int_0^1 \left(\frac{1}{2} - \cos(2k\pi x)\right)dx = \frac{1}{2}\left[1 - \frac{1}{k\pi}\sin(k\pi x)\right]_0^1 = \frac{1}{2}.$$

■

The proof above utilizes special identities for trigonometric functions. However, the orthogonality of the functions $\{u_k\}$ can also be derived directly from the fact that these functions are eigenfunctions of the symmetric operator L on the space $C_0^2\big((0,1)\big)$. From the symmetry of L it follows directly that

$$\lambda_k\langle u_k, u_m\rangle = \langle Lu_k, u_m\rangle = \langle u_k, Lu_m\rangle = \lambda_m\langle u_k, u_m\rangle,$$

or

$$(\lambda_k - \lambda_m)\langle u_k, u_m\rangle = 0.$$

Hence, since $\lambda_k \neq \lambda_m$ when $k \neq m$, we must have $\langle u_k, u_m\rangle = 0$.

2.4.2 The Discrete Eigenvalue Problem

We will now consider the discrete analog of the continuous eigenvalue problem (2.36). A real number μ is said to be an eigenvalue associated with the difference method (2.13) if

$$L_h v = \mu v \tag{2.44}$$

for a suitable nonzero[11] discrete function $v \in D_{h,0}$. We recall that the difference operator L_h was defined in Section 2.3.1.

It follows directly from the definition of L_h that if μ is an eigenvalue of (2.44), then μ is also an eigenvalue of the matrix

$$A = \frac{1}{h^2}\begin{pmatrix} 2 & -1 & 0 & \dots & 0 \\ -1 & 2 & -1 & \ddots & \vdots \\ 0 & \ddots & \ddots & \ddots & 0 \\ \vdots & \ddots & -1 & 2 & -1 \\ 0 & \dots & 0 & -1 & 2 \end{pmatrix}.$$

On the other hand, if $\mu \in \mathbb{R}$ is an eigenvalue of A, then μ is an eigenvalue of (2.44). Furthermore, since A is a symmetric matrix, any eigenvalue of A is real, i.e. there are no complex eigenvalues (see Project 1.2). Therefore, the eigenvalue problem (2.44) corresponds exactly to the eigenvalue problem associated with the matrix A. In particular, this means that there are, at most, n eigenvalues for (2.44).

Since the eigenfunctions of (2.36) are of the form $\sin(\beta x)$, it is reasonable to check whether functions of the form $v(x_j) = \sin(\beta x_j)$ are solutions of the finite difference equation (2.44). From the trigonometric identity

$$\sin(x+y) + \sin(x-y) = 2\cos(y)\sin(x)$$

we obtain

$$(L_h v)(x) = \frac{2}{h^2}[1 - \cos(\beta h)]v(x).$$

Furthermore, from the identity

$$1 - \cos(y) = 2\sin^2(y/2)$$

this can be written

$$(L_h v)(x) = \mu v(x),$$

where $\mu = \frac{4}{h^2}\sin^2(\frac{\beta h}{2})$. Also, note that the function $v(x_j) = \sin(\beta x_j)$ is in $D_{h,0}$ if $\beta = k\pi$, where k is an integer. Therfore, if k is an integer, we conclude that

$$\mu_k = \frac{4}{h^2}\sin^2(\frac{k\pi h}{2})$$

is an eigenvalue for (2.44), with corresponding discrete eigenfunction $v_k \in D_{h,0}$ given by

$$v_k(x_j) = \sin(k\pi x_j), \qquad j = 1, 2, \dots, n.$$

[11] A function $v \in D_{h,0}$ is referred to as nonzero if it is $\neq 0$ in at least one grid point.

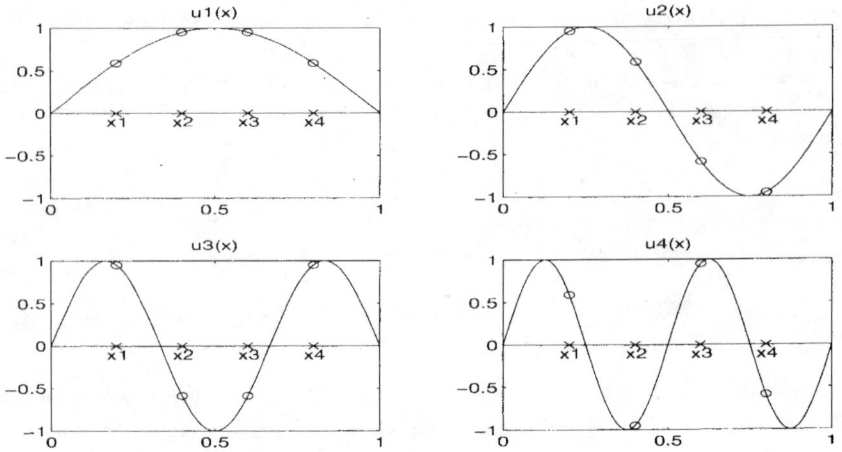

FIGURE 2.5. *The plots show the discrete and continuous eigenfunctions in the case of* $n = 4$. *Note that the discrete eigenfunctions interpolate the corresponding continuous eigenfunction.*

Hence, it seems as if the eigenvalue problem (2.44) has infinitely many eigenvalues. This contradicts our claim above that this problem has at most n eigenvalues. However, from the periodicity of $\sin(x)$ it follows that

$$v_{n+1}(x_j) = 0, \qquad j = 1, 2, \dots, n$$

and therefore μ_{n+1} is not an eigenvalue. In a similar way we also derive that

$$\mu_{n+1+k} = \mu_{n+1-k}, \qquad k = 1, 2, \dots, n,$$

and

$$\mu_{2(n+1)+k} = \mu_k.$$

Therefore, the n eigenvalues of (2.44) are given by

$$0 < \mu_1 < \mu_2 < \cdots < \mu_n < \frac{4}{h^2}.$$

The discrete eigenfunctions, when $n = 4$, are plotted in Fig. 2.5.

We summarize the properties of the eigenvalue problem (2.44):

Lemma 2.9 *The eigenvalues* μ_k *of the problem (2.44) are given by*

$$\mu_k = \frac{4}{h^2} \sin^2(k\pi h/2) \quad \text{for} \quad k = 1, \dots, n. \tag{2.45}$$

The corresponding discrete eigenfunctions $v_k \in D_{h,0}$, $k = 1, \dots, n$, *are given by*

$$v_k(x_j) = \sin(k\pi x_j) \quad \text{for} \quad j = 1, \dots, n.$$

Furthermore, the discrete eigenfunctions are orthogonal and satisfy

$$\langle v_k, v_m \rangle_h = \begin{cases} 0 & k \neq m, \\ 1/2 & k = m. \end{cases} \tag{2.46}$$

Proof: We have already derived the eigenvalues μ_k and eigenvectors v_k. As in the continuous case above, the orthogonality property can be derived directly from the symmetry of the operator L_h. Assume that $k \neq m$ and consider $\langle v_k, v_m \rangle_h$. It follows from Lemma 2.3 that

$$\mu_k \langle v_k, v_m \rangle_h = \langle L_h v_k, v_m \rangle_h = \langle v_k, L_h v_m \rangle_h = \mu_m \langle v_k, v_m \rangle_h,$$

and therefore, since $\mu_k \neq \mu_m$, $\langle v_k, v_m \rangle_h = 0$. An alternative proof of the orthogonality property is given in Exercise 2.30. In the same exercise $\langle v_k, v_k \rangle_h = 1/2$ is also established. ■

Before we complete this chapter we will discuss an important consequence of this final result. The functions $v_1, v_2, \dots, v_n \in D_{h,0}$ are orthogonal, and, more specifically, they are linearly independent. Therefore, since $D_{h,0}$ is a linear space of dimension n, the set $\{v_1, v_2, \dots, v_n\}$ forms a basis for $D_{h,0}$. Hence, any function $g \in D_{h,0}$ can be written in the form

$$g = \sum_{k=1}^{n} c_k v_k,$$

where $c_1, c_2, \dots, c_n$ are real coefficients. Furthermore, by Lemma 2.9,

$$\langle g, v_m \rangle_h = \sum_{k=1}^{n} c_k \langle v_k, v_m \rangle_h = \frac{c_m}{2},$$

which implies that

$$c_m = 2\langle g, v_m \rangle_h.$$

Hence, since $v_k(x_j) = \sin(k\pi x_j)$, we obtain that any function $g \in D_{h,0}$ can be expressed in the form

$$g(x_j) = \sum_{k=1}^{n} 2\langle g, v_k \rangle_h \sin(k\pi x_j), \qquad j = 1, \dots, n. \tag{2.47}$$

This representation of g as a finite sum of sine functions is referred to as a finite Fourier series.

Note that as the number of mesh points n tends to infinity, the distance h between the mesh points will go to zero. Hence, in the limit we can guess that any function f defined on $[0, 1]$ can be written in the form

$$f(x) = \sum_{k=1}^{\infty} c_k u_k = \sum_{k=1}^{\infty} c_k \sin(k\pi x), \tag{2.48}$$

where the coefficients c_k are given by

$$c_k = 2\langle f, u_k\rangle = 2\int_0^1 f(x)u_k(x)dx.$$

Here the eigenfunctions u_k are given by (2.41). As we shall see later, this hypothesis is nearly correct, but, not surprisingly, we need some assumptions on f in order to guarantee a representation of the form (2.48).

An infinite series of the form (2.48) is referred to as a Fourier series or a sine series. Such series will be used frequently in later chapters, and their properties will be studied carefully below.

2.5 Exercises

EXERCISE 2.1 Compute the sup-norm, on the unit interval, for the following functions:

(a) $f(x) = \sqrt{x(1-x)}$,

(b) $f(x) = \sin(100x)$,

(c) $f(x) = x\ln(x)$,

(d) $f(x) = 200890x/(x + 230187)$.

EXERCISE 2.2 Find the solution of the two-point boundary value problem

$$-u''(x) = f(x), \quad x \in (0,1), \quad u(0) = u(1) = 0.$$

where the right-hand side f is given by

(a) $f(x) = x^2$,

(b) $f(x) = e^x$,

(c) $f(x) = \cos(ax)$, where a is a given real number.

EXERCISE 2.3 Consider the boundary value problem

$$-u''(x) = f(x), \quad x \in (a,b), \quad u(a) = u(b) = 0,$$

where $a < b$ are given real numbers. Find the Green's function for this problem; i.e. find a function $G = G(x,y)$ such that the solution of the boundary value problem can be written in the familiar way,

$$u(x) = \int_a^b G(x,y)f(y)\,dy.$$

Use this representation to compute the solution for the following right-hand sides:

(a) $f(x) = 1$

(b) $f(x) = x$

(c) $f(x) = x^2$

EXERCISE 2.4 Construct a solution of the following two-point boundary value problem:

$$-u''(x) = f(x), \quad x \in (0,1), \quad u(0) = \alpha, \quad u(1) = \beta,$$

where α and β are given real numbers.

EXERCISE 2.5 Find a Green's function for the following two-point boundary value problem:

$$-u''(x) = f(x), \quad x \in (0,1), \quad u(0) = 0, \quad u'(1) = 0.$$

Is there a unique solution of this problem?

EXERCISE 2.6 Consider Poisson's equation with Neumann-type boundary values, i.e.

$$-u''(x) = f(x), \quad x \in (0,1), \quad u'(0) = 0, \quad u'(1) = 0.$$

(a) Show that the condition

$$\int_0^1 f(x)\,dx = 0, \tag{2.49}$$

is necessary in order for this problem to have a solution.

(b) Assume that u is a solution and define $v(x) = u(x) + c$, where c is some given constant. Is v a solution of the problem? Is the solution of this problem unique?

(c) Assume that the condition (2.49) is satisfied. Show that the problem then always has a solution. Furthermore, show that the solution is uniquely determined by the extra condition

$$\int_0^1 u(x)\,dx = 0, \tag{2.50}$$

EXERCISE 2.7 Repeat Exercise 2.6 for the following problem involving periodic boundary conditions:

$$-u''(x) = f(x), \quad x \in (0,1), \quad u'(0) = u'(1), \quad u(0) = u(1).$$

EXERCISE 2.8 Consider the boundary value problem

$$-(u''(x) + u(x)) = f(x), \quad x \in (0,1), \quad u(0) = u(1) = 0.$$

Show that the solution of this problem can be written in the form (2.9) where the Green's function is given by

$$G(x,y) = \begin{cases} c\sin(y)\sin(1-x) & \text{if } 0 \le y \le x, \\ c\sin(x)\sin(1-y) & \text{if } x \le y \le 1, \end{cases}$$

where $c = 1/\sin(1)$.

EXERCISE 2.9 The purpose of this exercise is to study the stability of the solution of our standard problem:

$$-u''(x) = f(x), \quad x \in (0,1), \quad u(0) = u(1) = 0.$$

Assume that this equation models some physical phenomena and that the function f is obtained from certain measurements. Hence, inaccuracies can be introduced into our model due to small errors in the measurements. It is therefore important to study how such inaccuracies may effect the solution of the problem. Suppose that $F = F(x)$ is the exact data that we are trying to measure, whereas $f = f(x)$ is the function representing the data that we have actually measured. Let U denote the solution of the problem

$$-U''(x) = F(x), \quad x \in (0,1), \quad U(0) = U(1) = 0,$$

and show that

$$\|U - u\|_\infty \le (1/8)\|F - f\|_\infty.$$

Thus, if the measurements are fairly accurate, so is the solution of our model. This property is referred to as *stability with respect to perturbation in the data;* a concept that is of fundamental importance in the use of mathematical models.

EXERCISE 2.10 Consider the boundary value problem of Example 2.5, and compute, by hand, the numerical approximation described in the example for $n = 1, 2, 3$, and compare your results with the exact solution.

EXERCISE 2.11 Write a procedure based on Algorithm 2.1 that solves the linear system

$$Av = b,$$

where A is given by (2.18). The size of the system, n, and the vectors containing α, β, γ, and b should be input and the solution v should be output. Do not store more vectors than necessary.

In order to debug the procedure, define the vector $z \in \mathbb{R}^n$ by

$$z_j = j, \qquad j = 1, \ldots, n,$$

and the vector $b \in \mathbb{R}^n$ by

$$b = Az,$$

where the matrix $A \in \mathbb{R}^{n,n}$ is given by (2.14). Use your procedure to solve the system

$$Av = b.$$

If v is not very close to z, there is something wrong in your code.

EXERCISE 2.12 Use the procedure implemented in the exercise above to compute the numerical solutions of the boundary value problems of Exercise 2.2. Compare the exact and numerical solutions by computing the error E_h defined in (2.16).

EXERCISE 2.13 In many applications we want to solve a series of linear systems with different right-hand sides, but where the matrix is kept fixed. In such cases the computational effort can be reduced by changing Algorithm 2.1 slightly.

Suppose we want to solve

$$Av_\ell = b_\ell$$

for $\ell = 1, \ldots, N$. Here the matrix $A \in \mathbb{R}^{n,n}$ is given by (2.18) and does not depend on ℓ. The right-hand side $b_\ell \in \mathbb{R}^n$ is given for each value of ℓ.

(a) Modify Algorithm 2.1 by introducing the following three steps:

1. Compute the factors m_k, δ_k for $k = 1, \ldots, n$.
2. Compute c_k for $k = 1, \ldots, n$.
3. Compute v_k for $k = 1, \ldots, n$.

Observe that for the system above, the first step can be done once and for all, whereas steps 2 and 3 must be performed for each ℓ.

(b) Use the modified algorithm to generate numerical solutions of the following problems:

$$-u''(x) = e^{x/\ell}, \qquad u(0) = u(1) = 0,$$

for $\ell = 1, 2, \ldots, 10$. By doing further experiments, try to guess the limit solution as ℓ tends to infinity.

EXERCISE 2.14 Consider the boundary value problem

$$-u''(x) = f(x), \quad u(0) = 0, \quad u'(1) = 1. \tag{2.51}$$

(a) Define two finite difference schemes, S_1 and S_2, approximating the solution of this problem. The differential equation and the left boundary condition can be handled as usual, but the two schemes differ at the approximation of the second boundary condition. In the scheme S_1, we use the approximation

$$\frac{u_{n+1} - u_n}{h} = 1,$$

and in S_2 we introduce an auxiliary unknown u_{n+2}, and approximate the boundary condition by

$$\frac{u_{n+2} - u_n}{2h} = 1.$$

For both these schemes, find the corresponding matrices A_1 and A_2, and the right-hand sides b_1 and b_2, such that the two approximations defined by the schemes S_1 and S_2 can be found by solving the linear systems

$$A_1 v_1 = b_1 \quad \text{and} \quad A_2 v_2 = b_2.$$

(b) Are the matrices A_1 and A_2

1. symmetric and positive definite?
2. diagonal dominant?

(c) Let

$$f(x) = -e^{x-1},$$

and show that

$$u(x) = e^{-1}(e^x - 1)$$

is the exact solution of (2.51). Compare the numerical approximations generated by the schemes S_1 and S_2 for this example by computing the error given by (2.16) for both schemes. What can you say about the rate of convergence[12] of the two approximations?

EXERCISE 2.15 In the Gaussian elimination Algorithm 2.1, the computation of δ_k is the most critical part. Consider the matrix A given by (2.14), and show that for this particular matrix we have

$$\delta_k = \frac{k+1}{k}, \qquad k = 1, 2, \ldots, n.$$

EXERCISE 2.16 The purpose of this exercise is to show that in some particular cases, the approximate solution defined by the finite difference scheme (2.13) gives the exact solution of the boundary value problem evaluated at the grid points.

(a) Consider the boundary value problem

$$-u''(x) = 1, \quad x \in (0,1), \qquad u(0) = u(1) = 0,$$

with the exact solution given by $u(x) = x(1-x)/2$. Show that for this problem, the finite difference solution defined by (2.13) is given by

$$v_j = x_j(1-x_j)/2, \quad j = 1, \ldots, n,$$

which coincides with the exact solution at the grid points x_j. Discuss this result in view of Theorem 2.2.

(b) Assume that the solution of the problem

$$-u''(x) = f(x), \quad x \in (0,1), \qquad u(0) = u(1) = 0, \tag{2.52}$$

is a polynomial of degree less than or equal to three. Show that the finite difference solution then coincides with the exact solution at the grid points.

(c) Describe the class of functions f such that the assumption in (b) is valid.

---

[12]In Project 1.1, we discuss how to estimate the rate of convergence numerically.

EXERCISE 2.17 Consider the problem (2.26) with

$$f(x) = 100e^{-10x}.$$

The exact solution is given by

$$u(x) = 1 - (1 - e^{-10})x - e^{-10x}.$$

We want to use this problem to investigate the sharpness of the error estimate given in Theorem 2.2. Make a table with three columns: one for h, one for the actual error, and one for the error estimate provided by the theorem. Fill in the table for $h = 1/10, 1/20, 1/40, 1/80, 1/160$, and use the result to comment on the sharpness of the estimate.

EXERCISE 2.18 We want to solve the problem (2.1) numerically for a right-hand side satisfying

$$||f''||_\infty \leq 1705,$$

and we want an approximation for which the error measured in absolute values at the grid points is less than 1/80545.

(a) How large do we have to choose n in order to be sure that the approximate solution defined by (2.27) is sufficiently accurate?

(b) It turns out that we are only interested in the solution at $x = 1/10$ and at $x = 9/10$. Is this information of any help? Can we reduce the number of grid points computed above?

EXERCISE 2.19 Consider the differential equation

$$-u''(x) + u(x) = f(x)$$

and the difference approximation

$$-\frac{v_{j-1} - 2v_j + v_{j+1}}{h^2} + v_j = f(x_j).$$

(a) Identify the differential operator L and the difference operator L_h.

(b) Define and compute the truncation error τ_h.

(c) Show that the scheme is consistent provided that the solution u is sufficiently smooth.

Exercise 2.20 Let $u, v \in C_0^2\big((0,1)\big)$. Prove that

$$|\langle u, v\rangle - \langle u, v\rangle_h| \leq \frac{h^2}{12}||(uv)''||_\infty.$$

Exercise 2.21 Show that a matrix $A \in \mathbb{R}^{n,n}$ is symmetric, i.e.

$$A^T = A,$$

if and only if

$$(Ax, y) = (x, Ay)$$

for all vectors x and y in $\mathbb{R}^n$. Here, $(\cdot,\cdot)$ is the Euclidean inner product on R^n, defined by

$$(x, y) = \sum_{j=1}^{n} x_j y_j.$$

Exercise 2.22 Show that the matrix A given by (2.14) on page 47 is positive definite.

Exercise 2.23 In this exercise we shall define an approximation to the solution of the two-point boundary value problem (2.26) for all x in the unit interval. The approximation is based on a piecewise linear extension of the solution v of the discrete problem (2.27).

For $j = 0, 1, \ldots, n$, we let

$$v_\Delta(x) = v_j + \frac{x - x_j}{h}(v_{j+1} - v_j) \quad \text{for} \quad x \in [x_j, x_{j+1}].$$

This approximation simply draws a straight line between the points defined by $\{x_j, v_j\}$. Show that

$$||u - v_\Delta||_\infty = O(h^2),$$

where the norm covers the whole unit interval and not only the grid points.

Exercise 2.24 Consider the eigenvalue problem

$$-u'' + \alpha u = \lambda u, \quad x \in (0,1), \quad u(0) = u(1) = 0,$$

where $\alpha \in \mathbb{R}$ is a constant. Find all eigenvalues and eigenvectors.

EXERCISE 2.25 Consider the eigenvalue problem

$$-u'' = \lambda u, \quad x \in (a,b), \quad u(a) = u(b) = 0,$$

where $a < b$ are given real numbers. Find all eigenvalues and eigenvectors.

EXERCISE 2.26 Find all the eigenvalues of the matrix

$$A = \begin{pmatrix} 2 & -1 \\ -1 & 2 \end{pmatrix}.$$

Use the result to verify Lemma 2.9 when $n = 2$.

EXERCISE 2.27 Let $\mu_1 = \frac{4}{h^2}\sin^2(\pi h/2)$ be the smallest eigenvalue of the problem (2.44). Hence, $\mu_1 = \mu_1(h)$, i.e. μ_1 can be considered as a function of h.

(a) Show that $\lim_{h\to 0}\mu_1(h) = \lambda_1 = \pi^2$.

(b) Show that $4 \le \mu_1(h) \le \pi^2$ for $0 < h \le 1$.

EXERCISE 2.28 The purpose of this exercise is to show that all eigennvalues of the problem (2.36) are real. Assume more generally that $Lu = \lambda u$, where

$$u(x) = v(x) + iw(x) \quad \text{and} \quad \lambda = \alpha + i\beta.$$

Here $i = \sqrt{-1}$, $v, w \in C_0^2((0,1))$ and $\alpha, \beta \in \mathbb{R}$. In addition u should not be the zero function.

(a) Show that

$$Lv = \alpha v - \beta w \quad \text{and} \quad Lw = \beta v + \alpha w.$$

(b) Use the symmetry of the operator L (see Lemma 2.2) to show that

$$\beta(\langle v, v\rangle + \langle w, w\rangle) = 0.$$

(c) Explain why $\beta = 0$ and why the real eigenvalue $\lambda = \alpha$ has a real eigenfunction.

EXERCISE 2.29 In this problem we shall derive some properties for finite Fourier series. Such series occur frequently for example in signal processing. Consider finite Fourier series of the form

$$g(x) = \sum_{k=1}^{n} c_k \sin(k\pi x),$$

where $c_1, c_2, \ldots, c_n$ are real coefficients. Furthermore, let, as usual, x_j denote the grid points $x_j = j/(n+1)$ for $j = 1, \ldots, n$.

(a) Let $z_1, z_2, \ldots, z_n$ be arbitrary real numbers. Show that the interpolation conditions

$$g(x_j) = z_j \quad for \quad j = 1, \ldots, n$$

are satisfied if and only if

$$c_k = 2h \sum_{j=1}^{n} z_j \sin(k\pi x_j) \quad \text{for} \quad k = 1, \ldots, n.$$

(b) Let $T \in \mathbb{R}^{n,n}$ be the matrix with coefficients $t_{j,k} = \sin(k\pi x_j)$. Show that T is symmetric and nonsingular.

(c) Show that $T^2 = \frac{1}{2h} I$, where $I \in \mathbb{R}^{n,n}$ is the identity matrix.

(d) Write a program which computes the coefficients $c_1, c_2, \ldots, c_n$ when the values $z_1, z_2, \ldots, z_n$ are given.

(e) Let

$$f(x) = \begin{cases} x & \text{if } 0 \le x \le 1/2, \\ 1-x & \text{if } 1/2 \le x \le 1, \end{cases}$$

and let $z_j = f(x_j)$. Make plots of the functions $g(x)$ and $f(x)$ for different values of n. Does g approach f as n grows?

EXERCISE 2.30 The purpose of this exercise is to complete the proof of Lemma 2.9 by showing that

$$\langle v_k, v_k \rangle_h = 1/2.$$

In addition we will derive an alternative proof of the orthogonality property. The argument here is a discrete analog of the proof of Lemma 2.8. Recall that the complex exponential function e^{ix}, where $i = \sqrt{-1}$ and x is real, is given by

$$e^{ix} = \cos(x) + i \sin(x).$$

As a consequence of this we obtain

$$\cos(x) = \frac{1}{2}(e^{ix} + e^{-ix}).$$

(a) Use the representation of the cosine function above to show that

$$S_k := \sum_{j=0}^{n} \cos(k\pi x_j) = \frac{1}{2} \sum_{j=0}^{n} (e^{ik\pi x_j} + e^{-ik\pi x_j}),$$

and use the formula for summation of a finite geometric series to show that $S_k = 0$ for k even and $S_k = 1$ if k is odd.

(b) Use the trigonometric formula

$$\sin(\alpha)\sin(\beta) = \frac{1}{2}(\cos(\alpha-\beta) - \cos(\alpha+\beta))$$

to show that

$$\langle v_k, v_m \rangle_h = 0$$

for $k \neq m$.

(c) Show that

$$\langle v_k, v_k \rangle_h = 1/2.$$

2.6 Projects

Project 2.1 *A Numerical Method*

Both in the text and in the exercises above, we have seen the usefulness of the formula (2.9) on page 42, based on Green's function, for the exact solution of two-point boundary value problems. Of course, when the integral involved in (2.9) can be evaluated explicitly, we know everything about the solution of the problem. But, as we know from basic calculus courses, we are not always able to carry out the process of integrating elementary functions. Furthermore, if the right-hand side of our problem, i.e. the function f, is given to us through some kind of measurement, the function is simply not known at every point, and a straightforward application of the solution formula is impossible. In this project we shall use the solution formula to derive a numerical method that can be applied in these cases.

Consider our standard problem

$$-u''(x) = f(x), \quad x \in (0,1), \quad u(0) = u(1) = 0. \tag{2.53}$$

We will find it convenient to write the solution formula (2.9) in the following form:

$$u(x) = x\int_0^1 (1-y)f(y)\,dy - \int_0^x (x-y)f(y)\,dy; \tag{2.54}$$

see (2.7) on page 41. Since we assume that the integrals involved here cannot be computed directly, we will have to provide numerical approximations of the integrals involved.

(a) Find any elementary book in numerical analysis[13] and read about the trapezoidal rule for numerical integration. Explain the derivation of the following approximation:

$$\int_a^b F(x)\,dx \approx h(F(a)/2 + \sum_{i=1}^{n} F(x_i) + F(b)/2). \qquad (2.55)$$

Here, $n > 0$ is a given integer, $h = (b-a)/(n+1)$, and $x_i = a + ih$.

(b) Write a procedure that, for given $a, b, F(x)$, and n, computes the approximation defined by (2.55).

(c) Put $F(x) = x^5$ and $G(x) = \sqrt{|x - \frac{1}{2}|}$. Compute the integrals

$$\int_0^1 F(x)\,dx \qquad \text{and} \qquad \int_0^1 G(x)\,dx$$

analytically and provide numerical approximations by using the trapezoidal rule for some values of n, say $n = 10, 20, 40, 80, 160$. Use the technique derived in Project 1.1 above to estimate the rate of convergence for the approximations of these integrals. Discuss your results in the light of the theory for numerical integration by the trapezoidal rule.

(d) Next we consider how to use this type of numerical integration in order to define an approximation to the solution $u(x)$ of (2.53) given by (2.54). Define the functions

$$\alpha(x) = \int_0^x f(y)\,dy \quad \text{and} \quad \beta(x) = \int_0^x y f(y)\,dy,$$

and show that $u(x)$ is given by

$$u(x) = x(\alpha)(1) - \beta(1)) + \beta(x) - x\alpha(x).$$

(e) We define an approximation of $u(x)$ by integrating α and β numerically. Let $f_i = f(x_i)$ for $i = 0, \ldots, n+1$ where we recall that $x_i = ih = i/(n+1)$ for a given integer $n \geq 1$. Similarly, we define $f_{i+1/2} = f(x_{i+1/2}) = f(x_i + h/2)$. Set $\alpha_0 = \beta_0 = 0$, and define

$$\alpha_{i+1} = \alpha_i + \frac{h}{2}(f_i + 2f_{i+1/2} + f_{i+1}),$$

$$\beta_{i+1} = \beta_i + \frac{h}{2}(x_i f_i + 2(x_{i+1/2} f_{i+1/2} + x_{i+1} f_{i+1}),$$

for $i = 0, \ldots, n$. Explain why $\alpha_i \approx \alpha(x_i)$ and $\beta_i \approx \beta(x_i)$.

---

[13]See, e.g., Conte and de Boor [7], Isaacson and Keller [14], Dahlquist and Bjørk [8], or Burlisch and Stoer [24].

(f) Define

$$u_i = x_i(\alpha_{n+1} - \beta_{n+1}) + \beta_i - x_i\alpha_i$$

for $i = 1, \dots, n$, and put $u_0 = u_{n+1} = 0$. Implement this approximation on a computer and test your procedure by applying it to the exact solutions computed in Example 2.1, Example 2.2, and in Exercise 2.2. Discuss how the accuracy of your approximation varies with the value of n. Do these observations fit the theory of the trapezoidal integration scheme?

(g) In the problems considered in this chapter, the function f has been given explicitly, and the only reason for introducing numerical integration is to be able to solve problems outside the realm of directly integrable functions. But, as we mentioned above, another motivation for introducing numerical integration is to be able to deal with problems where the function f is obtained through measurements of some kind, meaning that f is not known at each point in the interval concerned. In such a case one cannot simply increase the value of n in order to get a more accurate estimate for u, simply because f is not available at more than a fixed number of points. Assuming that the function that is measured is sufficiently smooth, how would you then go about to increase the accuracy of the approximations? We suggest that you once again return to the elementary books of numerical analysis.

Project 2.2 *Error Analysis: A Case Study*

The purpose of this project is to carefully analyze the error of a finite difference approximation for the two-point boundary value problem

$$-u'' + \alpha^2 u = \alpha^2, \qquad x \in (0,1), \qquad u(0) = 1, \qquad u(1) = 0, \tag{2.56}$$

where $\alpha > 0$ is a given real number. The exact solution of this problem is given by

$$u(x) = 1 - \frac{\sinh(\alpha x)}{\sinh \alpha}.$$

The problem is approximated by the following finite difference scheme:

$$-\frac{v_{j+1} - 2v_j + v_{j-1}}{h^2} + \alpha^2 v_j = \alpha^2, \qquad j = 1, \dots, n, \tag{2.57}$$

with the boundary conditions $v_0 = 1$ and $v_{n+1} = 0$.

(a) Find a tridiagonal matrix $A \in \mathbb{R}^{n,n}$ and a vector $b \in \mathbb{R}^n$ such that the linear system (2.57) takes the form $Av = b$.

(b) Prove that the system $Av = b$ can be solved by the Algorithm 2.1.

(c) Write a computer program that solves the system $Av = b$, and use this program to generate plots of the numerical and exact solutions for $\alpha = 1, 5, 100$ using $n = 100$.

(d) Use this program to estimate β such that

$$||u - v||_{h,\infty} = O(h^\beta).$$

A technique for computing such estimates is discussed in Project 1.1.

The rest of this project is devoted to investigating the quality of this estimate. We will do this for $\alpha = 1$.

(e) Define $\theta > 0$ by requiring

$$\cosh(\theta) = 1 + \frac{1}{2}h^2 , \tag{2.58}$$

and show that

$$v_j = 1 - \frac{\sinh(j\theta)}{\sinh((n+1)\theta)}$$

solves (2.57).

(f) Use the Taylor series of $\cosh(\theta)$ to show that

$$\theta < h \quad \text{for } h > 0. \tag{2.59}$$

(g) Define the error

$$e_j = u(x_j) - v_j , \qquad j = 0, 1, \dots , n+1,$$

and show that

$$e_j = \frac{\sinh(j\theta)\sinh((n+1)h) - \sinh(jh)\sinh((n+1)\theta)}{\sinh((n+1)\theta)\sinh((n+1)h)}.$$

(h) Show that

$$\frac{\sinh(j\theta) - \sinh(jh)}{\sinh(1)} \leq e_j \leq \frac{\sinh((n+1)h) - \sinh((n+1)\theta)}{\sinh(1)} .$$

(Hint: $\sinh(x) \leq \sinh(y)$ for $x \leq y$.)

(i) Show that

$$| e_j | \leq \frac{1}{\sinh(1)} \left[\sinh((n+1)h) - \sinh((n+1)\theta)\right]. \tag{2.60}$$

(j) Show that there is a finite constant $\tilde{c}$ that is independent of h such that

$$0 < h - \theta \leq \tilde{c}h^3. \tag{2.61}$$

(k) Show that there is a finite constant c that is independent of h such that

$$||u - v||_{h,\infty} \leq ch^2.$$

(l) Discuss how this result relates to the result of your numerical experiments in (d).

3
The Heat Equation

The historical paths of mathematical physics, mathematical analysis, and methods for solving partial differential equations are strongly interlaced, and it is often difficult to draw boundaries between them. In particular, this is the case in the field of Fourier analysis. This field was initiated by Joseph Fourier (1768–1830), a French physicist who studied heat conduction. In his analysis of this problem, he invented the most influential method for solving partial differential equations to this day. For over 200 years his work has been the foundation of certain areas of mathematical analysis. Any student of engineering or of the natural sciences has to master his techniques.

After two centuries of polishing, Fourier's theory is very elegant and comprehensible. But that has not always been the case. It took brilliant mathematicians years to agree upon the validity of Fourier series. We highly recommend that the interested student read about this story in the book by Davis and Hersh [9].

In this chapter, we will introduce Fourier's method for solving partial differential equations. The method will be applied to the heat equation. We will demonstrate how the solution of such problems can be expressed in terms of an infinite series. To prove that we actually have a solution, we must face the question of convergence of Fourier series. This issue is discussed in Chapter 9.

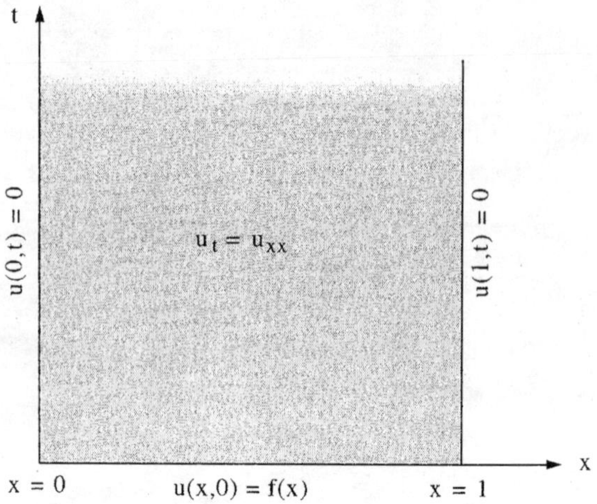

FIGURE 3.1. *The initial-boundary value problem.*

3.1 A Brief Overview

Before we start deriving the details of Fourier's method, let us take a brief look at the basic principles involved. The problem is to find a solution of the following partial differential equation

$$u_t = u_{xx}, \quad \text{for} \quad x \in (0,1) \quad t > 0, \tag{3.1}$$

subject to the boundary conditions

$$u(0,t) = u(1,t) = 0, \qquad t > 0 \tag{3.2}$$

and the initial condition

$$u(x,0) = f(x), \qquad x \in (0,1), \tag{3.3}$$

see Fig. 3.1. Here $f = f(x)$ is a given function.

In order to relate Fourier's method to something familiar, let us consider a linear system of ordinary differential equations of the form

$$v_t = Av, \qquad v(0) = v^0,$$

where $A \in \mathbb{R}^{n,n}$ and $v^0 \in \mathbb{R}^n$ are given, and where the unknown function $v(t) \in \mathbb{R}^n$. It is obvious to anybody who has a background in ordinary differential equations that the key to finding the general solution of this problem is the eigenvalue problem for the matrix A. Fourier's method generalizes this principle to linear partial differential equations. For the problem (3.1)–(3.3) the general solution will be derived from the eigenvalue problem (2.36). The similarity between Fourier's method and the eigenvalue/eigenvector method for linear systems of ordinary differential equations will be clarified in Project 3.1.

The first step in Fourier's method is to find several particular solutions of the problem defined by (3.1) and (3.2). The initial condition (3.3) will be taken into account later. In order to find a family of particular solutions $\{u_k(x,t)\}$, we simply guess that such solutions can be separated into their x and t dependency. Thus, we make the ansatz

$$u_k(x,t) = X_k(x)\,T_k(t), \tag{3.4}$$

from which a set of ordinary differential equations can be derived. Luckily, these ordinary differential equations can be solved explicitly, and hence formulas for the family of particular solutions $\{u_k(x,t)\}$ are available. The method of splitting the x and t dependency as in (3.4) is called *separation of variables*; a very appropriate term which should help you remember the idea of the method for the rest of your life. Summarizing this step, we have

Step 1: Find a family $\{u_k(x,t)\}$ of solutions satisfying the differential equation (3.1) and the boundary condition (3.2).

Next, we appeal to the *principle of superposition*. This principle is far from being as mysterious as it sounds; it simply states that the particular solutions $\{u_k(x,t)\}$ can be added to get new solutions of (3.1) and (3.2). Actually, we can form any linear combination of particular solutions

$$u(x,t) = \sum_k c_k u_k(x,t), \tag{3.5}$$

and the result is still a solution of (3.1) satisfying (3.2). Thus the collection of particular solutions forms a vector space spanned by the basis $\{u_k(x,t)\}$. Summarizing the second step, we have

Step 2: Any linear combination of the form (3.5) is a new solution of (3.1) and (3.2).

The next problem is to determine the coefficients $\{c_k\}$ of (3.5) such that the initial condition (3.3) is satisfied. More precisely, we want to determine the coefficients $\{c_n\}$ such that

$$f(x) = \sum_k c_k u_k(x,0). \tag{3.6}$$

This is exactly what the Fourier series is about; determining coefficients such that a given function is expanded as a series of particular functions. The problem of deriving formulas for $\{c_k\}$ is quite straightforward, although some nasty integrals may be introduced. We summarize this part as

Step 3: Find the coefficients $\{c_k\}$ such that the initial condition (3.3) is satisfied.

Here, one important question arises. When can a function f be expanded in the way indicated in (3.6)? Obviously this is not possible for a completely general family of functions $\{u_k(x,t)\}$. Informally speaking, the family must contain sufficiently many different functions in order to span a wide class of functions. This is referred to as the problem of *completeness*, which will be discussed in Chapters 8 and 9. Here we simply state that the family $\{u_k(x,t)\}$ has an infinite number of members, and that it spans a very large class of functions f

The observation that we need infinite series to expand initial functions as in (3.6), has some serious implications. The arguments outlined above assume finite linear combinations. When dealing with an infinite series, we have to verify that this series converges towards a well-defined function u, and that u is a solution of our problem. These tasks can be summarized as follows:

Step 4:

(a) Verify that the series in (3.5) converges toward a well-defined function $u = u(x,t)$.

(b) Verify that the limit u solves the differential equation (3.1).

(c) Verify that the limit u satisfies the boundary condition (3.2).

(d) Verify that the limit u satisfies the initial condition (3.3).

The rest of this section will be devoted to the steps 1, 2, and 3. Here we will simply leave the questions of convergence open, and just derive formal solutions of our problems. When we refer to a solution as formal, it means that not every step in the derivation of the solution is rigorously justified. Formal solutions are often used in preliminary studies of problems, leaving the justification to a later stage. This is often is a fruitful way of working.

3.2 Separation of Variables

Now we return to step 1 above, and the task is to find particular solutions $\{u_k(x,t)\}$ of the form (3.4) satisfying the differential equation

$$(u_k)_t = (u_k)_{xx} \quad \text{for} \quad x \in (0,1), \quad t > 0, \tag{3.7}$$

subject to the boundary conditions

$$u_k(0,t) = u_k(1,t) = 0. \tag{3.8}$$

By inserting the ansatz

$$u_k(x,t) = X_k(x)\,T_k(t) \tag{3.9}$$

into (3.7), we get

$$X_k(x)T_k'(t) = X_k''(x)T_k(t).$$

Next, we divide both sides of this identity by $X_k(x)T_k(t)$ and obtain

$$\frac{T_k'(t)}{T_k(t)} = \frac{X_k''(x)}{X_k(x)}. \tag{3.10}$$

Here we notice that the left-hand side only depends on t, whereas the right-hand side only depends on x. Hence, both expressions must be equal to a common constant, i.e.,

$$\frac{T_k'(t)}{T_k(t)} = \frac{X_k''(x)}{X_k(x)} = -\lambda_k. \tag{3.11}$$

As will become clear below, the minus sign here is introduced for reasons of convenience. For the time being, we just note that $(-\lambda_k)$ is some constant for each pair of functions X_k and T_k.

From (3.11), we get the following two ordinary differential equations:

$$X_k''(x) + \lambda_k X_k(x) = 0, \tag{3.12}$$

$$T_k'(t) + \lambda_k T_k(t) = 0. \tag{3.13}$$

We first consider (3.12). It follows from the boundary condition (3.8) that we must have

$$X_k(0) = X_k(1) = 0. \tag{3.14}$$

Hence, the functions $X_k(x)$ are eigenfunctions of the problem (2.36), with corresponding eigenvalues λ_k. Therefore, from the discussion in Section 2.4 we can conclude that

$$\lambda_k = (k\pi)^2 \quad \text{for} \quad k = 1, 2, \ldots, \tag{3.15}$$

and

$$X_k(x) = \sin(k\pi x) \quad \text{for} \quad k = 1, 2, \ldots. \tag{3.16}$$

Having solved the second-order problem (3.12), we turn our attention to the first-order problem (3.13), i.e.,

$$T_k'(t) + \lambda_k T_k(t) = 0.$$

This problem has a solution of the form

$$T_k(t) = e^{-\lambda_k t} = e^{-(k\pi)^2 t}, \tag{3.17}$$

where we have chosen the constant to be equal to one. Now it follows by (3.16) and (3.17) that

$$u_k(x,t) = e^{-(k\pi)^2 t}\sin(k\pi x) \quad \text{for} \quad k = 1, 2, \ldots. \tag{3.18}$$

This is the family $\{u_k\}$ of particular solutions we have been looking for.

3.3 The Principle of Superposition

In step 1 we found that the functions $\{u_k(x,t)\}$ given by (3.18) solve the following problems:

$$\begin{aligned} (u_k)_t &= (u_k)_{xx} \quad \text{for} \quad x \in (0,1), \quad t > 0, \\ u_k(0,t) &= u_k(1,t) = 0, \\ u_k(x,0) &= \sin(k\pi x), \end{aligned} \tag{3.19}$$

for $k = 1, 2, \ldots$. Now, we want to use these solutions to solve more general problems of the form

$$\begin{aligned} u_t &= u_{xx} \quad \text{for} \quad x \in (0,1), \quad t > 0, \\ u(0,t) &= u(1,t) = 0, \\ u(x,0) &= f(x). \end{aligned} \tag{3.20}$$

Suppose first that the initial function f can be written as a finite linear combination of the eigenfunctions $\{\sin(k\pi x)\}$. Thus, there exist constants $\{c_k\}_{k=1}^N$ such that

$$f(x) = \sum_{k=1}^{N} c_k \sin(k\pi x). \tag{3.21}$$

Then, by linearity, it follows that the solution of (3.20) is given by

$$u(x,t) = \sum_{k=1}^{N} c_k e^{-(k\pi)^2 t} \sin(k\pi x). \tag{3.22}$$

You can easily check that this is a solution by explicit differentiation.

EXAMPLE 3.1 Let us look at one simple example showing some typical features of a solution of the heat equation. Suppose

$$f(x) = 3\sin(\pi x) + 5\sin(4\pi x);$$

then the solution of (3.20) is given by

$$u(x,t) = 3e^{-\pi^2 t}\sin(\pi x) + 5e^{-16\pi^2 t}\sin(4\pi x).$$

This solution is graphed, as a function of x, in Fig. 3.2 for $t = 0$, 0.01, 0.1. Notice here that the maximum value of the solution is attained at $t = 0$, and that the entire solution becomes smaller as t increases. We easily see, both from the figure and from the formulas, that this solution approaches zero as t tends to infinity.

■

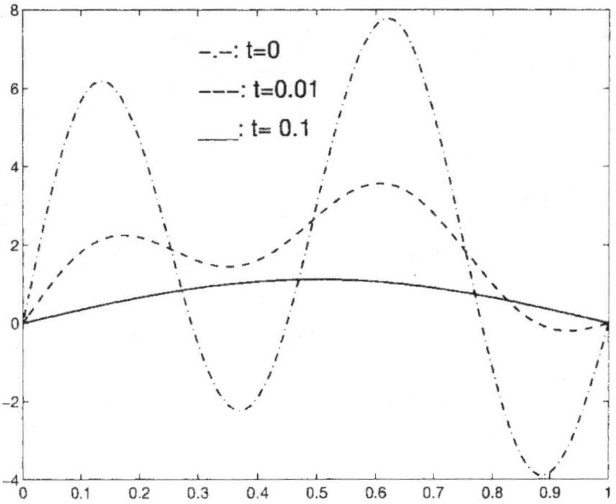

FIGURE 3.2. *The solution of the heat equation with* $f(x) = 3\sin(\pi x) + 5\sin(4\pi x)$ *for* $t = 0,\ 0.01,\ 0.1$.

Now we are able to solve the heat equation for all initial data that can be written in the form (3.21). By varying the coefficients c_k and allowing a large value of N, we can of course cover quite a large class of functions. However, it turns out that this class is not wide enough. Let us look at another example.

EXAMPLE 3.2 Consider a uniform rod of length 1 with initial temperature u of the entire rod equal to 1. Then, at $t = 0$, we start cooling the rod at the endpoints $x = 0$ and $x = 1$. By an appropriate choice of scales, the heat equation (3.20) with $f(x) = 1$ models the temperature distribution in the rod for $t > 0$. In order to find the temperature by following the steps outlined above, we have to represent the function $f(x) = 1$ as a finite sum of sine functions. However, this is impossible and the procedure fails at the simplest possible initial condition! On the other hand, if we allow infinite linear combinations, it can be shown that[1]

$$1 = \frac{4}{\pi} \sum_{k=1}^{\infty} \frac{1}{2k-1} \sin((2k-1)\pi x) \tag{3.23}$$

for x in the unit interval. In Fig. 3.3, we have plotted the Nth partial sum of this series for $N = 3, 10$, and 100. We easily see that the series converge towards $f(x) = 1$ within the unit interval, and we notice that the convergence is very slow near the boundaries.

[1]Here we have to embark on a major detour; the simplest possible function is expressed by an infinite series. It is essential that you understand the reason for this detour.

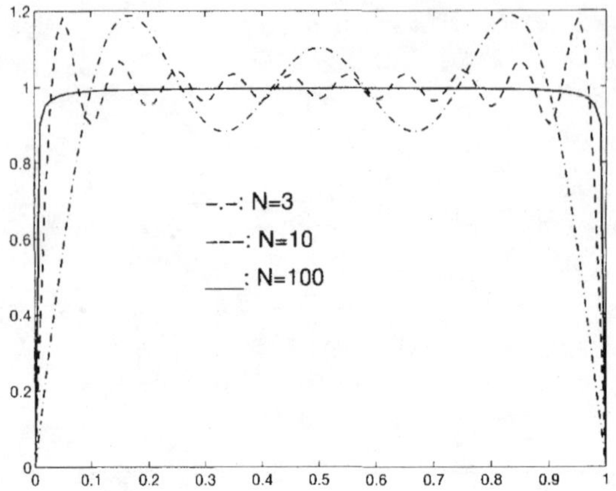

FIGURE 3.3. *The first 3, 10, and 100 terms of the sine-series approximation of* $f(x) = 1$.

When allowing infinite series in the initial data, the solution given by (3.22) also becomes an infinite series. For the present example, we get the following formal solution:

$$u(x,t) = \frac{4}{\pi}\sum_{k=1}^{\infty}\frac{1}{2k-1}e^{-((2k-1)\pi)^2 t}\sin\left((2k-1)\pi x\right). \tag{3.24}$$

Recall here that this solution is referred to as being formal since we have not proved that the series and its derivatives converge and satisfy all the requirements of the heat equation (3.20).

We have plotted the formal solution of this problem as a function of x at $t = 0, 0.01, 0.1$ in Fig. 3.4. Note that the observations concerning the qualitative behavior of the solution stated in Example 3.1 also apply to the present solution.

■

The key observation of the example above is that finite linear combinations of eigenfunctions are not sufficient to cover all interesting initial functions $f(x)$. Thus we are led to allow infinite linear combinations of the form

$$f(x) = \sum_{k=1}^{\infty} c_k \sin(k\pi x). \tag{3.25}$$

By letting N tend to infinity in (3.22), we obtain the corresponding formal solution of the problem (3.20),

$$u(x,t) = \sum_{k=1}^{\infty} c_k e^{-(k\pi)^2 t}\sin(k\pi x). \tag{3.26}$$

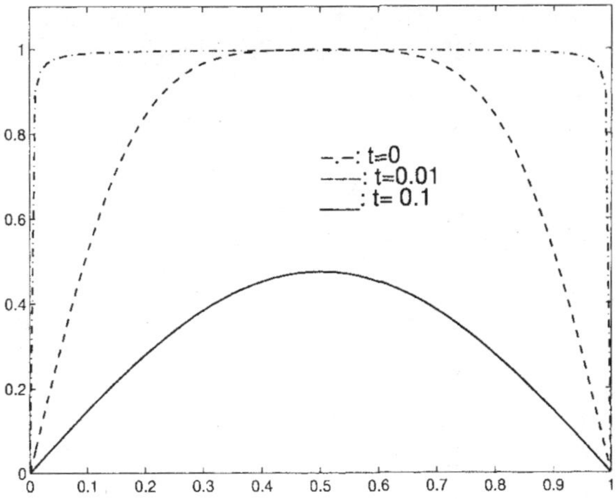

FIGURE 3.4. *The solution of the heat equation with* $f(x) = 1$ *for* $t = 0, 0.01, 0.1$.

It will be established in Chapter 10 that this is not only a formal solution, but a rigorous solution in a strict mathematical sense. In the next section we will show how the coefficients $\{c_k\}$ can be computed from the initial function f, and we will use these values to provide formal solutions for some examples.

3.4 Fourier Coefficients

In this section we show how to compute the coefficients $\{c_k\}$ in (3.25). This approach is identical to what was done for finite Fourier series in Section 2.4. The basic property we will use is that eigenfunctions $\{\sin(k\pi x)\}_{k\geq 1}$ are orthogonal with respect to the inner product $\langle \cdot, \cdot \rangle$ defined by

$$\langle f, g \rangle = \int_0^1 f(x)g(x)dx.$$

More precisely, we have from Lemma 2.8 on page 68 that

$$\langle \sin(k\pi x), \sin(m\pi x) \rangle = \begin{cases} 0 & k \neq m, \\ 1/2 & k = m. \end{cases} \tag{3.27}$$

By using this property of the eigenfunctions, we can easily find formulas for the coefficients $\{c_k\}$ such that

$$f(x) = \sum_{k=1}^{\infty} c_k \sin(k\pi x). \tag{3.28}$$

For any index $m \geq 1$, we take the inner product of this expression with the mth eigenfunction, i.e. with $\sin(m\pi x)$. Then by (3.27) we get

$$\langle f(x), \sin(m\pi x)\rangle = c_k \langle \sin(m\pi x), \sin(m\pi x)\rangle = \frac{c_m}{2}.$$

Hence we have

$$c_k = 2\langle f(x), \sin(k\pi x)\rangle \quad \text{for} \quad k = 1, 2, \ldots . \tag{3.29}$$

These coefficients are referred to as *Fourier coefficients*, and the corresponding series is called a *Fourier series*, or more specifically a *Fourier sine series*. Fourier cosine series will be developed later.

In principle, having these coefficients we are able to express any function in terms of the basis provided by the eigenfunctions $\{\sin(k\pi x)\}$. For most trivial functions this procedure works fine, but more complicated functions may lead to problems related to convergence of the series defined by these coefficients. Also, it might be difficult to find explicit formulas for the integrals involved.

As discussed above, a formal solution of the initial-boundary value problem (3.20) is now given by

$$u(x,t) = \sum_{k=1}^{\infty} c_k e^{-(k\pi)^2 t} \sin(k\pi x), \tag{3.30}$$

where the coefficients $\{c_k\}$ are given by (3.29).

Let us look at some examples of Fourier series and corresponding solutions of the heat equation.

EXAMPLE 3.3 Going back to Example 3.2 above, we want to express the function $f(x) = 1$ in terms of a Fourier sine series. Using (3.29) above, we get

$$c_k = 2\int_0^1 \sin(k\pi x)\,dx = \frac{2}{k\pi}(1 - \cos(k\pi)).$$

Hence

$$c_k = \begin{cases} \frac{4}{k\pi} & \text{for} \quad k = 1, 3, 5 \ldots, \\ 0 & \text{for} \quad k = 2, 4, 6 \ldots, \end{cases}$$

and we have

$$1 = \frac{4}{\pi}\sum_{k=1}^{\infty} \frac{1}{2k-1} \sin((2k-1)\pi x).$$

This verifies the coefficients used in (3.23) above. ■

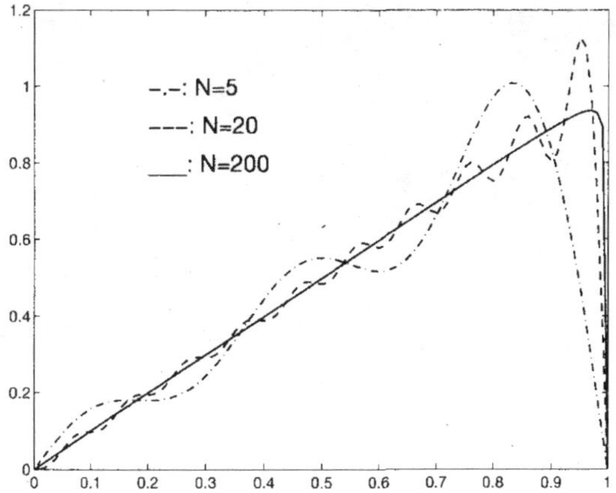

FIGURE 3.5. *The first 5, 20, and 200 terms of the sine-series approximation of* $f(x) = x$.

EXAMPLE 3.4 Next we want to compute the Fourier sine series of $f(x) = x$. Using (3.29), we get

$$c_k = 2\int_0^1 x \sin(k\pi x)\,dx = \left[\frac{2}{(k\pi)^2}\sin(k\pi x) - \frac{2x}{k\pi}\cos(k\pi x)\right]_0^1 = \frac{2}{k\pi}(-1)^{k+1}.$$

Hence the Fourier sine series of $f(x) = x$ on the unit interval is given by

$$x = \frac{2}{\pi}\sum_{k=1}^{\infty}\frac{(-1)^{k+1}}{k}\sin(k\pi x). \tag{3.31}$$

The Nth partial sums for $N = 5, 20$, and 200 are graphed in Fig. 3.5.

Having this expansion, it follows from the discussion above that a formal solution of the heat equation with initial data given by $f(x) = x$ is given by

$$u(x,t) = \frac{2}{\pi}\sum_{k=1}^{\infty}\frac{(-1)^{k+1}}{k}e^{-(k\pi)^2 t}\sin(k\pi x). \tag{3.32}$$

■

3.5 Other Boundary Conditions

So far, we have been concerned with very simple boundary conditions for the heat equation. We have only considered problems where the solution vanishes at the boundary. More generally, if the solution u is given specific values at the boundaries, say

$$u(0,t)=a, \qquad u(1,t)=b, \tag{3.33}$$

with a and b given, we have a *Dirichlet*-type boundary condition. In many applications, other types of boundary conditions appear. If the derivatives rather than the function itself are specified, we have a *Neumann*-type boundary condition. Generally, Neumann conditions can be written in the following form:

$$u_x(0,t)=a, \qquad u_x(1,t)=b, \tag{3.34}$$

where a and b are given.

By combining the Dirichlet- and Neumann-type boundary conditions we get a *Robin*-type boundary condition, which can be written in the form

$$au_x(0,t)+bu(0,t)=c, \qquad \alpha u_x(1,t)+\beta u(1,t)=\gamma, \tag{3.35}$$

for given constants a, b, c, α, β, and γ.

Finally, we have the *periodic* boundary condition

$$u(0,t)=u(1,t), \qquad u_x(0,t)=u_x(1,t). \tag{3.36}$$

We will not give detailed presentations of how to solve the model problems for all these different boundary conditions. Some of them will be addressed in the exercises, and in the next section we will derive a formal solution of a Neumann-type problem along the lines sketched above.

3.6 The Neumann Problem

The purpose of this section is to illustrate the techniques discussed above for another type of boundary conditions. Although the ideas remain more or less the same, it might be useful to see the Fourier method applied to a different type of boundary data. In this way we get a feeling of how the method can be generalized to other more challenging problems.

Our aim is to derive a formal solution of the following problem:

$$\begin{aligned} u_t &= u_{xx} \quad \text{for} \quad x\in(0,1), \quad t>0, \\ u_x(0,t) &= u_x(1,t)=0, \quad t>0, \\ u(x,0) &= f(x), \quad x\in(0,1). \end{aligned} \tag{3.37}$$

We notice that this initial-boundary value problem is identical to the problem (3.20), except for the choice of boundary values.

As for the Dirichlet problem, we start in step 1 by searching for particular solutions of the following problem:

$$(u_k)_t = (u_k)_{xx} \quad \text{for} \quad x \in (0,1), \quad t > 0, \tag{3.38}$$

subject to the boundary conditions

$$(u_k)_x(0,t) = (u_k)_x(1,t) = 0. \tag{3.39}$$

The particular solutions $\{u_k\}$ are found by separation of variables. Inserting the ansatz

$$u_k(x,t) = X_k(x)T_k(t) \tag{3.40}$$

into (3.38), we derive the following ordinary differential equations:

$$X_k''(x) + \lambda_k X_k(x) = 0, \tag{3.41}$$

$$T_k'(t) + \lambda_k T_k(t) = 0. \tag{3.42}$$

3.6.1 The Eigenvalue Problem

We start the analysis of these equations by considering (3.41). This is an eigenvalue/eigenfunction problem; we want to find eigenvalues λ_k and corresponding eigenfunctions X_k such that

$$X_k''(x) + \lambda_k X_k(x) = 0, \qquad X_k'(0) = X_k'(1) = 0. \tag{3.43}$$

Here the boundary conditions stem from (3.39). Before we solve this problem, it is useful to determine what kind of values the eigenvalue λ_k can attain. To this end, we multiply the differential equation by X_k and integrate over the unit interval. Taking the boundary condition into account, this gives

$$\lambda_k = \frac{\langle X_k', X_k' \rangle}{\langle X_k, X_k \rangle}.$$

Hence we have shown that

$$\lambda_k \geq 0. \tag{3.44}$$

Suppose that $\lambda_k = 0$; then we seek a solution to the problem

$$X_k''(x) = 0, \qquad X_k'(0) = X_k'(1) = 0.$$

We easily see that any constant function satisfies these requirements, and we therefore define

$$\lambda_0 = 0 \quad \text{and} \quad X_0(x) = 1. \tag{3.45}$$

Next, we turn to the case of $\lambda_k > 0$, and define

$$\beta_k = \sqrt{\lambda_k}.$$

This leads to the equation

$$X_k''(x) + \beta_k^2 X_k(x) = 0,$$

which has a general solution of the form

$$X_k(x) = c_1 \cos(\beta_k x) + c_2 \sin(\beta_k x). \tag{3.46}$$

Now the boundary conditions $X_k'(0) = 0$ forces $c_2 = 0$, while the other boundary condition implies that

$$c_1 \sin(\beta_k) = 0.$$

Thus we have to choose

$$\beta_k = k\pi \quad \text{for} \quad k = 0, 1, \dots . \tag{3.47}$$

We summarize these results as follows:

Lemma 3.1 *The eigenvalues and eigenfunctions of the problem (3.43) are given by*

$$\lambda_k = (k\pi)^2 \quad \textit{for} \quad k = 0, 1, 2, \dots, \tag{3.48}$$

and

$$X_k(x) = \cos(k\pi x) \quad \textit{for} \quad k = 0, 1, 2, \dots . \tag{3.49}$$

It should be noted here that this result differs from the Dirichlet case in that $k = 0$ is allowed. As we observed above, a zero eigenvalue in the Neumann case gives a nonzero eigenfunction. This is different from the Dirichlet case, where all eigenvalues are strictly positive.[2]

3.6.2 Particular Solutions

Next we solve the first-order problem (3.42). The solution of this problem is

$$T_k(t) = e^{-(k\pi)^2 t} \quad \text{for} \quad k = 0, 1, 2, \dots . \tag{3.50}$$

Using (3.49) and (3.50), it follows that the family of particular solutions is given by

$$u_k(x,t) = e^{-(k\pi)^2 t} \cos(k\pi x) \quad \text{for} \quad k = 0, 1, 2, \dots . \tag{3.51}$$

[2]In terms of operators, we refer to the differential operator in the Dirichlet case as positive definite, whereas the operator in the Neumann case is positive semidefinite.

3.6.3 A Formal Solution

From this collection of particular solutions, we proceed by deriving a formal solution of (3.37). The formal solution is defined through an infinite linear combination of particular solutions[3]

$$u(x,t) = \frac{c_0}{2} + \sum_{k=1}^{\infty} c_k e^{-(k\pi)^2 t} \cos(k\pi x). \tag{3.52}$$

Here the coefficients $\{c_k\}$ have to be determined by the initial data. This means that we have to expand the initial function $f(x)$ in a *Fourier cosine series*, i.e. we have to determine the coefficients $\{c_k\}$ such that

$$f(x) = \frac{c_0}{2} + \sum_{k=1}^{\infty} c_k \cos(k\pi x). \tag{3.53}$$

In order to find these coefficients, we have to apply the following properties of the eigenfunctions:

Lemma 3.2 *The functions* $\{\cos(k\pi x)\}$ *satisfy*

$$\langle \cos(k\pi x), \cos(m\pi x)\rangle = \begin{cases} 0 & k \neq m, \\ 1/2 & k = m \geq 1, \\ 1 & k = m = 0. \end{cases} \tag{3.54}$$

These relations can be verified by direct integration; this task is left to the reader in Exercise 3.14.

Given these orthogonality properties, we can derive the Fourier coefficients by taking the inner products of the eigenfunction on both sides of (3.53). This gives

$$c_k = 2\langle f(x), \cos(k\pi x)\rangle \quad \text{for} \quad k = 0, 1, 2, \ldots . \tag{3.55}$$

Let us look at some examples of solutions to the heat equation with Neumann-type boundary conditions.

Example 3.5 We want to solve (3.37) with the initial data

$$f(x) = 9 + 3\cos(\pi x) + 5\cos(4\pi x).$$

Since this function is written in the form (3.53), the Fourier coefficients are easily found, and the solution of (3.20) is given by

$$u(x,t) = 9 + 3e^{-\pi^2 t}\cos(\pi x) + 5e^{-16\pi^2 t}\cos(4\pi x).$$

This solution is graphed, as a function of x, in Fig. 3.6 for $t = 0, 0.01, 0.1$. You can observe from the figure that the Neumann-type boundary conditions are satisfied.

■

[3] Putting $\frac{1}{2}$ in front of c_0 helps us in deriving a formula for c_k which holds for all $k \geq 0$.

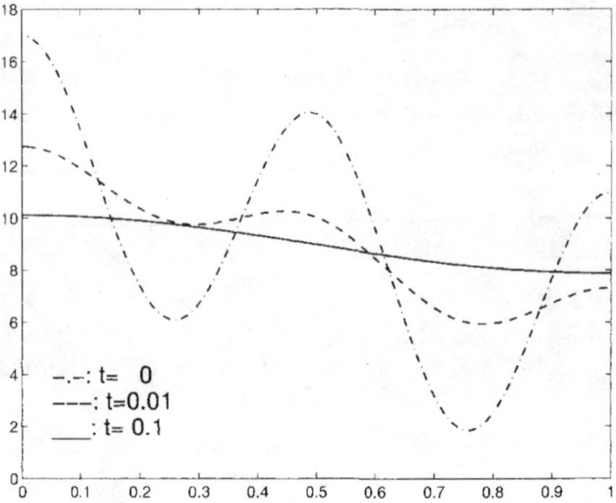

FIGURE 3.6. *The solution of the heat equation for* $t = 0,\ 0.01,\ 0.1$.

EXAMPLE 3.6 Next we want to solve the problem (3.37) with the initial data given by $f(x) = x$.

We start by finding the Fourier cosine series of $f(x) = x$. Observe that

$$c_0 = 2\int_0^1 x\,dx = 1,$$

and then, using integration by parts, we find

$$c_k = 2\int_0^1 x\cos(k\pi x)\,dx = 2\left[\frac{x\sin(k\pi x)}{k\pi} + \frac{\cos(k\pi x)}{(k\pi)^2}\right]_0^1 = 2\frac{(-1)^k - 1}{(k\pi)^2},$$

for $k = 1, 2, \ldots$. Hence, a formal solution of this problem is given by

$$u(x,t) = \frac{1}{2} + \frac{2}{\pi^2}\sum_{k=1}^{\infty}\left(\frac{(-1)^k - 1}{k^2}\right)e^{-(k\pi)^2 t}\cos(k\pi x).$$

The solution, as a function of x, is plotted in Fig. 3.7 for $t = 0,\ 0.05,\ 0.2$. ■

3.7 Energy Arguments

So far in this chapter we have studied a technique, referred to as Fourier's method, which has enabled us to find a formula, or a representation, of the solution u of the initial and boundary value problem (3.1)–(3.3). However, it is often possible to derive certain properties of the solution of a differential equation without knowing the solution in detail. Such techniques are particularly important in the analysis of nonlinear problems, where an

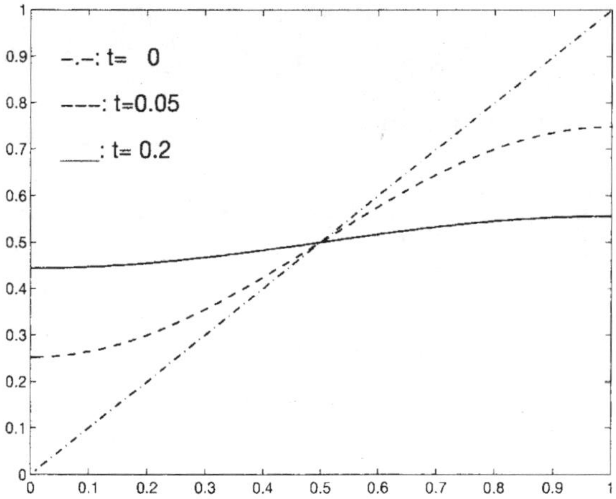

FIGURE 3.7. *The solution of the heat equation with Neumann-type boundary conditions and initial data given by $f(x) = x$.*

analytical representation of the solution is usually impossible to derive. Energy arguments are typical examples of such techniques,[4] and here we will illustrate how such arguments can be applied to the problem (3.1)–(3.3):

Find a function $u = u(x,t)$ which solves the differential equation

$$u_t = u_{xx} \quad \text{for} \quad x \in (0,1), \quad t > 0, \tag{3.56}$$

subject to the boundary conditions

$$u(0,t) = u(1,t) = 0, \qquad t > 0 \tag{3.57}$$

and the initial condition

$$u(x,0) = f(x), \qquad x \in (0,1). \tag{3.58}$$

Throughout this section we assume that $u = u(x,t)$ is a function such that

- $u, u_t, u_{xx} \in C\big([0,1] \times [0,\infty)\big)$,
- u satisfies (3.56)–(3.58).

For each $t \geq 0$ let

$$E(t) = \int_0^1 u^2(x,t)dx.$$

[4] Maximum principles are another set of properties that can be derived without analytical formulas for the solution. This is studied in Chapter 6.

We now consider how $E(t)$, which is a scalar variable, evolves in time. We consider

$$E'(t) \equiv \frac{d}{dt}\int_0^1 u^2(x,t)dx.$$

For smooth functions u we can interchange the order of differentiation and integration such that for $t > 0$

$$E'(t) = \int_0^1 \frac{\partial}{\partial t}u^2(x,t)dx. \tag{3.59}$$

In this case we then derive from equations (3.56)–(3.57) and integration by parts that

$$\begin{aligned} E'(t) &= 2\int_0^1 u(x,t)u_t(x,t)dx \\ &= 2\int_0^1 u(x,t)u_{xx}(x,t)dx \\ &= 2[u(x,t)u_x(x,t)]_0^1 - 2\int_0^1 (u_x(x,t))^2\,dx \\ &= -2\int_0^1 (u_x(x,t))^2\,dx \leq 0. \end{aligned}$$

Hence, $E(t)$ is a nonincreasing function, i.e.

$$E(t) \leq E(0).$$

As pointed out above, the derivation of this inequality requires that we can interchange the order of differentiation and integration such that the identity (3.59) holds. This will in fact follow from Proposition 3.1, given in the next section.

We summarize the result above as follows:

Theorem 3.1 *If u is a solution of (3.56)–(3.58) such that $u, u_t, u_{xx} \in C([0,1] \times [0,\infty))$, then*

$$\int_0^1 u^2(x,t)dx \leq \int_0^1 f^2(x)dx, \qquad t \geq 0. \tag{3.60}$$

An inequality of the form (3.60) is frequently referred to as a *stability estimate*, since it expresses that the size of the solution, measured by the integral $E(t)$ can be bounded by the corresponding size of the initial data f. A consequence of this result is also that small perturbations of the initial function lead to small perturbations of the solution. In order to see this, we assume that there are two solutions $u_1(x,t)$ and $u_2(x,t)$ of (3.56)–(3.58) with initial functions f_1 and f_2. Let $w = u_1 - u_2$. Then

$$w(0,t) = w(1,t) = 0 \quad \text{and} \quad w(x,0) = f_1 - f_2.$$

Furthermore,

$$w_t = (u_1)_t - (u_2)_t = (u_1)_{xx} - (u_2)_{xx} = w_{xx}.$$

Therefore w is a solution of (3.56)–(3.58) with initial function $f_1 - f_2$. From (3.60) we therefore obtain that

$$\int_0^1 (u_1 - u_2)^2(x,t)dx = \int_0^1 w^2(x,t)dx \leq \int_0^1 (f_1 - f_2)^2(x,t)dx. \quad (3.61)$$

Therefore, the size of the difference of the solutions at time t is bounded by the size of the difference of the initial functions.

The estimate (3.61) implies in particular that if $f_1 = f_2$, then $u_1(x,t) = u_2(x,t)$. Hence, for each initial function there is at most one solution of the problem (3.56)–(3.58).

Corollary 3.1 *Two solutions u_1 and u_2 of (3.56)–(3.58), of the form described in Theorem 3.1, satisfy the stability estimate (3.61). In particular, for each initial function f there is at most one solution.*

At the beginning of this section, we claimed that energy arguments can also be used for nonlinear problems since these arguments do not rely on a representation of the solution. In order to illustrate this, consider instead of (3.56)–(3.58) the nonlinear problem

$$u_t = u_{xx} - u^3 \quad \text{for} \quad x \in (0,1), \quad t > 0, \quad (3.62)$$

subject to the boundary conditions

$$u(0,t) = u(1,t) = 0 \quad (3.63)$$

and the initial condition,

$$u(x,0) = f(x). \quad (3.64)$$

Because of the appearance of the nonlinear term u^3, it is not possible to apply Fourier's method[5] to this problem. However, as above let

$$E(t) = \int_0^1 u^2(x,t)dx.$$

[5] Try!

We then obtain

$$
\begin{aligned}
E'(t) &= 2\int_0^1 u(x,t)u_t(x,t)dx \\
&= 2\int_0^1 u(x,t)(u_{xx}(x,t) - u^3(x,t))dx \\
&= -2\int_0^1 (u_x(x,t))^2\,dx - 2\int_0^1 u^4(x,t)dx \le 0.
\end{aligned}
$$

Hence, even if the problem (3.62)-(3.64) is nonlinear, any solution u of the problem satisfies[6]

$$\int_0^1 u^2(x,t)dx \le \int_0^1 f^2(x)dx, \qquad t \ge 0.$$

This energy estimate does not, however, directly imply stability in the way we observed for the linear case.

3.8 Differentiation of Integrals

Above we encountered the problem of computing the derivative of the energy $E(t)$ given by

$$E(t) = \int_0^1 u^2(x,t)dx$$

with respect to the time t. The problem of differentiating an integral with respect to a parameter occurs frequently in the analysis of differential equations. We will therefore take the time to discuss the problem in a general setting. Let

$$F(y) = \int_a^b f(x,y)dx. \tag{3.65}$$

Our aim is to give a proper condition which guarantees that F is differentiable and that

$$F'(y) = \int_a^b f_y(x,y)dx.$$

where $f_y = \frac{\partial}{\partial y} f$.

---

[6] A sharper estimate is derived in Project 11.2 on page 362.

Proposition 3.1 *Let F be given by (3.65) and assume that f and f_y both are continuous on $[a,b] \times [c,d]$. Then $F'(y)$ exists for all $y \in (c,d)$ and*

$$F'(y) = \int_a^b f_y(x,y)dx.$$

Proof: By the definition of $F'(y)$ we must show that

$$\lim_{h \to 0} \left| \frac{F(y+h) - F(y)}{h} - \int_a^b f_y(x,y)dx \right| = 0.$$

Let us first recall that since f_y is continuous on the compact set $[a,b] \times [c,d]$, it follows that f_y is uniformly continuous.[7] In particular, this implies that

$$\begin{aligned} \lim_{h \to 0} \; & \| f_y(\cdot, y+h) - f_y(\cdot, y) \|_\infty \\ & = \lim_{h \to 0} \sup_{x \in [a,b]} |f_y(x, y+h) - f_y(x,y)| \;\; = \;\; 0 \end{aligned} \tag{3.66}$$

for any $y \in (c,d)$. From the mean value theorem it follows that

$$\begin{aligned} \frac{1}{h}\left(F(y+h) - F(y)\right) &= \int_a^b \frac{f(x,y+h) - f(x,y)}{h} dx \\ &= \int_a^b f_y(x, y+\delta)dx, \end{aligned}$$

where δ is between 0 and h and depends on x, y, and h. Hence, we have

$$\begin{aligned} \left| \frac{F(y+h) - F(y)}{h} - \int_a^b f_y(x,y)dx \right| &= \left| \int_a^b \left(f_y(x, y+\delta) - f_y(x,y)\right) dx \right| \\ &\leq (b-a) \sup_{|\delta| \leq |h|} \|f_y(\cdot, y+\delta) - f_y(\cdot, y)\|_\infty. \end{aligned}$$

However, by (3.66) the right-hand side of this inequality tends to zero as h tends to zero. ■

It is straightforward to check that Proposition 3.1 justifies the formula (3.59). We have assumed that $\frac{\partial}{\partial t}u^2 = uu_t$ is continuous on $[0,1] \times [0,T]$ for arbitrary $T > 0$. Therefore, by Proposition 3.1, formula (3.59) holds for all $t \in (0,T)$. Since $T > 0$ is arbitrary, this mean that (3.59) holds for all $t > 0$.

---

[7] Check your calculus book for the definition of uniform continuity and make sure you understand why (3.66) follows.

3.9 Exercises

EXERCISE 3.1 Find the Fourier sine series on the unit interval for the following functions:

(a) $f(x) = 1 + x$,

(b) $f(x) = x^2$,

(c) $f(x) = x(1 - x)$.

EXERCISE 3.2 Find the Fourier cosine series on the unit interval for the functions given in Problem (3.1).

EXERCISE 3.3 Write a computer program that computes the Nth partial sums of Fourier series. Use the program to plot the function $f(x) = x$ and the corresponding Fourier sine and Fourier cosine series on the unit interval. Then use the program to plot the series for $x \in [-3, 3]$.

EXERCISE 3.4 Find the formal solution of the problem

$$\begin{aligned} u_t &= u_{xx} \quad \text{for} \quad x \in (0,1), \quad t > 0 \\ u(0,t) &= u(1,t) = 0 \\ u(x,0) &= f(x), \end{aligned}$$

for the initial functions

(a) $f(x) = \sin(14\pi x)$,

(b) $f(x) = x(1 - x)$,

(c) $f(x) = \sin^3(\pi x)$.

EXERCISE 3.5 Find a formal solution of the problem

$$\begin{aligned} u_t &= u_{xx} \quad \text{for} \quad x \in (0,1), \quad t > 0, \\ u_x(0,t) &= u_x(1,t) = 0, \\ u(x,0) &= f(x), \end{aligned}$$

for the initial functions

(a) $f(x) = \cos(14\pi x)$,

(b) $f(x) = \sin(\pi x)$,

(c) $f(x) = x^3$.

EXERCISE 3.6 Verify, by direct calculation, that the functions $\{u_n(x,t)\}$ given by (3.18) are solutions of (3.7),(3.8).

EXERCISE 3.7 Verify, by a direct calculation, that the functions $\{u_n(x,t)\}$ given by (3.51) are solutions of (3.38),(3.39).

EXERCISE 3.8 Find a family of particular solutions to the following problem:

$$\begin{aligned} u_t &= u_{xx} - u \quad \text{for} \quad x \in (0,1), \quad t > 0, \\ u(0,t) &= u(1,t) = 0. \end{aligned}$$

EXERCISE 3.9 Find a family of particular solutions to the following problem:

$$\begin{aligned} u_t &= u_{xx} + u_x \quad \text{for} \quad x \in (0,1), \quad t > 0, \\ u(0,t) &= u(1,t) \ = 0. \end{aligned}$$

EXERCISE 3.10 Find a formal solution of the following problem:

$$\begin{aligned} u_t &= u_{xx} \quad \text{for} \quad x \in (0,\ell), \quad t > 0, \\ u(0,t) &= u(\ell,t) = 0, \\ u(x,0) &= f(x), \end{aligned} \tag{3.67}$$

where ℓ is a given constant greater than zero.

EXERCISE 3.11 Find a formal solution of the following problem:

$$\begin{aligned} u_t &= u_{xx} \quad \text{for} \quad x \in (0,1), \quad t > 0, \\ u(0,t) &= a, \quad u(1,t) = b, \\ u(x,0) &= f(x), \end{aligned} \tag{3.68}$$

for given constants a and b. Here, you may find it helpful to introduce $v(x,t) = u(x,t) - (a + (b-a)x)$, and use the formal solution derived for the problem (3.20) above.

EXERCISE 3.12 Find a formal solution of the following problem:

$$\begin{aligned} u_t &= u_{xx} + 2x \quad \text{for} \quad x \in (0,1), \quad t > 0, \\ u(0,t) &= 0, \quad u(1,t) = 0, \\ u(x,0) &= f(x). \end{aligned} \tag{3.69}$$

Here, you may find it helpful to introduce $v(x,t) = u(x,t) + w(x)$ for a suitable w which is only a function x.

EXERCISE 3.13 Consider a nonhomogeneous problem of the form

$$u_t = u_{xx} + g \quad \text{for} \quad x \in (0,1), \quad t > 0, \tag{3.70}$$
$$u(0,t) = u(1,t) = 0, \tag{3.71}$$
$$u(x,0) = f(x), \tag{3.72}$$

where $g = g(x,t)$ is a given function. Assume that f and g can be represented by Fourier sine series of the form

$$f(x) = \sum_{k=1}^{\infty} a_k \sin(k\pi x),$$
$$g(x,t) = \sum_{k=1}^{\infty} b_k(t) \sin(k\pi x).$$

(a) Derive a formal solution of the problem (3.70)–(3.72) of the form

$$u(x,t) = \sum_{k=1}^{\infty} T_k(t) \sin(k\pi x).$$

Let $T > 0$ be given. A T-periodic solution of the problem (3.70)–(3.72) is a function $u = u(x,t)$ which satisfies (3.70) and (3.71), and where the initial function f is chosen such that

$$u(x,T) = u(x,0) = f(x).$$

(b) Show that when g is given, there is a unique formal T-periodic solution.

(c) Let $g(x,t) = t\sin(\pi x)$ and $T = 1$. Compute the unique T-periodic solution in this case.

EXERCISE 3.14.

(a) Prove the result stated in Lemma 3.2.

(b) Show that the functions $\{\cos(k\pi x)\}_{k=0}^{\infty}$ and $\{\sin(k\pi x)\}_{k=1}^{\infty}$ satisfy the following orthogonality relations on $[-1,1]$:

$$\int_{-1}^{1} \sin(k\pi x)\sin(m\pi x)dx = \begin{cases} 0 & k \neq m, \\ 1 & k = m. \end{cases}$$

$$\int_{-1}^{1} \cos(k\pi x)\cos(m\pi x)dx = \begin{cases} 0 & k \neq m, \\ 1 & k = m \geq 1, \\ 2 & k = m = 0. \end{cases}$$

$$\int_{-1}^{1} \sin(k\pi x)\cos(m\pi x)dx = 0.$$

EXERCISE 3.15

(a) Consider the eigenvalue problem

$$\begin{aligned} -X''(x) &= \lambda X(x), \qquad x \in (-1,1), \\ X(-1) &= X(1), \qquad X'(-1) = X'(1). \end{aligned}$$

These boundary conditions are referred to as periodic. Show that the eigenfunctions of this problem are given by $\{\cos(k\pi x)\}_{k=0}^{\infty}$ and $\{\sin(k\pi x)\}_{k=1}^{\infty}$. (cf. Exercise 3.14 (b)).

(b) Let f be a function defined on the interval $[-1,1]$ and assume that f can be expanded in series of the form

$$f(x) = \frac{a_0}{2} + \sum_{k=1}^{\infty} \left(a_k \cos(k\pi x) + b_k \sin(k\pi x)\right).$$

Show that

$$a_k = \int_{-1}^{1} f(x)\cos(k\pi x)dx, \qquad k = 0,1,2,\dots,$$

and

$$b_k = \int_{-1}^{1} f(x)\sin(k\pi x)dx, \qquad k = 1,2,\dots.$$

A series of the form above where f is expressed as a sum of sine and cosine functions will be referred to as a full Fourier series.

(c) Find a formal solution of the following problem:

$$\begin{aligned} u_t &= u_{xx} \quad \text{for} \quad x \in (-1,1), \quad t > 0, \\ u(-1,t) &= u(1,t), \quad u_x(-1,t) = u_x(1,t), \\ u(x,0) &= f(x). \end{aligned} \tag{3.73}$$

Note that these boundary data are periodic.

EXERCISE 3.16 Assume that $u(x,t)$ is a solution of the Neumann problem (3.37). Use energy arguments to show that

$$\int_0^1 u^2(x,t)dx \le \int_0^1 f^2(x)dx, \qquad t \ge 0.$$

EXERCISE 3.17 Let $g = g(u)$ be a function u such that $ug(u) \leq 0$ for all u. Use energy arguments to show that any solution of the (possibly nonlinear) problem

$$\begin{aligned} u_t &= u_{xx} + g(u) \quad \text{for} \quad x \in (0,1), \quad t > 0, \\ u(0,t) &= u(1,t) = 0, \\ u(x,0) &= f(x). \end{aligned}$$

satisfies the estimate

$$\int_0^1 u^2(x,t)dx \leq \int_0^1 f^2(x)dx, \qquad t \geq 0.$$

EXERCISE 3.18 Let $a = a(x,t,u)$ be a strictly positive function. Use energy arguments to show that any solution of the (possibly nonlinear) problem

$$\begin{aligned} u_t &= (a(x,t,u)u_x)_x \quad \text{for} \quad x \in (0,1), \quad t > 0, \\ u(0,t) &= u(1,t) = 0, \\ u(x,0) &= f(x), \end{aligned}$$

satisfies the estimate

$$\int_0^1 u^2(x,t)dx \leq \int_0^1 f^2(x)dx, \qquad t \geq 0.$$

EXERCISE 3.19 Consider the nonlinear initial and boundary value problem (3.62)–(3.64). Assume that u_1 and u_2 are two solutions with initial functions f_1 and f_2, respectively. Let $w = u_1 - u_2$.

Show that w solves a linear problem of the form

$$w_t = w_{xx} + aw \tag{3.74}$$

with boundary conditions

$$w(0,t) = w(1,t)$$

and initial condition

$$w(x,0) = f_1(x) - f_2(x),$$

where $a = a(x,t)$ depends on u_1 and u_2.

You should compare the results here with the arguments used to prove Corollary 3.1. In the present case, we observe a typical effect of nonlinear problems. The differential equation (3.74) for the difference w is different from the original equation (3.62).

EXERCISE 3.20 Consider the problem

$$\begin{aligned} u_t &= u_{xx} + u \quad \text{for} \quad x \in (0,1), \qquad t > 0, \\ u(0,t) &= u(1,t) = 0, \\ u(x,0) &= f(x). \end{aligned}$$

Show that

$$\frac{d}{dt}\left[e^{-2t}\int_0^1 u^2(x,t)dx\right] \leq 0$$

and conclude that

$$\int_0^1 u^2(x,t)dx \leq e^{2t}\int_0^1 f^2(x)dx, \qquad t \geq 0.$$

Use this estimate to bound the difference between two solutions in terms of the difference between the initial functions. Does this problem have a unique solution for each initial function f?

3.10 Projects

Project 3.1 *Semidiscrete Approximation.*

The purpose of this project is to illustrate the close relation between the initial-boundary value problem (3.1)–(3.3) and systems of ordinary differential equations of the form

$$v_t = Av, \qquad v(0) = v^0. \tag{3.75}$$

Here the matrix $A \in \mathbb{R}^{n,n}$ and the initial vector $v^0 \in \mathbb{R}^n$ are given, and the solution $v(t)$ is a vector in $\mathbb{R}^n$ for each t.

(a) Let $\mu \in \mathbb{R}$ be an eigenvalue of A, with corresponding eigenvector $w \in \mathbb{R}^n$. Verify that

$$v(t) = e^{\mu t}w$$

satisfies (3.75) with $v^0 = w$.

(b) Assume that the matrix A has n eigenvalues, $\mu_1, \mu_2, \ldots, \mu_n \in \mathbb{R}$, with corresponding linearly independent eigenvectors $w_1, w_2, \ldots, w_n$. Show that a vector-valued function $v(t)$ of the form

$$v(t) = \sum_{k=1}^{n} c_k e^{\mu_k t} w_k, \tag{3.76}$$

where the coefficients $c_1, c_2, \dots, c_n \in \mathbb{R}$, is a solution of (3.75) with initial vector

$$v^0 = \sum_{k=1}^{n} c_k w_k.$$

(c) Assume that the matrix A has n real eigenvalues $\mu_1, \mu_2, \dots, \mu_n$ as above. Explain why all solutions of (3.75) are of the form (3.76).

The solution procedure for linear systems of ordinary differential equations outlined above is often referred to as the *eigenvalue/eigenvector method.* The method can also be extended to the case where some of the eigenvalues μ_k of A are complex. This is simply done by considering solutions of the form (3.76), but where also the coefficients c_k and eigenvectors w_k are allowed to be complex. The discussion below will illustrate that Fourier's method is a generalization of the eigenvalue/eigenvector method to linear partial differential equations.

Recall that the problem (3.1)–(3.3) can be written in the form

$$\begin{aligned} u_t(x,t) &= -(Lu)(x,t) \quad \text{for} \quad x \in (0,1), \quad t > 0, \\ u(x,0) &= f(x), \end{aligned}$$

where for each $t > 0$, $u(\cdot,t) \in C_0^2$. Here, as in Chapter 2, $Lu = -u_{xx}$. From our discussion in Chapter 2 it seems reasonable to approximate this problem with the semidiscrete system

$$\begin{aligned} v_t(x_j,t) &= -(L_h v)(x_j,t) \qquad \text{for} \quad j = 1,2,\dots,n, \\ v(x_j,0) &= f(x_j) \qquad \text{for} \quad j = 1,2,\dots,n, \end{aligned} \tag{3.77}$$

where we assume that for each $t \ge 0$, $v(\cdot,t) \in D_{h,0}$. Here $h = 1/(n+1)$ and $x_j = jh$. We refer to Section 2.3.1 for the definition of the difference operator L_h and the discrete space $D_{h,0}$. The system (3.77) is referred to as a *semidiscrete system* since it is discrete with respect to x, but continuous with respect to t.

(d) Assume first that $n = 2$. Show that (3.77) is equivalent to a system of the form (3.75), where the matrix A is given by

$$A = -9 \begin{pmatrix} 2 & -1 \\ -1 & 2 \end{pmatrix}.$$

(e) Assume that $f(x) = \sin(\pi x) - 3\sin(2\pi x)$. Find the solution of (3.77) when $n = 2$.

(f) Explain why the problem (3.77) in general is equivalent to a system of the form (3.75). In particular, identify the matrix A in (3.75).

(g) Explain why any solution of (3.77) can be written in the form

$$v(x_j,t)=\sum_{k=1}^{n} c_k e^{-\mu_k t}\sin(k\pi x_j) \quad for \quad j=1,2,\ldots,n, \tag{3.78}$$

where $\mu_k=\frac{4}{h^2}\sin^2(k\pi h/2)$.

The formula (3.78) can be used to define the semidiscrete solution $v(x,t)$ for all $x\in[0,1]$.

(h) Consider the initial function

$$f(x)=3\sin(\pi x)+5\sin(4\pi x)$$

used in Example 3.1 on page 92. Find the semidiscrete solution $v(x,t)$ when $n=2$ and $n\geq 4$. Compare the semidiscrete solution $v(x,t)$ and the analytical solution $u(x,t)$ by plotting $v(x,0.01)$, for $n=2,4,6$, and $u(x,0.01)$.

(i) If v is a solution of (3.77), define the corresponding discrete energy by

$$E_h(t)=\langle v(\cdot,t),v(\cdot,t)\rangle_h \quad for \quad t\geq 0.$$

Here, the discrete inner product $\langle\cdot,\cdot\rangle_h$ is defined in Section 2.3.1. Use energy arguments, together with the result of Lemma 2.4, to show that

$$E_h(t)\leq E_h(0) \quad for \quad t\geq 0.$$

In order to obtain a fully discrete finite difference method, where no differential equation has to be solved, the semidiscrete system (3.77) must be discretized with respect to time. The simplest time discretization is Euler's method. If we apply this to the system (3.77) we obtain, for $m\geq 0$ and $j=1,2,\ldots,n$,

$$\begin{aligned}\frac{v(x_j,t_{m+1})-v(x_j,t_m)}{\Delta t}&=-(L_h v)(x_j,t_m)\\ v(x_j,0)&=f(x_j)\end{aligned} \tag{3.79}$$

where $t_m=m\Delta t$. Here $\Delta t>0$ is the time step.

(j) Let the vectors $v^m\in\mathbb{R}^n$ be given by

$$v_j^m=v(x_j,t_m) \quad \text{for} \quad j=1,2,\ldots,n.$$

Show that

$$v^{m+1}=(I+\Delta tA)v^m,$$

where $I\in\mathbb{R}^{n,n}$ is the identity matrix and $A\in\mathbb{R}^{n,n}$ is as above.

(k) Show that any solution of the finite difference method (3.79) can be written in the form

$$v(x_j, t_m) = \sum_{k=1}^{n} c_k (1 - \Delta t \mu_k)^m \sin(k\pi x_j),$$

where the coefficients $c_1, c_2, \ldots, c_n \in \mathbb{R}$.

4

Finite Difference Schemes for the Heat Equation

In the previous chapter we derived a very powerful analytical method for solving partial differential equations. By using straightforward techniques, we were able to find an explicit formula for the solution of many partial differential equations of parabolic type. By studying these analytical solutions, we can learn a lot about the qualitative behavior of such models. This qualitative insight will also be useful in understanding more complicated equations.

In this section we will turn our attention to numerical methods for solving parabolic equations. Having spent quite some time on deriving elegant analytical formulas for solving such equations, you may wonder why we need numerical methods. There are several reasons. We can summarize the main difficulties of the Fourier method as follows:

Nonlinear problems. The Fourier method derived above cannot handle nonlinear equations. Both separation of variables and the principle of superposition will in general fail for such equations. This is quite an important drawback, since many applications give rise to nonlinear problems. There is a strong tradition for linearizing such equations in order to apply linear techniques like the Fourier method discussed above. But this approximation is in some cases very crude. Nonlinear equations can be handled adequately using finite difference methods. The basic ideas are exactly the same as for linear problems, but difficulties may arise in, for example, solving the nonlinear algebraic equations involved.

Variable coefficients. Even linear problems with variable coefficients may be hard to solve using the Fourier method. In particular this is the case for discontinuous coefficients. Variable coefficients are fairly easy to handle with finite difference schemes.

Integrals. As mentioned above, some of the integrals involved in computing the Fourier coefficients can be difficult or even impossible to evaluate analytically. In such cases, numerical integration is needed.

Infinite series. In order to actually graph the Fourier solution of a problem, we have to compute the sum of the series. If the series is infinite, we have to rely on an approximation based on a truncation of the series. Furthermore, except for some trivial cases, the partial sum has to be computed numerically, i.e. on a computer.

We conclude that there are a lot of interesting problems that cannot be solved by the Fourier method. And for most problems that can be solved, we are dependent on some sort of numerical procedure in order to actually graph the solution. These observations clearly motivate the study of numerical methods in a more general setting. However, you should not be misled into believing that numerical methods solve every problem. There are a lot of dangers in using these methods. The most important difficulties will be pointed out here.

Although there are a lot of different numerical methods available for solving parabolic equations, we focus on finite difference schemes. The reason for this is that they are very simple to understand and easy to generalize for quite complicated problems. Furthermore, they are very easy to implement on a computer.

In numerical analysis, much effort is spent on the search for efficient schemes in the sense of optimizing accuracy and minimizing CPU time and memory requirements.[1] No claim is made here that finite difference schemes are optimal in this respect. Often, the most powerful techniques for solving partial differential equations use as much analytical information as possible. The so-called spectral methods are illustrative examples of this phenomenon. These methods are carefully constructed in order to take advantage of all the analytical insight we have gained through the development of Fourier techniques. You may consult the notes by Gottlieb and Orszag [12] to read more about this topic.

The most popular method in applied areas is probably the finite element method. Although in many respects it is quite similar to the finite

[1]For scalar computers, i.e., for computers where you only have access to one single processor, it is fairly easy to rank different schemes with respect to these criteria. Just compute, either a priori or run-time, the number of arithmetic operations needed for the different schemes. In the era of parallel computing this issue is more complicated. On such machines, the quality of a certain method has to be related to how well it exploits the access to numerous processors.

difference method, the finite element method is preferable when it comes to complicated geometries in several space dimensions. You can find a friendly introduction to this topic in the book by C. Johnson [15]. A mathematically more advanced approach is given by Brenner and Scott [4], and engineers seem to prefer the book by Zienkiewicz [31]. Further references can be found in these books.

4.1 An Explicit Scheme

In this section we will derive a finite difference approximation of the following initial-boundary value problem:

$$\begin{aligned} u_t &= u_{xx} \quad \text{for} \quad x \in (0,1), \quad t > 0, \\ u(0,t) &= u(1,t) = 0, \\ u(x,0) &= f(x). \end{aligned} \tag{4.1}$$

The way we derive the finite difference scheme for (4.1) is very similar to the way we derived a scheme for the two-point boundary value problem in Section 2.2. The basic idea is again to replace the derivatives involved in (4.1) by finite differences. But for this problem we have to approximate both the space and the time derivatives.

Let $n \geq 1$ be a given integer, and define the grid spacing in the x-direction by $\Delta x = 1/(n+1)$. The grid points in the x-direction are given by $x_j = j\Delta x$ for $j = 0, 1, \ldots, n+1$. Similarly, we define $t_m = m\Delta t$ for integers $m \geq 0$, where Δt denotes the time step. Finally, we let v_j^m denote an approximation of $u(x_j, t_m)$.

Before we define the scheme, let us recall that we have the following approximations[2]

$$u_t(x,t) = \frac{u(x,t+\Delta t) - u(x,t)}{\Delta t} + O(\Delta t),$$

and

$$u_{xx}(x,t) = \frac{u(x-\Delta x,t) - 2u(x,t) + u(x+\Delta x,t)}{\Delta x^2} + O(\Delta x^2).$$

These approximations motivate the following scheme:

$$\frac{v_j^{m+1} - v_j^m}{\Delta t} = \frac{v_{j-1}^m - 2v_j^m + v_{j+1}^m}{\Delta x^2} \quad \text{for } j = 1, \ldots, n, \quad m \geq 0.$$

By using the boundary conditions of (4.1), we put

$$v_0^m = 0 \quad \text{and} \quad v_{n+1}^m = 0$$

[2] See Project 1.1, page 28.

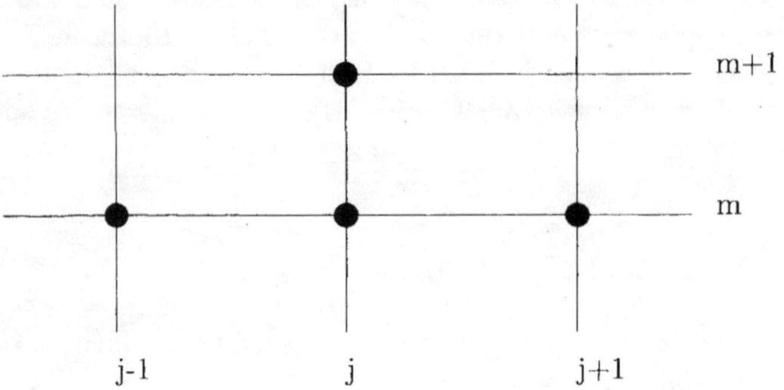

FIGURE 4.1. *The computational molecule of the explicit scheme.*

for all $m \geq 0$. The scheme is initialized by

$$v_j^0 = f(x_j) \quad \text{for} \quad j = 1, \ldots, n.$$

Let $r = \Delta t/\Delta x^2$; then the scheme can be rewritten in a more convenient form

$$v_j^{m+1} = rv_{j-1}^m + (1-2r)v_j^m + rv_{j+1}^m, \qquad j = 1, \ldots, n, \quad m \geq 0. \tag{4.2}$$

When the scheme is written in this form, we observe that the values on the time level t_{m+1} are computed using only the values on the previous time level t_m. Therefore we refer to this scheme as *explicit*. This is in contrast to *implicit* schemes where we have to solve a tridiagonal system of linear equations in order to pass from one time level to the next. Such schemes will be discussed below.

In Fig. 4.1, we have sketched the basic structure of this scheme. We often refer to such illustrations as the *computational molecule*[3] of the scheme.

Before we start analyzing the properties of the scheme, we present some examples illustrating how this scheme works.

EXAMPLE 4.1 In the first example we look at how well this scheme approximates one of the particular solutions derived in Section 3.2 above. Thus, we let

$$f(x) = \sin(2\pi x),$$

and recall from (3.18) that the exact solution is given by

$$u(x,t) = e^{-4\pi^2 t} \sin(2\pi x).$$

[3]Sometimes the term "stencil" is used for such illustrations.

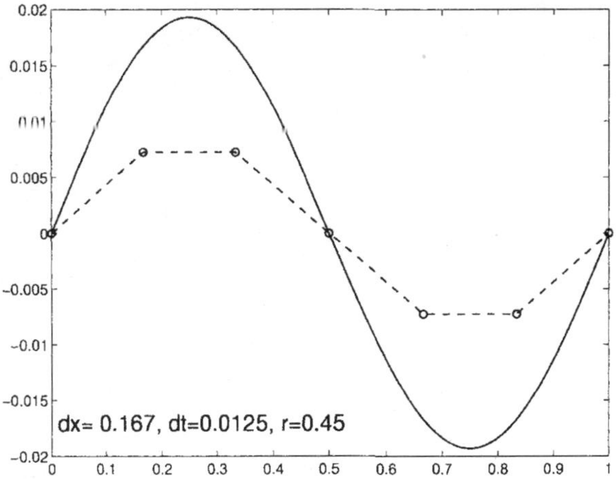

FIGURE 4.2. *The finite difference solution (dashed line) and the Fourier-based solutions (solid line) of the heat equation. For the finite difference scheme we used a relatively coarse mesh;* $\Delta x = 1/6$ *and* $\Delta t = 1/80$.

We choose $\Delta x = 1/6$ and $\Delta t = 1/80$ and compute the numerical solution for $0 \leq t_m \leq 1/10$. The numerical and analytical solution at $t = 1/10$ is plotted in Fig. 4.2. As usual we have used piecewise linear interpolation between the grid points.

In Fig. 4.3 we have used a finer mesh, $\Delta x = 1/20$ and $\Delta t = 1/800$. Note that the approximation is much better in this case.

■

EXAMPLE 4.2 In our next example we consider the following initial data for the problem (4.1):

$$f(x) = \begin{cases} 2x & x \leq 1/2, \\ 2(1-x) & x \geq 1/2. \end{cases}$$

A formal solution of this problem can be derived using Fourier's method. The solution is

$$u(x,t) = \frac{8}{\pi^2} \sum_{k=1}^{\infty} \left(\frac{\sin(k\pi/2)}{k^2} \right) e^{-(k\pi)^2 t} \sin(k\pi x). \tag{4.3}$$

In Fig. 4.4 we have plotted the Fourier solution given by (4.3) as a function of x for $t = 0.1$. The series is truncated after 200 terms. We also plot the numerical solution generated by the scheme (4.2) using $\Delta x = 1/50$ and $\Delta t = 1/5000$, hence $r = 1/2$. In Fig. 4.5, we have plotted the Fourier solution and the numerical solution again, but we have increased the time step slightly; $\Delta t = 0.000201$. This gives $r = 0.5025$ and we observe from the plot that something is wrong; the numerical solution oscillates, whereas the

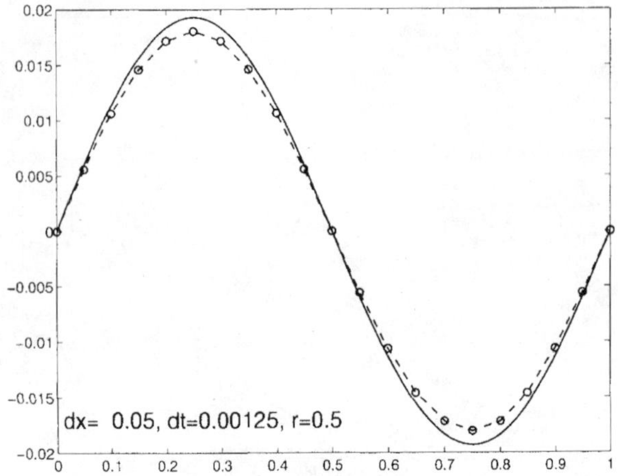

FIGURE 4.3. *The finite difference solution (dashed line) and the Fourier-based solutions (solid line) of the heat equation. For the finite difference scheme we used the mesh parameters* $\Delta x = 1/20$ *and* $\Delta t = 1/800$.

analytical solution is smooth and very well behaved. This behavior of the scheme is referred to as an *instability problem*. Much effort will be invested below in deriving precise conditions to avoid such problems.

■

4.2 Fourier Analysis of the Numerical Solution

In our experiments above, we observed two interesting features of the numerical solutions. First we noticed that the scheme may generate very good approximations of the analytical solutions. And secondly, we saw that by changing the grid parameters slightly, severe oscillations appeared in the numerical solution. It is quite clear that such oscillations are unacceptable. The analytical solution is smooth and well behaved, and we look for an approximation which shares these properties.

Our aim in this section is to understand these observations. We want to know when oscillations can appear and we want to know how to prevent such behavior. Furthermore, we want to gain some insight into why the scheme generates very good approximations for proper grid sizes. It is not the purpose of this section to derive any rigorous error analysis for the discrete solutions. The problem of determining the convergence rate of the scheme will be addressed later. Here we will use our insight in the analytical solution to try to understand in a qualitative way why the scheme works.

Our analysis of the numerical method will, step by step, follow the procedure outlined in Section 3.1. By using a discrete version of the Fourier method, we will derive an explicit formula for the discrete solution. This

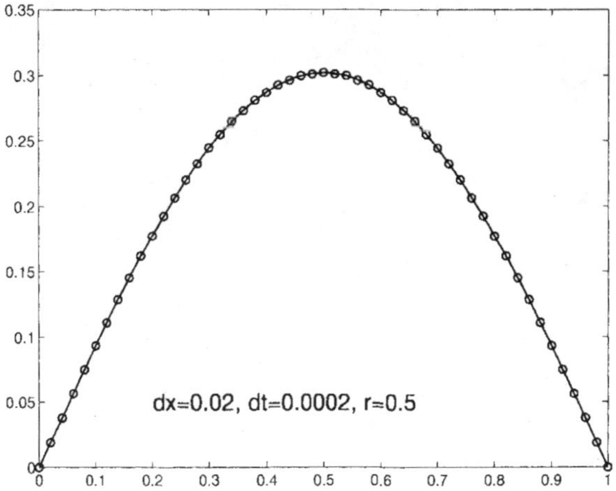

FIGURE 4.4. *The numerical (dashed line) and Fourier-based solution (solid line) of the heat equation. For the numerical method we have used* $r = 1/2$.

formula enables us to compare the analytical and the numerical solutions term by term.

In the next section, we will present von Neumann's stability analysis. This is a versatile tool which is commonly used to investigate the stability of numerical methods for solving time-dependent partial differential equations. The basis of von Neumann's method is the term-by-term analysis mentioned above. Both analytical and numerical methods can be decomposed into linear combinations of particular solutions. Thus, in order to compare the solutions, it is sufficient to compare the particular solutions. In this section we will do such a comparison thoroughly in order to prepare for the next section. So if you feel that the present section is a bit lengthy and full of details, we promise you that the von Neumann technique we arrive at in the next section is very simple and powerful.

4.2.1 Particular Solutions

The first step in our discrete Fourier analysis is to derive a family of particular solutions of the following problem:

$$\frac{v_j^{m+1} - v_j^m}{\Delta t} = \frac{v_{j-1}^m - 2v_j^m + v_{j+1}^m}{\Delta x^2} \quad \text{for} \quad j = 1, \ldots, n, \quad m \geq 0, \tag{4.4}$$

with the boundary conditions

$$v_0^m = 0 \text{ and } v_{n+1}^m = 0, \quad m \geq 0. \tag{4.5}$$

The initial data will be taken into account later.

We are looking for particular solutions of the problem (4.4) and (4.5). In the continuous case, we derived the particular solutions by guessing that

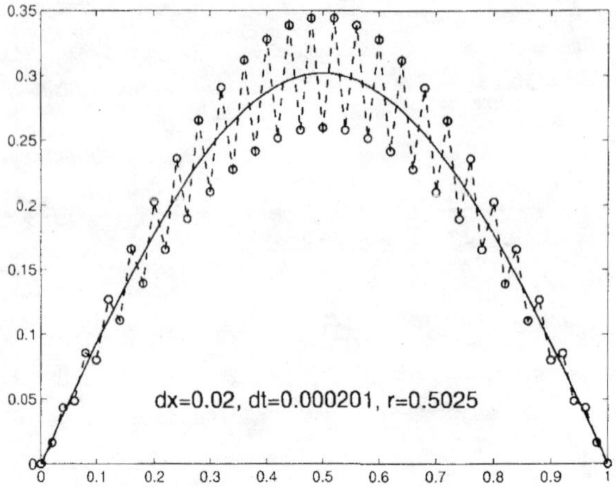

FIGURE 4.5. *The numerical (dashed line) and Fourier-based solution (solid line) of the heat equation. For the numerical method we have used* $r = 0.5025$. *Note that the numerical solution for this value of* r *oscillates.*

the space and time dependency could be separated. Thus, we inserted an ansatz of the form

$$u(x,t) = X(x)T(t)$$

into the equation. Exactly the same procedure will be applied in the discrete case. Hence we seek particular solutions of the form

$$w_j^m = X_j T_m \quad \text{for} \quad j = 1, \ldots, n, \quad m \geq 0. \tag{4.6}$$

Here X is a vector of n components, independent of m, while $\{T_m\}_{m\geq 0}$ is a sequence of real numbers. By inserting (4.6) into (4.4), we get

$$\frac{X_j T_{m+1} - X_j T_m}{\Delta t} = \frac{X_{j-1}T_m - 2X_j T_m + X_{j+1}T_m}{(\Delta x)^2}.$$

Since we are looking only for nonzero solutions, we assume that $X_j T_m \neq 0$, and thus we obtain

$$\frac{T_{m+1} - T_m}{\Delta t T_m} = \frac{X_{j-1} - 2X_j + X_{j+1}}{(\Delta x)^2 X_j}.$$

The left-hand side only depends on m and the right-hand side only depends on j. Consequently, both expressions must be equal to a common constant, say $(-\mu)$, and we get the following two difference equations:

$$\frac{X_{j-1} - 2X_j + X_{j+1}}{(\Delta x)^2} = -\mu X_j \quad \text{for} \quad j = 1, \ldots, n, \tag{4.7}$$

$$\frac{T_{m+1} - T_m}{\Delta t} = -\mu T_m \quad \text{for} \quad m \geq 0. \tag{4.8}$$

We also derive from the boundary condition (4.5) that

$$X_0 = X_{n+1} = 0. \tag{4.9}$$

We first consider the equation (4.8). We define[4] $T_{k,0} = 1$, and consider the difference equation

$$T_{m+1} = (1 - \Delta t\,\mu)T_m \quad \text{for} \quad m \geq 0. \tag{4.10}$$

Some iterations of (4.10),

$$T_{m+1} = (1 - \Delta t\,\mu)T_m = (1 - \Delta t\,\mu)^2 T_{m-1} \dots ,$$

clearly indicate that the solution is

$$T_m = (1 - \Delta t\,\mu)^m \qquad \text{for} \quad m \geq 0. \tag{4.11}$$

This fact is easily verified by induction on m.

Next we turn our attention to the problem (4.7) with boundary condition (4.9). In fact this is equivalent to the eigenvalue problem (2.44). Hence, from Lemma 2.9 we obtain that the n eigenvalues $\mu_1, \mu_2, \dots , \mu_n$ are given by

$$\mu_k = \frac{4}{(\Delta x)^2} \sin^2 (k\pi\Delta x/2) \quad \text{for} \quad k = 1, \dots , n \tag{4.12}$$

and the corresponding eigenvectors $X_k = (X_{k,1}, X_{k,2}, \dots , X_{k,n}) \in \mathbb{R}^n$, $k = 1, \dots , n$, have components given by

$$X_{k,j} = \sin (k\pi x_j) \quad \text{for} \quad j = 1, \dots , n.$$

Hence, we obtain particular solutions $w_{k,j}^m$ of the form

$$w_{k,j}^m = (1 - \Delta t\,\mu_k)^m \sin (k\pi x_j). \tag{4.13}$$

So far we have derived a family of particular solutions $\{w_k\}_{k=1}^n$ with values $w_{k,j}^m$ at the grid point (x_j, t_m). Next, we observe that any linear combination of particular solutions

$$v = \sum_{k=1}^{n} \gamma_k w_k,$$

where the γ_ks are scalars, is also a solution of (4.4) and (4.5). This observation corresponds to the second step in Section 3.1. Finally, we determine the coefficients $\{\gamma_k\}$ by using the initial condition

$$v_j^0 = f(x_j) \qquad \text{for} \quad j = 1, \dots , n.$$

---

[4] We are free to choose any nonzero constant; cf. the similar discussion in the continuous case on page 91.

Since $w_k = X_k$ at $t = 0$, we want to determine $\{\gamma_k\}$ such that

$$\sum_{k=1}^{n} \gamma_k X_{k,j} = f(x_j) \quad \text{for } j = 1, \dots, n. \tag{4.14}$$

Hence, it follows from (2.47) that

$$\gamma_k = 2\Delta x \sum_{j=1}^{n} f(x_j) X_{k,j} \quad \text{for } k = 1, \dots, n. \tag{4.15}$$

There is an alternative procedure to derive the representation of the general solution of the finite difference scheme (4.4)–(4.5). Let $A \in \mathbb{R}^{n,n}$ be the matrix

$$A = \frac{1}{(\Delta x)^2} \begin{pmatrix} 2 & -1 & 0 & \dots & 0 \\ -1 & 2 & -1 & \ddots & \vdots \\ 0 & \ddots & \ddots & \ddots & 0 \\ \vdots & \ddots & -1 & 2 & -1 \\ 0 & \dots & 0 & -1 & 2 \end{pmatrix}, \tag{4.16}$$

and for $m \geq 0$ let $v^m \in \mathbb{R}^n$ be the vector

$$v^m = (v_1^m, v_2^m, \dots, v_n^m).$$

The difference scheme (4.4)–(4.5) can then be equivalently written in the form

$$v^{m+1} = (I - \Delta t A) v^m, \tag{4.17}$$

where I is the identity matrix. As a simple consequence of this, we obtain

$$v^m = (I - \Delta t A)^m v^0. \tag{4.18}$$

Recall from Section 2.4 that the eigenvalue problem (2.44) is equivalent to the eigenvalue problem for the matrix A. Hence, the vectors X_k introduced above are eigenvectors for the matrix A with corresponding eigenvalue μ_k. Hence, X_k is also an eigenvector for the matrix $(I - \Delta t A)^m$ with eigenvalue $(1 - \Delta t\, \mu_k)^m$. Hence, if $v^0 = X_k$, we derive from (4.18) that

$$v_j^m = (1 - \Delta t\, \mu_k)^m \sin(k\pi x_j)$$

is a particular solution of (4.4)–(4.5). This solution corresponds exactly to (4.13) above. In the same way as above, the general solution is obtained by taking linear combinations of these solutions.

4.2.2 Comparison of the Analytical and Discrete Solution

We now have explicit formulas of the solutions for both the continuous problem (4.1) and the discrete problem (4.2). For ease of reference, we repeat the formulas here. The solution[5] of the continuous problem is given by

$$u(x,t) = \sum_{k=1}^{\infty} c_k e^{-\lambda_k t} \sin(k\pi x), \tag{4.19}$$

where $\lambda_k = (k\pi)^2$ and

$$c_k = 2\int_0^1 f(x) \sin(k\pi x)\, dx. \tag{4.20}$$

We want to compare this analytical solution with the discrete solution given by

$$v_j^m = \sum_{k=1}^{n} \gamma_k (1 - \Delta t\, \mu_k)^m \sin(k\pi x_j), \tag{4.21}$$

where

$$\mu_k = \frac{4}{(\Delta x)^2} \sin^2(k\pi \Delta x/2),$$

and

$$\gamma_k = 2\Delta x \sum_{j=1}^{n} f(x_j) \sin(k\pi x_j).$$

for $k = 1, \ldots, n$. In order to compare the analytical and numerical solution at a grid point (x_j, t_m), we define $u_j^m = u(x_j, t_m)$, i.e.

$$u_j^m = \sum_{k=1}^{\infty} c_k e^{-\lambda_k t_m} \sin(k\pi x_j). \tag{4.22}$$

Our aim is to give a rough, but instructive, argument for the fact that

$$v_j^m \approx u_j^m,$$

under appropriate conditions on the mesh parameters Δx and Δt. To avoid technicalities, we consider a fixed grid point (x_j, t_m) where $t_m \geq \bar{t}$ for $\bar{t} > 0$ independent of the mesh parameters. Furthermore, we assume that the initial function f is smooth and satisfies the boundary conditions, i.e.

[5] ... still formal.

$f(0) = f(1) = 0$. Finally, we assume that the mesh parameters Δt and Δx are sufficiently small.

In order to compare u_j^m and v_j^m, we note that

$$\begin{aligned} u_j^m &= \sum_{k=1}^{\infty} c_k e^{-\lambda_k t_m} \sin(k\pi x_j) \\ &= \sum_{k=1}^{n} c_k e^{-\lambda_k t_m} \sin(k\pi x_j) + \sum_{k=n+1}^{\infty} c_k e^{-\lambda_k t_m} \sin(k\pi x_j). \end{aligned}$$

Here, we want to show that

$$\sum_{k=n+1}^{\infty} c_k e^{-\lambda_k t_m} \sin(k\pi x_j) \approx 0. \tag{4.23}$$

To do this we make the following observations:

- Since f is smooth, it is also bounded, and then the Fourier coefficients c_k given by (4.20) are bounded[6] for all k. Hence there is a finite constant c such that $|c_k| \leq c$ for all k.
- Obviously, we have $|\sin(k\pi x_j)| \leq 1$.

By using these observations, we get

$$\begin{aligned} \left| \sum_{k=n+1}^{\infty} c_k e^{-\lambda_k t_m} \sin(k\pi x_j) \right| &\leq \max_k |c_k| \sum_{k=n+1}^{\infty} e^{-(k\pi)^2 t_m} \\ &\leq c \sum_{k=n+1}^{\infty} (e^{-\pi^2 \bar{t}})^k \\ &= c(e^{-\pi^2 \bar{t}})^{n+1} \frac{1}{1 - e^{-\pi^2 \bar{t}}} \approx 0, \end{aligned}$$

for large values of n. Here we have used the summation formula for a geometric series, and we have exploited the fact that $t_m \geq \bar{t}$. Since we have verified (4.23), it follows that

$$u_j^m \approx \sum_{k=1}^{n} c_k e^{-\lambda_k t_m} \sin(k\pi x_j). \tag{4.24}$$

Now we want to compare the finite sums (4.24) and (4.21). Motivated by the derivation of the solutions, we try to compare the two sums termwise. Thus, we keep k fixed, and we want to compare

$$c_k e^{-\lambda_k t_m} \sin(k\pi x_j)$$

[6]The Fourier coefficients actually converge towards zero as k tends to infinity. This is a consequence of Bessel's inequality, which is discussed in Chapter 8 below. Here, boundedness is sufficient.

and

$$\gamma_k(1 - \Delta t\, \mu_k)^m \sin(k\pi x_j).$$

Since the sine part here is identical, it remains to compare the Fourier coefficients c_k and γ_k, and the time-dependent terms $e^{-\lambda_k t_m}$ and $(1 - \Delta t\, \mu_k)^m$.

We start by considering the Fourier coefficients, and note that γ_k is a good approximation of c_k because

$$2\Delta x \sum_{j=1}^{n} f(x_j) \sin(k\pi x_j)$$

is the trapezoidal-rule[7] approximation of

$$2\int_0^1 f(x) \sin(k\pi x)\, dx.$$

In fact, by Exercise 2.1 on page 82, we have

$$|c_k - \gamma_k| = O((\Delta x)^2).$$

for f sufficiently smooth.

4.2.3 Stability Considerations

Finally, we must compare the time-dependent terms $e^{-\lambda_k t_m}$ and $(1-\Delta t \mu_k)^m$. Before we compare the actual values of these expressions, let us briefly consider the magnitudes involved. Since $\lambda_k t_m$ is positive, we have

$$|e^{-\lambda_k t_m}| \leq 1$$

for all $k = 1, \ldots, n$; it is reasonable to require that also

$$|1 - \Delta t\, \mu_k| \leq 1$$

for all $k = 1, \ldots, n$. From this requirement, it follows that $\Delta t \mu_k \leq 2$, or equivalently

$$\frac{4\Delta t}{(\Delta x)^2} \sin^2(k\pi \Delta x/2) \leq 2$$

for all $k = 1, \ldots, n$. This is always the case if the condition

[7] The trapezoidal method of numerical integration is discussed in Project 2.1 on page 82.

$$\frac{\Delta t}{(\Delta x)^2} \leq 1/2 \tag{4.25}$$

is satisfied.

Recall now that in Example 4.2 on page 121 we observed that by not obeying the stability condition (4.25), severe oscillations appeared in our solution. Now we see the reason; if $|1 - \Delta t\, \mu_k| > 1$ for some index k, the term $(1 - \Delta t\, \mu_k)^m$ blows up as m becomes large. Since such behavior cannot appear in the analytical solution, we conclude that the condition (4.25) must be satisfied in order to expect reliable numerical results. This stability condition will be derived again later using other techniques.

4.2.4 The Accuracy of the Approximation

Let us now return to the accuracy considerations initiated above. Of course, we assume from now on that the mesh sizes are chosen such that (4.25) is satisfied. The remaining problem is to discuss how well the term $(1 - \Delta t \mu_k)^m$ approximates the term $e^{-\lambda_k t_m}$. In order to study this question, we simplify the problem a bit by choosing a fixed time t_m, say $t_m = 1$, and we assume that $\Delta t = (\Delta x)^2/2$. As a consequence, we want to compare the terms

$$\alpha_k = e^{-\lambda_k}$$

and

$$\beta_k = (1 - \Delta t \mu_k)^{1/\Delta t} = (1 - 2\sin^2(k\pi\sqrt{\Delta t/2}))^{1/\Delta t}.$$

Obviously, α_k is very small for large values of k. The first three terms are given by

$$\alpha_1 \approx 5.172 \cdot 10^{-5}, \quad \alpha_2 \approx 7.157 \cdot 10^{-18}, \quad \alpha_3 \approx 2.65 \cdot 10^{-39}.$$

The values of β_k depends both on k and the mesh parameters. By choosing a relatively fine mesh, say $\Delta x = 1/100$, we get

$$\beta_1 \approx 5.164 \cdot 10^{-5}, \quad \beta_2 \approx 6.973 \cdot 10^{-18}, \quad \beta_3 \approx 2.333 \cdot 10^{-39}.$$

These computations clearly indicate that both α_k and β_k become very small when k is large. Furthermore, we observe that β_k seems to approximate α_k adequately. We will consider this problem a bit more closely. Since both α_k and β_k are very small for large values of k, it is sufficient to compare them for small k.

In order to compare α_k and β_k for small values of k, we start by recalling that

$$\sin(y) = y + O(y^3).$$

Thus, we get

$$2\sin^2(k\pi\sqrt{\Delta t/2}) \approx (k\pi)^2\Delta t.$$

Furthermore, we have in general that

$$e^y \approx (1+\epsilon y)^{1/\epsilon},$$

for ϵ sufficiently small. By using these facts we derive

$$\begin{aligned}\beta_k &= (1 - 2\sin^2(k\pi\sqrt{\Delta t/2}))^{1/\Delta t}\\ &\approx (1-(k\pi)^2\Delta t)^{1/\Delta t}\\ &\approx e^{-(k\pi)^2}\\ &= \alpha_k.\end{aligned}$$

This shows that also the time dependent term $e^{-\lambda_k t_m}$ is well approximated by its discrete counterpart, $(1-\Delta t\,\mu_k)^m$.

4.2.5 Summary of the Comparison

In order to summarize our analysis, we go back to the exact solution given by

$$u(x,t) = \sum_{k=1}^{\infty} c_k e^{-\lambda_k t}\sin(k\pi x),$$

and the representation of the discrete solution given by

$$v_j^m = \sum_{k=1}^{n} \gamma_k (1-\Delta t\,\mu_k)^m \sin(k\pi x_j).$$

We have seen that

$$(i)\quad u(x_j,t_m) \approx \sum_{k=1}^{n} c_k e^{-\lambda_k t_m}\sin(k\pi x_j)$$

by truncating the Fourier series. In (i), the Fourier coefficients satisfy

$$(ii)\quad c_k = 2\int_0^1 f(x)\sin(k\pi x)\,dx \approx 2\Delta x\sum_{j=1}^{n} f(x_j)\sin(k\pi x_j) = \gamma_k$$

by the trapezoidal rule of numerical integration. Furthermore, if the mesh parameters satisfy

$$\frac{\Delta t}{(\Delta x)^2} \leq 1/2,$$

then

$$(iii) \quad e^{-\lambda_k t_m} \approx (1 - \Delta t\, \mu_k)^m,$$

by the properties of the eigenvalues.

The observations $(i), (ii)$, and (iii) imply that

$$u(x_j, t_m) \approx \sum_{k=1}^{n} \gamma_k (1 - \Delta t\, \mu_k)^m \sin(k\pi x_j) = v_j^m.$$

This explains why we get good approximations for appropriate choices of Δx and Δt.

We have derived a stability condition (4.25) which has to be satisfied in order to get well-behaved numerical approximations. Secondly, we have utilized a discrete version of the Fourier method to show that each significant term in the analytical solution is well approximated by a similar term in the discrete solution. Although this analysis is rough and does not supply a precise error estimate, it explains very well what is going on in our computations. And furthermore, it is useful in order to prepare for the von Neumann stability analysis that we will present in the next section. The basic idea of this technique is exactly the same as in the present section; stability of the entire approximation is studied by analyzing particular solutions. It is important to note that this way of analyzing a scheme is purely linear; no similar method is found for general nonlinear problems. Therefore, we will come back to other ways of investigating the stability of numerical methods later.

4.3 Von Neumann's Stability Analysis

In the section above, we saw that a discrete version of the Fourier method enabled us to understand important features of a numerical scheme. The basic observation is that questions regarding stability and convergence can be analyzed by comparing the analytical and numerical solutions term by term. Thus, particular solutions of both problems become very important. In this section we will continue along the same lines. Our aim is to generalize this way of investigating the stability of a scheme to cover a wider class of equations and boundary conditions.

4.3.1 Particular Solutions: Continuous and Discrete

Let us start by recalling some particular solutions of the heat equation,

$$u_t = u_{xx} \quad \text{for} \quad x \in (0,1), \quad t > 0.$$

In the presence of Dirichlet boundary conditions,

$$u(0,t) = u(1,t) = 0, \quad t \geq 0,$$

the particular solutions are given by

$$F_D = \{T_k(t) \sin(k\pi x)\}_{k=1}^{\infty},$$

where

$$T_k(t) = e^{-(k\pi)^2 t}.$$

For Neumann data,

$$u_x(0,t) = u_x(1,t) = 0, \quad t \geq 0,$$

the particular solutions are given by

$$F_N = \{T_k(t) \cos(k\pi x)\}_{k=0}^{\infty};$$

see Section 3.6. And finally, for periodic boundary conditions

$$u(-1,t) = u(1,t) \quad \text{and} \quad u_x(-1,t) = u_x(1,t), \quad t \geq 0,$$

where the space variable $x \in (-1,1)$, we have

$$F_P = F_D \cup F_N;$$

cf. Exercise 3.15 on page 111.

In order to be able to handle all these particular solutions in a uniform manner, it is convenient to write them in a slightly different form. We know from calculus that

$$e^{ix} = \cos(x) + i \sin(x), \tag{4.26}$$

where i is the imaginary unit. Hence, we have

$$\cos(x) = \frac{1}{2}(e^{ix} + e^{-ix})$$

and

$$\sin(x) = \frac{1}{2i}(e^{ix} - e^{-ix}).$$

Using these formulas, all the functions in the families F_D, F_N, and F_P can be expressed as linear combinations of the following functions[8]

$$F = \{T_k(t)e^{ik\pi x}\}_{k=-\infty}^{\infty}.$$

In a similar way, we can argue that the corresponding discrete problems have a family of particular solutions of the form

$$F_\Delta = \{(a_k)^m e^{ik\pi x_j}\}_{k=-\infty}^{\infty},$$

where a_k represents the time dependency of the discrete solutions. In the explicit scheme for the heat equation, it is given by $a_k = 1 - \Delta t \mu_k$; see (4.21). This term is often referred to as the *amplification factor* of the scheme.

4.3.2 Examples

The basic idea of von Neumann's method is to compare the growth of the analytical and discrete particular solutions. More precisely, we want to derive conditions on the mesh parameters Δx and Δt such that the growth of the discrete solutions are bounded by the growth of the analytical solutions. Let us look at two examples to clarify this procedure.

EXAMPLE 4.3 We consider the heat equation

$$u_t = u_{xx}. \tag{4.27}$$

By inserting a particular solution of the form

$$u_k(x,t) = T_k(t)e^{ik\pi x}$$

into (4.27), we get

$$T_k'(t) = -(k\pi)^2 T_k(t).$$

Hence,

$$T_k(t) = e^{-(k\pi)^2 t}, \tag{4.28}$$

where we as usual have defined $T_k(0) = 1$. The solution of the partial differential equation is approximated by the scheme

$$\frac{v_j^{m+1} - v_j^m}{\Delta t} = \frac{v_{j-1}^m - 2v_j^m + v_{j+1}^m}{\Delta x^2}.$$

---

[8]Why do we suddenly need complex functions? So far, everything has been real. Do not worry too much about this; the introduction of complex variables here is merely a tool to simplify our calculations. We can do this directly by using the sine and cosine functions, or we can handle both at once by using the complex exponentials.

By inserting a particular solution of the form

$$(a_k)^m e^{ik\pi x_j},$$

we get

$$\frac{(a_k)^{m+1} - (a_k)^m}{\Delta t} e^{ik\pi x_j} = \frac{e^{ik\pi x_{j-1}} - 2e^{ik\pi x_j} + e^{ik\pi x_{j+1}}}{(\Delta x)^2} (a_k)^m.$$

Since $x_j = j\Delta x$, this implies

$$\begin{aligned} \frac{a_k - 1}{\Delta t} &= \frac{e^{-ik\pi\Delta x} - 2 + e^{ik\pi\Delta x}}{(\Delta x)^2} \\ &= 2\frac{\cos(k\pi\Delta x) - 1}{(\Delta x)^2} \\ &= -\frac{4}{(\Delta x)^2}\sin^2(k\pi\Delta x/2). \end{aligned}$$

Hence, we have

$$a_k = 1 - \frac{4\Delta t}{(\Delta x)^2}\sin^2(k\pi\Delta x/2).$$

Since T_k, given by (4.28), satisfies

$$|T_k(t)| \leq 1$$

for all k, we also require that

$$|(a_k)^m| \leq 1$$

for all k. As above, cf. (4.25), this inequality holds if the following condition is satisfied:

$$\frac{\Delta t}{(\Delta x)^2} \leq 1/2. \tag{4.29}$$

Thus, for both Dirichlet, Neumann, and periodic boundary conditions, this condition has to be satisfied in order to get reasonable numerical results. ■

EXAMPLE 4.4 Let us apply the procedure to the following equation:

$$u_t = u_{xx} + u. \tag{4.30}$$

Again, by inserting the particular solution

$$u_k(x,t) = T_k(t)e^{ik\pi x}$$

into the equation (4.30), we get

$$T_k'(t) = (1 - (k\pi)^2)T_k(t).$$

Hence

$$T_k(t) = e^{(1-(k\pi)^2)t}$$

and thus, for all k, we have

$$|T_k(t)| \leq e^t, \qquad t \geq 0. \tag{4.31}$$

The solution of equation (4.30) is approximated by the scheme

$$\frac{v_j^{m+1} - v_j^m}{\Delta t} = \frac{v_{j-1}^m - 2v_j^m + v_{j+1}^m}{\Delta x^2} + v_j^m.$$

By inserting

$$(a_k)^m e^{ik\pi x_j},$$

we get

$$\frac{a_k - 1}{\Delta t} = \frac{e^{-ik\pi\Delta x} - 2 + e^{ik\pi\Delta x}}{(\Delta x)^2} + 1,$$

or

$$a_k = 1 + \Delta t - \frac{4\Delta t}{(\Delta x)^2}\sin^2(k\pi\Delta x/2). \tag{4.32}$$

Based on the bound (4.31) for the analytical solution, it is reasonable to require that the numerical solution satisfies

$$|(a_k)^m| \leq e^{t_m}$$

for all k. Suppose that the usual condition

$$\frac{\Delta t}{(\Delta x)^2} \leq 1/2 \tag{4.33}$$

is satisfied. Then by (4.32) we get

$$\begin{aligned}
|(a_k)^m| &\leq |a_k|^m \\
&\leq \left(|1 - \frac{4\Delta t}{(\Delta x)^2}\sin^2(k\pi\Delta x/2)| + \Delta t\right)^m \\
&\leq (1 + \Delta t)^m \\
&\leq e^{m\Delta t} = e^{t_m},
\end{aligned}$$

where we have employed the useful inequality

$$(1+y)^m \leq e^{my}, \qquad m \geq 0, \tag{4.34}$$

which holds for all $y \geq -1$. The proof of this fact is left to the reader in Exercise 4.27. The conclusion of this example is that, again, the condition (4.33) must be satisfied in order to get stable results. ■

We summarize our discussion so far by stating that a numerical solution is said to be *stable in the sense of von Neumann* if the growth of the discrete particular solutions can be bounded by the growth of the continuous particular solutions. More precisely, if we let

$$T(t) = \max_k |T_k(t)|,$$

then we say that the scheme is stable in the sense of von Neumann if

$$\max_k |(a_k)^m| \leq T(t_m)$$

for all $t_m \geq 0$.

4.3.3 A Nonlinear Problem

As mentioned above, the method of von Neumann is only valid for linear problems with constant coefficients. That, of course, is a major drawback, because linear problems with constant coefficients are about the only problems we can solve analytically. So, under the circumstances where we badly need numerical methods, e.g. for nonlinear problems and problems with variable coefficients, the method of von Neumann cannot be applied directly. However, in practical computations, the method of von Neumann is applied far beyond the family of problems where we have actually shown that the method works. It is frequently applied to both linear problems with variable coefficients and to nonlinear problems. The procedure is roughly as follows: Given a nonlinear problem, we linearize the equation and freeze[9] the coefficients by considering the problem locally. In this manner, we derive a linear problem with constant coefficients. For this problem, the method of von Neumann can be applied, and a stability condition can be derived. Certainly, this condition will depend on the frozen coefficients involved. The trick is then to choose a conservative time step, covering all possible values of the frozen coefficient. More precisely, we try to find coefficient values that lead to the most restrictive time step. In some cases, this can be quite difficult, since we do not know any bounds for the coefficients. One

[9] Freezing the coefficient means to approximate the coefficients by a constant. Of course, this can only be valid locally, and thus freezing of coefficients often leads to a family of problems with different constant coefficients.

practical solution of this problem, is to introduce a variable time step, and update the bounds on the coefficients at each time step.[10]

Let us illustrate this procedure by an example.

EXAMPLE 4.5 Consider the following nonlinear heat equation,

$$\begin{aligned} u_t &= (\alpha(u)u_x)_x \quad \text{for} \quad x \in (0,1), \quad t > 0, \\ u(0,t) &= u(1,t) = 0, \\ u(x,0) &= f(x), \end{aligned}$$

where

$$\alpha(u) = \frac{1+3u^2}{1+u^2}.$$

An explicit finite difference scheme for this problem is given by

$$\frac{v_j^{m+1} - v_j^m}{\Delta t} = \frac{\alpha_{j+1/2}^m (v_{j+1}^m - v_j^m) - \alpha_{j-1/2}^m (v_j^m - v_{j-1}^m)}{\Delta x^2}, \tag{4.35}$$

where $\alpha_{j+1/2}^m = (\alpha(v_{j+1}^m) + \alpha(v_j^m))/2$. The derivation of this scheme will be considered in Exercise 4.20.

Consider this problem locally, i.e. close to some fixed location (x_0, t_0). If u is smooth, we can approximate the function $\alpha(u)$ by a constant value $\alpha_0 = \alpha(u(x_0, t_0))$ close to (x_0, t_0). This approximation leads to the equation

$$u_t = \alpha_0 u_{xx},$$

and the associated scheme

$$\frac{v_j^{m+1} - v_j^m}{\Delta t} = \alpha_0 \frac{v_{j-1}^m - 2v_j^m + v_{j+1}^m}{\Delta x^2} \quad \text{for} \quad j = 1, \ldots, n, \quad m \geq 0.$$

The particular solutions of the linearized equation are given by

$$T_k(t)e^{ik\pi x},$$

where

$$T_k(t) = e^{-(k\pi)^2 \alpha_0 t}$$

satisfies the usual bound

$$|T_k(t)| \leq 1, \qquad t \geq 0,$$

[10]The technique of linearizing the equation and freezing the coefficients can be carried out in order to analyze amazingly complicated problems. An excellent introduction to this procedure is given in the book by Kreiss and Lorenz [17]. A thorough discussion of the practical use of von Neumann's method for complicated problems can be found in the book by Godunov and Ryabenkii [10].

for all k since $\alpha_0 \geq 1$. Consequently, we require that the particular solutions of the corresponding finite difference scheme,

$$(a_k)^m e^{ik\pi x_j},$$

satisfy the bound

$$|a_k| \leq 1$$

for all k. By inserting the discrete particular solution into the scheme, we find that

$$a_k = 1 - \frac{4\alpha_0 \Delta t}{(\Delta x)^2} \sin^2 (k\pi \Delta x/2).$$

Thus we require that the mesh parameters satisfy the bound

$$\alpha_0 \Delta t/(\Delta x)^2 \leq 1/2.$$

This heuristic argument indicates that for mesh parameters satisfying this bound, the scheme is stable, at least locally. In order to derive a global condition, we observe that

$$\alpha(u) = \frac{1+3u^2}{1+u^2} \leq 3$$

for all u. Thus any frozen coefficient α_0 is less than 3, and consequently the most restrictive requirement on the time step is given by

$$\Delta t \leq \frac{(\Delta x)^2}{6}. \tag{4.36}$$

In Fig. 4.6, we have plotted the numerical solution when the initial data is given by $f(x) = \sin(3\pi x)$, using the mesh parameters $\Delta t = 0.00005$ and $\Delta x = 0.02$. We observe that the solution seems to be well behaved. The reader is encouraged to do further experiments using this scheme; see Exercise 4.19.

We will return to the problem considered in this example later[11] and actually prove that the requirement (4.36) is a sufficient condition for the numerical solutions to be stable. As indicated earlier, we will derive techniques that are more suitable for nonlinear problems. However, the present example indicates that by doing some rough arguments, the von Neumann method can be used to derive reasonable time-step requirements even for nonlinear problems. Of course, time steps derived by this procedure must be applied with great care. ■

Further examples of stability analysis based on von Neumann's method will be given in the exercises.

---

[11] See Section 6.3.2 on page 190.

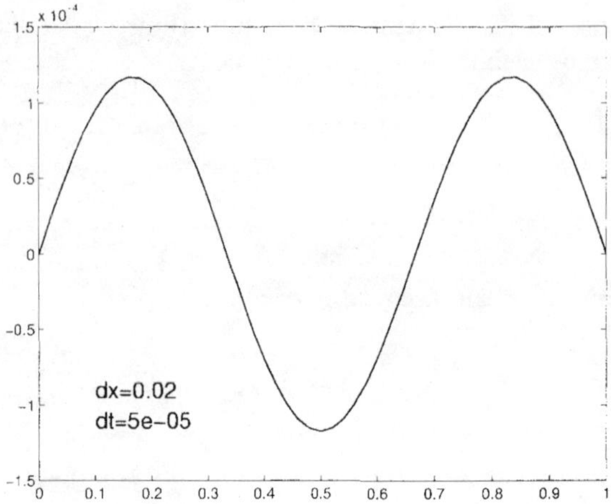

FIGURE 4.6. *The finite difference solution of the nonlinear heat equation using* $\Delta x = 0.02$ *and* $\Delta t = 0.00005$.

4.4 An Implicit Scheme

We have studied one particular numerical method for solving the heat equation. The method is given by (4.2) and is referred to as explicit. Obviously, the scheme is very simple to implement on a computer, and we have seen that it has some nice properties. Further properties will be derived below. However, the explicit method suffers from one major drawback; it requires very small time steps due to the stability condition.

Let us look a bit closer on the consequences of the stability conditions (4.25), i.e.

$$\frac{\Delta t}{(\Delta x)^2} \leq 1/2. \tag{4.37}$$

Suppose we want to compute a numerical solution of the heat equation at time $t = 1$, and that accuracy requirements force us to choose a fairly fine mesh, say $n = 100$. Then by the stability condition we must have

$$\Delta t \leq \frac{1}{20402}.$$

Since we want the solution at time $t_M = 1$, we must take $M = 20\,402$ time steps. Refining the mesh by choosing $n = 1000$, we have to compute $M = 2\,004\,002$ time steps. Clearly, even this very simple problem can put even our most powerful modern computers under strain. Of course, by turning our interest towards two or three space dimensions, this situation becomes even worse.

This unfortunate feature of the explicit scheme motivates us to search for alternatives with higher computational efficiency. Indeed there are a lot of methods available. In this section, we will present the simplest and probably most popular method: the standard implicit scheme. We do not want to go too deep into the discussion of different methods here; the topic is far too large and the interested reader may consult e.g. Thomee [27] and references given there for further discussions.

Before we start presenting the implicit scheme, let us remind ourselves of the basic difference between explicit and implicit schemes. We stated above, on page 120, that a scheme is called explicit if the solution at one time step can be computed directly from the solution at the previous time step. On the other hand, we call the scheme implicit if the solution on the next time level is obtained by solving a system of equations.

We want to derive an implicit scheme for the following equation:

$$\begin{aligned} u_t &= u_{xx} \quad \text{for} \quad x \in (0,1), \quad t > 0, \\ u(0,t) &= 0, \quad u(1,t) = 0, \\ u(x,0) &= f(x). \end{aligned} \tag{4.38}$$

Borrowing the notation from the explicit scheme, we apply the following approximations:

$$u_t(x, t+\Delta t) \approx \frac{u(x,t+\Delta t) - u(x,t)}{\Delta t},$$

and

$$u_{xx}(x, t+\Delta t) \approx \frac{u(x-\Delta x, t+\Delta t) - 2u(x,t+\Delta t) + u(x+\Delta x, t+\Delta t)}{\Delta x^2}.$$

This leads to the following scheme:

$$\frac{v_j^{m+1} - v_j^m}{\Delta t} = \frac{v_{j-1}^{m+1} - 2v_j^{m+1} + v_{j+1}^{m+1}}{\Delta x^2} \quad \text{for} \quad j = 1, \ldots, n, \quad m \geq 0.$$

The computational molecule of this scheme is depicted in Fig. 4.7.

The boundary conditions of (4.38) imply that

$$v_0^m = 0 \quad \text{and} \quad v_{n+1}^m = 0$$

for all $m \geq 0$, and the initial condition gives

$$v_j^0 = f(x_j) \quad \text{for} \quad j = 1, \ldots, n.$$

In order to write this scheme in a more convenient form, we introduce the vector $v^m \in \mathbb{R}^n$ with components $v^m = (v_1^m, \ldots, v_n^m)^T$. Then we observe that the scheme can be written as

$$(I + \Delta t A)v^{m+1} = v^m, \quad m \geq 0. \tag{4.39}$$

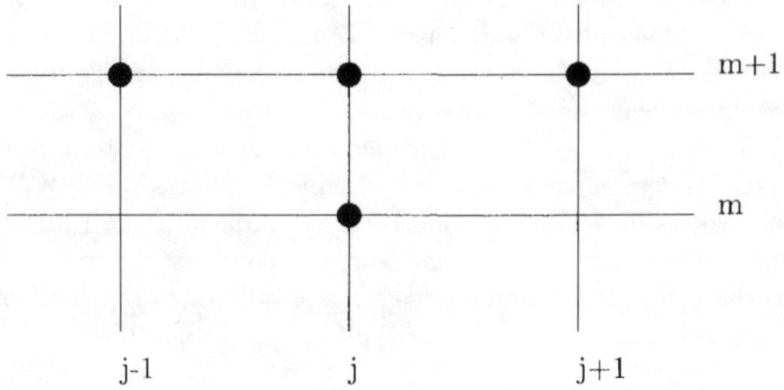

FIGURE 4.7. *The computational molecule of the implicit scheme.*

where $I \in \mathbb{R}^{n,n}$ is the identity matrix, and where the matrix $A \in \mathbb{R}^{n,n}$ is given by (4.16) above, i.e.

$$A = \frac{1}{(\Delta x)^2} \begin{pmatrix} 2 & -1 & 0 & \dots & 0 \\ -1 & 2 & -1 & \ddots & \vdots \\ 0 & \ddots & \ddots & \ddots & 0 \\ \vdots & \ddots & -1 & 2 & -1 \\ 0 & \dots & 0 & -1 & 2 \end{pmatrix}. \tag{4.40}$$

In order to compute numerical solutions based on this scheme, we have to solve linear systems of the form (4.39) at each time step. Hence, it is important to verify that the matrix $(I + \Delta t A)$ is nonsingular such that v^{m+1} is uniquely determined by v^m. In order to prove this, we use the properties of the matrix A derived in Lemma 2.9 on page 70.

Lemma 4.1 *The matrix $(I + \Delta t A)$ is symmetric and positive definite for all mesh parameters.*

Proof: The matrix $(I + \Delta t A)$ is obviously symmetric, since A is symmetric. Furthermore, the eigenvalues of $(I + \Delta t A)$ are of the form $1 + \Delta t \mu$, where μ corresponds to eigenvalues of A. However, the eigenvalues of A, which are given by (4.12), are all positive. Therefore, all the eigenvalues of $(I + \Delta t A)$ are positive, and hence this matrix is positive definite. ■

Since $(I + \Delta t A)$ is symmetric and positive definite, it follows from Proposition 2.4 that the system (4.39) has a unique solution that can be computed using the Gaussian elimination procedure given in Algorithm 2.1 on page 53. From a computational point of view, it is important to note that the coefficient matrix $(I + \Delta t A)$ in (4.39) does not change in time. This obser-

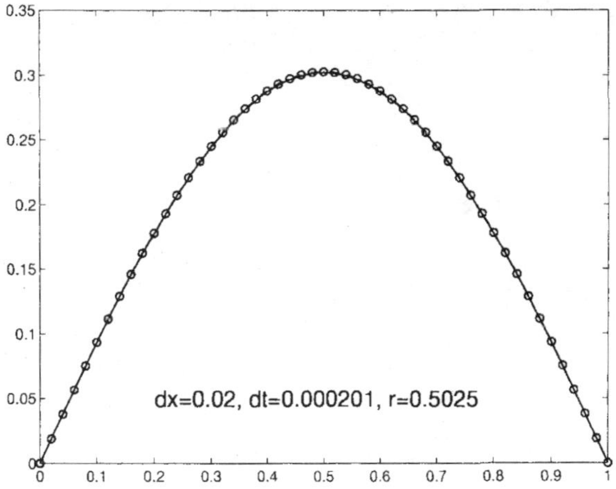

FIGURE 4.8. *The numerical solution (dashed line) computed by the implicit scheme and Fourier-based solution (solid line) of the heat equation. For the numerical method we have used* $r = 0.5025$.

vation can be used to reduce the total amount of computational effort in the scheme; see Exercise 2.13 on page 75.

Let us see how the implicit scheme works.

EXAMPLE 4.6 In Example 4.2 we observed that for one choice of grid parameters, the explicit scheme produces very good approximations. However, by increasing the timestep slightly, severe oscillations appear. Now we want to see how the implicit scheme handles this situation.

In Fig. 4.8 we have plotted the analytical solution, computed as in Example 4.2, and the numerical solution provided by the implicit scheme. The grid parameters are given by $\Delta x = 0.02$ and $\Delta t = 0.000201$, thus $r = \Delta t/(\Delta x)^2 = 0.5025$. We recall that these parameters gave an oscillatoric solution using the explicit scheme. From the figure, we observe that the numerical solution computed by the implicit scheme is very well behaved.

This observation leads us to believe that reliable solutions can be computed by this scheme without obeying the stability condition (4.37). Let us go one step further and choose $\Delta t = \Delta x = 0.02$, which gives $r = 50$. The results are given in Fig. 4.9, and we see that the numerical solution is still nice and smooth.

■

4.4.1 Stability Analysis

We observed in the previous example that the stability condition derived for the explicit scheme seems unnecessary for getting reasonable results using the implicit scheme. Let us look closer at this phenomenon by applying

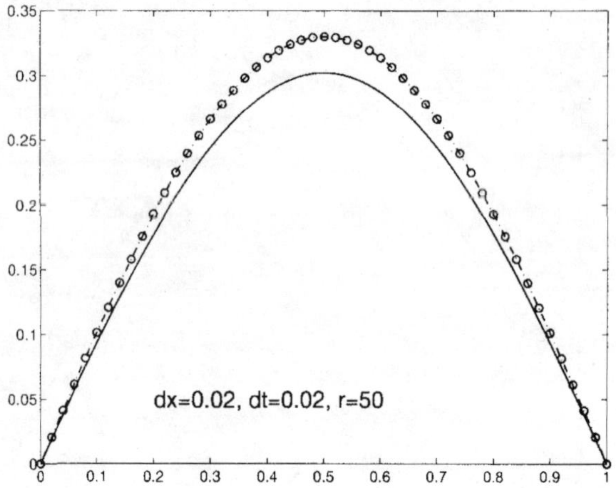

FIGURE 4.9. *The numerical (dashed line) and Fourier-based solution (solid line) of the heat equation. For the numerical method we have used* $r = 50$.

the von Neumann method derived above. Recall first that the particular solutions of the heat equation (4.38) are given by

$$u_k(x,t) = T_k(t)e^{ik\pi x},$$

where

$$T_k(t) = e^{-(k\pi)^2 t}.$$

By inserting a particular solution of the form

$$(a_k)^m e^{ik\pi x_j}$$

into the implicit scheme (4.39), we get

$$\begin{aligned}\frac{a_k - 1}{\Delta t} &= \frac{e^{-ik\pi\Delta x} - 2 + e^{ik\pi\Delta x}}{(\Delta x)^2} a_k \\ &= -\frac{4a_k}{(\Delta x)^2}\sin^2(k\pi\Delta x/2).\end{aligned}$$

Consequently,

$$a_k = \frac{1}{1 + \frac{4\Delta t}{(\Delta x)^2}\sin^2(k\pi\Delta x/2)}.$$

By observing that

$$|T_k(t)| \leq 1$$

for all k, we require

$$|(a_k)^m| \leq 1. \tag{4.41}$$

Since

$$a_k = \frac{1}{1 + \Delta t \mu_k},$$

where all the μ_ks are strictly positive, it follows that the requirement (4.41) is fulfilled for all mesh parameters. This explain why oscillations did not appear in the computations above.

Numerical methods that are well behaved for any choice of grid parameters are referred to as *unconditionally stable.* We have just seen that the implicit scheme deserves this label. The explicit scheme is analogously referred to as *conditionally stable.*

It should be noted here that although we are able to compute approximations using arbitrarily long time steps, the issue of accuracy has not yet been discussed. Obviously, by choosing Δt very large, the accuracy of the computation is poor. In numerical analysis this issue is a topic of lively discussion; should implicit or explicit schemes be used? The problem is of course how to compute good approximations using as little CPU time and memory resources as possible.[12]

4.5 Numerical Stability by Energy Arguments

In Section 3.7 we introduced energy arguments in order to derive a stability property for the solution of the heat equation. A similar analysis can also frequently be performed for finite difference solutions. It is possible to derive certain properties of the solution of the finite difference method without knowing the solution in detail. Here we shall illustrate these techniques by studying the solution of the explicit finite difference method (4.4) applied to the initial and boundary value problem (4.1). Recall that if $r = \Delta t/(\Delta x)^2$, this difference scheme has the form

$$v_j^{m+1} = v_j^m + r(v_{j-1}^m - 2v_j^m + v_{j+1}^m), \qquad j = 1, \ldots, n, \quad m \geq 0, \tag{4.42}$$

with boundary conditions

$$v_0^m = v_{n+1}^m = 0, \qquad m \geq 0. \tag{4.43}$$

[12]You might think that with the extremely powerful computers of today, such considerations are less important than earlier. This is not the case. We always strive for higher accuracy and more complicated models far beyond the capacity of any known computer.

Furthermore, we will assume throughout this section that the stability condition (4.25) holds, i.e.

$$1 - 2r \geq 0. \tag{4.44}$$

Along the same lines as in Section 3.7, we introduce, for each time level $m \geq 0$, the discrete energy

$$E^m = \Delta x \sum_{j=1}^{n} (v_j^m)^2, \tag{4.45}$$

and we are interested in the dynamics of this scalar variable. More precisely we want to show that E decreases with time, i.e.

$$E^{m+1} \leq E^m \quad m \geq 0. \tag{4.46}$$

Instead of computing the time derivative of the energy, as we did in Section 3.7, we consider the corresponding time difference,

$$\begin{aligned}
E^{m+1} - E^m &= \Delta x \sum_{j=1}^{n} \left((v_j^{m+1})^2 - (v_j^m)^2\right) \\
&= \Delta x \sum_{j=1}^{n} (v_j^{m+1} + v_j^m)(v_j^{m+1} - v_j^m) \\
&= r\Delta x \sum_{j=1}^{n} (v_j^{m+1} + v_j^m)(v_{j-1}^m - 2v_j^m + v_{j+1}^m) \qquad (4.47) \\
&= r\Delta x \left\{ \sum_{j=1}^{n} v_j^m (v_{j-1}^m - 2v_j^m + v_{j+1}^m) \right. \\
&\quad \left. -2\sum_{j=1}^{n} v_j^{m+1} v_j^m + \sum_{j=1}^{n} v_j^{m+1}(v_{j-1}^m + v_{j+1}^m) \right\},
\end{aligned}$$

where we have used the difference scheme (4.42).

We consider each of the three parts on the right-hand side separately. Observe first that from the boundary condition (4.43) and by summation by parts (cf. (2.31) on page 60) we obtain

$$\sum_{j=1}^{n} v_j^m (v_{j-1}^m - 2v_j^m + v_{j+1}^m) = -\sum_{j=1}^{n} (v_{j+1}^m - v_j^m)^2.$$

Furthermore, by applying the difference scheme (4.42) and summation by parts once more,

$$\begin{aligned}
-2\sum_{j=1}^{n} v_j^{m+1} v_j^m &= -2\sum_{j=1}^{n} \left((v_j^m)^2 + 2r(v_{j-1}^m - 2v_j^m + v_{j+1}^m)v_j^m\right) \\
&= -2\sum_{j=1}^{n} (v_j^m)^2 + 2r\sum_{j=1}^{n} (v_{j+1}^m - v_j^m)^2.
\end{aligned}$$

Finally, we use the inequality[13]

$$ab \leq \frac{1}{2}(a^2 + b^2)$$

to obtain

$$\begin{aligned}
\sum_{j=1}^{n} v_j^{m+1}(v_{j-1}^m + v_{j+1}^m) &\leq \sum_{j=1}^{n}((v_j^{m+1})^2 + \frac{1}{2}((v_{j-1}^m)^2 + (v_{j+1}^m)^2)) \\
&\leq \sum_{j=1}^{n}((v_j^{m+1})^2 + (v_j^m)^2).
\end{aligned}$$

Collecting these three inequalities, it follows from (4.47) that

$$\begin{aligned}
E^{m+1} - E^m &\leq r(E^{m+1} - E^m) - r(1-2r)\Delta x \sum_{j=1}^{n}(v_{j+1}^m - v_j^m)^2 \\
&\leq r(E^{m+1} - E^m),
\end{aligned}$$

where we have used the stability assumption (4.44). Hence,

$$(1-r)(E^{m+1} - E^m) \leq 0,$$

and by (4.44) this implies the desired inequality (4.46).

We summarize the result of the discussion above:

Theorem 4.1 *Let $\{v_j^m\}$ be a solution of the finite difference scheme (4.42)–(4.43) and let the corresponding energy $\{E^m\}$ be given by (4.45). If the stability condition (4.25) holds, then $\{E^m\}$ is nonincreasing with respect to m.*

Hence, we have seen that the stability condition (4.25), or (4.44), implies that the explicit difference scheme admits an estimate which is similar to the estimate (3.60) for the continuous problem. As for the continuous problem, this can be used to estimate the difference between two solutions, with different initial data. This follows since the difference of two solutions of the finite difference scheme is a new finite difference solution. We therefore obtain

Corollary 4.1 *Assume that the stability condition (4.25) holds and let $\{v_j^m\}$ and $\{w_j^m\}$ be two solutions of the finite difference scheme (4.42)–(4.43). Then, for all $m \geq 0$,*

$$\Delta x \sum_{j=1}^{n}(v_j^m - w_j^m)^2 \leq \Delta x \sum_{j=1}^{0}(v_j^0 - w_j^0)^2.$$

[13] See Exercise 4.24.

The interpretation of this result is that the difference scheme is a stable dynamical system in the sense that an error in the initial data bounds the error in the corresponding solutions. Energy arguments can also be performed for the implicit scheme (4.39). This is discussed in Exercise 4.25 below.

4.6 Exercises

EXERCISE 4.1 Verify, by direct calculation, that the discrete functions $\{w_n\}$ given by (4.13) are solutions of (4.4)–(4.5).

EXERCISE 4.2 Implement the scheme (4.2) for the heat equation and investigate the performance of the method by comparing the numerical results with the analytical solution given by

(a) Example 3.1 on page 92.

(b) Example 3.2 on page 93.

(c) Example 3.4 on page 97.

EXERCISE 4.3 Repeat Exercise 4.2 using the implicit scheme (4.39). Compare the numerical solutions provided by the explicit and the implicit schemes.

EXERCISE 4.4 In this exercise we want to study the rate of convergence of the explicit scheme by doing numerical experiments. Define the error by

$$e_\Delta(t_m) = \max_{j=0,\ldots,n+1} |u(x_j, t_m) - v_j^m|.$$

We want to estimate the rate of convergence for the scheme at time $t = 1/10$ for the problem considered in Exercise 4.2 (a).

(a) Estimate, using numerical experiments, α such that $e_\Delta(1/10) = O((\Delta t)^\alpha)$ for $\Delta t = (\Delta x)^2/2$.

(b) Repeat the experiments in (a) using $\Delta t = (\Delta x)^2/6$.

(c) Try to explain the difference in the rate of convergence encountered in the two cases above. Hint: Consider the truncation error discussed in Exercise 4.15 below.

EXERCISE 4.5 Consider the following initial-boundary value problem

$$\begin{aligned} u_t &= u_{xx} \quad \text{for} \quad x \in (0,1), \quad t > 0, \\ u(0,t) &= u_\ell(t), \quad u(1,t) = u_r(t), \\ u(x,0) &= f(x). \end{aligned}$$

Here $u_\ell(t)$, $u_r(t)$, and $f(x)$ are bounded functions satisfying $u_\ell(0) = f(0)$ and $u_r(0) = f(1)$.

(a) Derive an explicit scheme for this problem.

(b) Derive an implicit scheme for this problem and show that the linear system that arises can be solved by Gaussian elimination.

EXERCISE 4.6 Derive an explicit scheme for the following Neumann problem:

$$\begin{aligned} u_t &= u_{xx} \quad \text{for} \quad x \in (0,1), \quad t > 0, \\ u_x(0,t) &= u_x(1,t) = 0, \\ u(x,0) &= f(x). \end{aligned}$$

Use the analytical solution given in Example 3.5 on page 101 to check the quality of your approximations.

EXERCISE 4.7 Repeat Exercise 4.6 by deriving an implicit approximation of the problem. Compare the numerical solutions provided by the explicit and the implicit schemes.

EXERCISE 4.8 Consider the problem

$$\begin{aligned} u_t &= \alpha u_{xx} \quad \text{for} \quad x \in (-\ell,\ell), \quad t > 0, \\ u(-\ell,t) &= a, \quad u(\ell,t) = b, \\ u(x,0) &= f(x). \end{aligned}$$

where a, b and $\ell, \alpha > 0$ are given constants.

(a) Derive an explicit scheme.

(b) Derive an implicit scheme.

(c) Find the exact solution when $\alpha = 2$, $\ell = \pi$, $a = -\pi$, $b = \pi$, and $f(x) = x + \sin(3x)$.

(d) Implement the schemes derived in (a) and (b) and compare the results with the analytical solution derived in (c).

EXERCISE 4.9 Consider the problem

$$\begin{aligned} u_t &= 4u_{xx} - 10u + q(x,t) \quad \text{for} \quad x \in (\ell_1, \ell_2), \quad t > 0, \\ u(\ell_1, t) &= a(t), \quad u(\ell_2, t) = b(t), \\ u(x, 0) &= f(x), \end{aligned}$$

where $\ell_2 > \ell_1$ are given constants, and $a(t), b(t)$, and $q(x,t)$ are given functions.

(a) Derive an explicit scheme.

(b) Derive an implicit scheme.

(c) Suppose

$$\ell_1 = -2, \quad \ell_2 = 3, \quad a(t) = e^t - 2, \quad b(t) = e^t + 3, \quad f(x) = 1 + x,$$

and

$$q(x,t) = 11e^t + 10x.$$

Show that

$$u(x,t) = e^t + x$$

is an exact solution of the problem.

(d) Implement the schemes derived in (a) and (b) and compare the results with the analytical solution derived in (c).

EXERCISE 4.10 Consider the problem

$$\begin{aligned} u_t &= (\alpha(x,t)u_x)_x + c(x,t)u_x + q(x,t) \quad \text{for} \quad x \in (\ell_1, \ell_2), \quad t > 0, \\ u(\ell_1, t) &= a(t), \quad u(\ell_2, t) = b(t), \\ u(x, 0) &= f(x), \end{aligned}$$

where $\ell_2 > \ell_1$ are given constants, and $a(t), b(t), \alpha(x,t), c(x,t)$, and $q(x,t)$ are given smooth functions.

(a) Derive an explicit scheme.

(b) Derive an implicit scheme.

EXERCISE 4.11 Consider the problem

$$\begin{aligned} u_t &= (\alpha(x,t)u_x)_x + c(x,t)u_x + q(x,t)u \quad \text{for} \quad x \in (\ell_1, \ell_2), \quad t > 0, \\ u_x(\ell_1, t) &= a(t), \quad u_x(\ell_2, t) = b(t), \\ u(x,0) &= f(x), \end{aligned}$$

where $\ell_2 > \ell_1$ are given constants, and $a(t), b(t), \alpha(x,t), c(x,t)$, and $q(x,t)$ are given functions.

(a) Derive an explicit scheme.

(b) Derive an implicit scheme.

EXERCISE 4.12 Consider the equation

$$u_t = \alpha u_{xx},$$

with Dirichlet boundary conditions. Here $\alpha > 0$ is a given constant. We define an explicit scheme

$$\frac{v_j^{m+1} - v_j^m}{\Delta t} = \alpha \frac{v_{j-1}^m - 2v_j^m + v_{j+1}^m}{\Delta x^2} \quad \text{for} \quad j = 1, \ldots, n, \quad m \geq 0,$$

and an implicit scheme

$$\frac{v_j^{m+1} - v_j^m}{\Delta t} = \alpha \frac{v_{j-1}^{m+1} - 2v_j^{m+1} + v_{j+1}^{m+1}}{\Delta x^2} \quad \text{for} \quad j = 1, \ldots, n, \quad m \geq 0.$$

(a) Derive a stability condition for the explicit scheme using the von Neumann method.

(b) Show that the implicit scheme is unconditionally stable in the sense of von Neumann.

EXERCISE 4.13 Consider the equation

$$u_t = u_{xx},$$

with Neumann-type boundary conditions and the following explicit scheme

$$\frac{v_j^{m+1} - v_j^{m-1}}{2\Delta t} = \frac{v_{j-1}^m - 2v_j^m + v_{j+1}^m}{\Delta x^2} \quad \text{for} \quad j = 1, \ldots, n, \quad m \geq 0.$$

(a) Use the Taylor series to explain the derivation of this scheme.

(b) For what mesh sizes is this scheme stable in the sense of von Neumann?

EXERCISE 4.14 Consider the equation

$$u_t = u_{xx} - 9u,$$

with Dirichlet-type boundary conditions. Derive an explicit and an implicit scheme for this equation. Use the von Neumann method to investigate the stability of the methods.

EXERCISE 4.15 In Section 2.3.5 we introduced the concept of *truncation error* for a finite difference approximation for a two-point boundary value problem. Here we shall discuss a similar concept for difference approximations of the heat equation.

Observe that the scheme (4.4) can be written in the form

$$\frac{1}{\Delta t}(v^{m+1} - v^m) + Av^m = 0,$$

where $A \in \mathbb{R}^{n,n}$ is given by (4.16).

The truncation vector $\tau^m \in \mathbb{R}^n$ is given by

$$\tau^m = \frac{1}{\Delta t}(u^{m+1} - u^m) + Au^m,$$

where $u^m \in \mathbb{R}^n$ is given by $u_j^m = u(x_j, t_m)$ for a solution of the continuous problem (4.1).

(a) Show that under suitable smoothness assumptions on the solution u,

$$|\tau_j^m| = O(\Delta t) + O((\Delta x)^2). \tag{4.48}$$

In rest of this exercise we study a more general difference scheme of the form

$$\frac{1}{\Delta t}(v^{m+1} - v^m) + \theta(Av^{m+1}) + (1-\theta)Av^m = 0, \tag{4.49}$$

where $\theta \in [0, 1]$ is a parameter. Note that if $\theta = 0$, this corresponds to the explicit scheme (4.4), while if $\theta = 1$, it corresponds to the implicit scheme studied in Chapter 4.4 above.

(b) Sketch the computational molecule for the scheme when $\theta \in (0, 1)$.

(c) Show that for all $\theta \in [0, 1]$ the estimate (4.48) holds, and that the choice $\theta = 1/2$ leads to an improved estimate of the form

$$|\tau_j^m| = O((\Delta t)^2) + O((\Delta x)^2).$$

(Hint: Consider Taylor expansions at the point $(x_j, (t_{m+1} + t_m)/2)$.)

EXERCISE 4.16 Motivated by the result of Exercise 4.15, we study, in this exercise, the difference scheme 4.49 with $\theta = 1/2$. This difference scheme is usually referred to as the Crank-Nicholson scheme. In component form the scheme is given by

$$\frac{v_j^{m+1} - v_j^m}{\Delta t} = \frac{1}{2}\left(\frac{v_{j-1}^{m+1} - 2v_j^{m+1} + v_{j+1}^{m+1}}{(\Delta x)^2} + \frac{v_{j-1}^m - 2v_j^m + v_{j+1}^m}{(\Delta x)^2}\right)$$

for $j = 1, 2, \ldots, n$ and $m \geq 0$.

(a) Show that this implicit scheme is unconditionally stable in the sense of von Neumann.

(b) Discuss how the vectors $v^{m+1} \in \mathbb{R}^n$ can be computed from v^m.

(c) Show that the solution of the Crank-Nicholson scheme for the initial-boundary value problem (4.1) admits the representation

$$v_j^m = \sum_{k=1}^{n} \gamma_k \big(a(\mu_k)\big)^m \sin(k\pi x_j),$$

where $a(\mu) = \left(1 - \frac{\Delta t}{2}\mu\right)\left(1 + \frac{\Delta t}{2}\mu\right)^{-1}$ and $\gamma_k = 2\Delta x \sum_{j=1}^{n} v_j^0 \sin(k\pi x_j)$.

(d) Show that the amplification factor of the difference scheme, $a(\mu)$, satisfies

$$\left|a(\mu) - e^{-\mu\Delta t}\right| = O\big((\Delta t)^3\big).$$

How does this result relate to the corresponding result for the explicit scheme (4.4)? Compare your result with the conclusions you derived in Exercise 4.15.

(e) Implement the Crank-Nicholson scheme. Choose the initial function $f(x)$ as in Example 4.2 and try to verify that the scheme is unconditionally stable by varying the parameter $r = \frac{\Delta t}{(\Delta x)^2}$.

EXERCISE 4.17 Consider the general scheme (4.49). Use the von Neumann method to discuss the stability for any $\theta \in [0, 1]$.

EXERCISE 4.18 Consider the equation

$$u_t + cu_x = u_{xx},$$

with Dirichlet-type boundary conditions. Here $c \geq 0$ is given constant.

(a) Show that this problem has a family of particular solutions of the form

$$e^{-(ik\pi c + (k\pi)^2)t} e^{ik\pi x}.$$

(b) Show that

$$T_k(t) = e^{-(ik\pi c + (k\pi)^2)t}$$

satisfies the bound[14]

$$|T_k(t)| \leq 1 \qquad t \geq 0,$$

for all k.

Derive stability conditions for the following numerical methods by applying the von Neumann method.

(c)

$$\frac{v_j^{m+1} - v_j^m}{\Delta t} + c\frac{v_{j+1}^m - v_{j-1}^m}{2\Delta x} = \frac{v_{j-1}^m - 2v_j^m + v_{j+1}^m}{\Delta x^2}$$

(d)

$$\frac{v_j^{m+1} - v_j^m}{\Delta t} + c\frac{v_j^m - v_{j-1}^m}{\Delta x} = \frac{v_{j-1}^m - 2v_j^m + v_{j+1}^m}{\Delta x^2}$$

(e)

$$\frac{v_j^{m+1} - v_j^m}{\Delta t} + c\frac{v_{j+1}^m - v_j^m}{\Delta x} = \frac{v_{j-1}^m - 2v_j^m + v_{j+1}^m}{\Delta x^2}$$

(f)

$$\frac{v_j^{m+1} - v_j^m}{\Delta t} + c\frac{v_j^m - v_{j-1}^m}{\Delta x} = \frac{v_{j-1}^{m+1} - 2v_j^{m+1} + v_{j+1}^{m+1}}{\Delta x^2}$$

(g)

$$\frac{v_j^{m+1} - v_j^m}{\Delta t} + c\frac{v_j^{m+1} - v_{j-1}^{m+1}}{\Delta x} = \frac{v_{j-1}^{m+1} - 2v_j^{m+1} + v_{j+1}^{m+1}}{\Delta x^2}.$$

EXERCISE 4.19 In Example 4.5 on page 138, we derived a stability condition (see (4.36)) based on some rough considerations. The purpose of this exercise is to perform a numerical study of the quality of this condition. Consider the initial condition $f(x) = 100x(1-x)|x-1/2|$, and run the scheme (4.35) with several different grids. For this initial function, is the condition (4.36) sufficient in order to guarantee well-behaved numerical solutions?

---

[14]Recall that for a complex number $z = x + iy$, the absolute value, or the modulus, is given by $|z| = (x^2 + y^2)^{1/2}$. It is also useful to note that $|e^{i\theta}| = 1$ for any $\theta \in \mathbb{R}$.

EXERCISE 4.20 The purpose of this exercise is to derive a finite difference scheme for the following problem:

$$\begin{aligned} u_t &= (\alpha(u)u_x)_x \quad \text{for} \quad x \in (0,1), \quad t>0, \\ u(0,t) &= u(1,t) = 0, \\ u(x,0) &= f(x), \end{aligned}$$

where $\alpha(u)$ is a given strictly positive and smooth function.

(a) Put $v = \alpha(u)u_x$ and justify the following approximations:

$$\begin{aligned} u_t(x,t) &\approx \frac{u(x,t+\Delta t) - u(x,t)}{\Delta t}, \\ v_x(x,t) &\approx \frac{v(x+\Delta x/2,t) - v(x-\Delta x/2,t)}{\Delta x}. \end{aligned}$$

(b) Show that

$$v(x+\Delta x/2,t) \approx \frac{1}{2}(\alpha(u(x+\Delta x,t)) + \alpha(u(x,t)))\frac{u(x+\Delta x,t)-u(x,t)}{\Delta x}.$$

(c) Use these approximations to derive the scheme[15]

$$\frac{v_j^{m+1} - v_j^m}{\Delta t} = \frac{\alpha_{j+1/2}^m(v_{j+1}^m - v_j^m) - \alpha_{j-1/2}^m(v_j^m - v_{j-1}^m)}{\Delta x^2},$$

where $\alpha_{j+1/2}^m = (\alpha(v_{j+1}^m) + \alpha(v_j^m))/2$.

EXERCISE 4.21 Consider the initial-boundary value problem in Exercise 4.20. Derive an implicit scheme and investigate the stability of the method by the technique discussed in Example 4.5 on page 138.

EXERCISE 4.22 Consider the nonlinear heat equation

$$\begin{aligned} u_t &= (\alpha_\epsilon(u)u_x)_x \quad \text{for} \quad x \in (0,1), \quad t>0, \\ u(0,t) &= u(1,t) = 0, \\ u(x,0) &= \sin(3\pi x). \end{aligned}$$

Here $\alpha_\epsilon(u) = 2 + \epsilon\cos(u)$, where ϵ is a small parameter, $|\epsilon| \ll 1$.

(a) Derive an explicit finite difference scheme for this problem and find a stability condition using the technique discussed in Example 4.5.

[15]We will study the stability, or more precisely, a maximum principle for the numerical solutions generated by this scheme in Section 6.3.2 on page 190.

(b) Implement the scheme and plot the solution at time $t = 1/10$ for $\epsilon = 1/8,\ 1/16,\ 1/32,\ 1/64$.

(c) We want a rough estimate of the solution at time $t = 1/10$ for $\epsilon = 1/100$. Use the approximation $\alpha_\epsilon(u) \approx 2$ to derive an estimate. Can you use the results of the computations above to argue that the explicit formula you obtain is a good approximation of the exact solution?

EXERCISE 4.23 Consider the nonlinear heat equation

$$\begin{aligned} u_t &= (\alpha(u)u_x)_x \quad \text{for} \quad x \in (0,1), \quad 0 < t \leq 1, \\ u(0,t) &= a(t), \quad u(1,t) = b(t), \\ u(x,0) &= f(x). \end{aligned}$$

Here $\alpha(u)$ is a given strictly positive function, and the boundary conditions $a(t)$ and $b(t)$ are given functions.

(a) Implement the following explicit numerical method:

$$\frac{v_j^{m+1} - v_j^m}{\Delta t} = \frac{\alpha_{j+1/2}^m (v_{j+1}^m - v_j^m) - \alpha_{j-1/2}^m (v_j^m - v_{j-1}^m)}{\Delta x^2},$$

where $\alpha_{j+1/2}^m = (\alpha(v_{j+1}^m) + \alpha(v_j^m))/2$.

(b) Let

$$\alpha(u) = u, \quad a(t) = t, \quad b(t) = 1 + t, \quad \text{and} \quad f(x) = x.$$

Show that $u(x,t) = x + t$ is an exact solution of this problem.

(c) Show, by induction, that the explicit scheme gives the exact solution at each grid point, i.e. show that $v_j^m = x_j + t_m$, for any grid sizes.

(d) Compute the numerical solution at $t = 1$ using the scheme implemented in (a) for the problem defined in (b). Try the following grid parameters:

- $n = 4$ and $\Delta t = 1/65$.
- $n = 30$ and $\Delta t = 1/10$.

Discuss your observations in light of the result in (c).

(e) From the numerical results obtained in (d), it is clear that some kind of stability condition is needed. Use the procedure discussed in Example 4.5 on page 138 to derive a stability condition for this problem. Run some numerical experiments with mesh parameters satisfying this condition. Are the numerical solutions well-behaved if the condition on the mesh parameters is satisfied?

Exercise 4.24 Use the fact that $(a-b)^2 \geq 0$ to show that

$$ab \leq \frac{1}{2}(a^2+b^2)$$

for all real numbers a and b.

Exercise 4.25 Use an energy argument and the fact that the matrix A given by (4.40) is positive definite to show that the implicit difference scheme (4.39) satisfies an estimate of the form (4.46) for all mesh parameters.

Exercise 4.26 Similar to the discussion in Chapter 2 we introduce a discrete inner product which is an analog of the inner product $\langle \cdot,\cdot \rangle$ for continuous functions. For a vectors $v, w \in \mathbb{R}^n$ we define[16]

$$\langle v, w \rangle_\Delta = \Delta x \sum_{j=1}^{n} v_j w_j.$$

Hence, this inner product is just the ordinary Euclidean inner product multiplied by the scaling factor Δx. We recall from Chapter 2 that this inner product arises naturally when the vector v has the interpretation of a discrete function defined on the grid points $x_j = j\Delta x$. We let $||\cdot||_\Delta$ denote the corresponding norm, i.e.

$$||v||_\Delta^2 = \langle v, v \rangle_\Delta.$$

As above we let $X_k = (X_{k,1}, X_{k,2}, \ldots, X_{k,n}) \in \mathbb{R}^n, \quad k = 1, \ldots, n,$ be the vectors with components given by

$$X_{k,j} = \sin(k\pi x_j) \quad \text{for} \quad j = 1, \ldots, n.$$

Recall that these vectors are orthogonal with respect to the inner product $\langle \cdot,\cdot \rangle_\Delta$ and that $||X_k||_\Delta^2 = 1/2$ (see Lemma 2.30).

(a) Explain why any vector $v \in \mathbb{R}^n$ can be written in the form

$$v = \sum_{k=1}^{n} c_k X_k,$$

where

$$c_k = 2\langle v, X_k \rangle_\Delta.$$

[16]In Chapter 2 we used h to indicate the spacing in the x–variable, and hence we used this subscript to indicate the corresponding discrete inner product. Here, where we have two grid parameters Δx and Δt, we use Δ for the same purpose.

(b) Show that

$$||v||_\Delta^2 = \frac{1}{2}\sum_{k=1}^{n} c_k^2.$$

(c) Let $\{v^m\}_{m\geq 0}$ be a sequence of vectors generated by the finite difference scheme (4.39). As above, in Exercise 4.25, let

$$E^m = ||v^m||_\Delta^2.$$

Show that

$$E^m \leq \left(\frac{1}{1+\Delta t \mu_1}\right)^m E^0,$$

where $\mu_1 = \frac{4}{\Delta x^2}\sin^2(\pi \Delta x/2)$.

(d) Explain why

$$\lim_{m\to\infty} E^m = 0,$$

and compare this result with what you derived in Exercise 4.25 above.

EXERCISE 4.27 Show that

$$(1+y)^m \leq e^{my}$$

for all $m \geq 0$ and $y \geq -1$.

5 The Wave Equation

The purpose of this chapter is to study initial-boundary value problems for the wave equation in one space dimension. In particular, we will derive formal solutions by a separation of variables technique, establish uniqueness of the solution by energy arguments, and study properties of finite difference approximations.

The wave equation models the movement of an elastic, homogeneous string which undergoes relatively small transverse vibrations. The wave equation is of second order with respect to the space variable x and time t, and takes the form

$$u_{tt} = c^2 u_{xx}. \tag{5.1}$$

Here the constant c is called the wave speed. Since the equation is of second order with respect to time, an initial value problem typically needs two initial conditions. Hence, in addition to the differential equation (5.1) we specify two initial conditions of the form

$$u(x,0) = f(x) \quad \text{and} \quad u_t(x,0) = g(x). \tag{5.2}$$

If we study the pure initial value problem, i.e. where x varies over all of $\mathbb{R}$, then the solution of (5.1)–(5.2) is given by d'Alembert's formula

$$u(x,t) = \frac{1}{2}\left(f(x+ct) + f(x-ct)\right) + \frac{1}{2c}\int_{x-ct}^{x+ct} g(y)dy; \tag{5.3}$$

cf. page 16. However, in most practical applications, for example in modeling the movement of a guitar string, we are facing an initial and boundary value problem.

Throughout this chapter we shall consider the following initial and boundary value problem:

$$\begin{aligned} u_{tt} &= u_{xx} \quad \text{for} \quad x \in (0,1)\,,\ t > 0, \\ u(0,t) &= u(1,t) = 0, \quad t > 0, \\ u(x,0) &= f(x)\,, \quad u_t(x,0) = g(x)\,,\ x \in (0,1). \end{aligned} \tag{5.4}$$

We note that we have assumed that the wave speed c is set equal to 1. In fact, any problem with $c \neq 0$ can be transformed to a problem with $c = 1$ by introducing a proper time scale (see Exercise 5.2). Therefore, we set $c = 1$ for simplicity.

5.1 Separation of Variables

Let us try to find solutions of problem (5.4) of the form

$$u(x,t) = X(x)T(t).$$

By inserting this ansatz into the wave equation, we obtain

$$X(x)T''(t) = X''(x)T(t)$$

or

$$\frac{T''(t)}{T(t)} = \frac{X''(x)}{X(x)}. \tag{5.5}$$

As in Section 3.2 we can argue that since the left-hand side is independent of x and the right-hand side is independent of t, both expressions must be independent of x and t. Therefore,

$$\frac{T''(t)}{T(t)} = \frac{X''(x)}{X(x)} = -\lambda \tag{5.6}$$

for a suitable $\lambda \in \mathbb{R}$. In particular this means that the functions $X(x)$ satisfy the eigenvalue problem

$$\begin{aligned} -X''(x) &= \lambda X(x), \quad x \in (0,1), \\ X(0) &= X(1) = 0, \end{aligned} \tag{5.7}$$

where the boundary conditions follow from (5.4). Of course, this eigenvalue problem is by now familiar to us. From Lemma 2.7 we conclude that

$$\lambda = \lambda_k = (k\pi)^2 \quad \text{for} \quad k = 1, 2, \ldots \tag{5.8}$$

with corresponding eigenfunctions

$$X_k(x) = \sin(k\pi x) \quad \text{for} \quad k = 1, 2, \ldots\,. \tag{5.9}$$

On the other hand, the functions $T_k(t)$ must satisfy

$$-T_k''(t) = \lambda_k T_k(t) = (k\pi)^2 T_k(t).$$

This equation has two linearly independent solutions given by

$$T_k(t) = e^{ik\pi t} \quad \text{and} \quad T_k(t) = e^{-ik\pi t}. \tag{5.10}$$

The general *real* solution is therefore of the form

$$T_k(t) = a_k \cos(k\pi t) + b_k \sin(k\pi t),$$

where a_k, $b_k \in \mathbb{R}$ are arbitrary constants. Hence, we conclude that the functions

$$u_k(x,t) = \sin(k\pi x)\big(a_k \cos(k\pi t) + b_k \sin(k\pi t)\big) \tag{5.11}$$

satisfy the differential equation and the boundary values prescribed by the initial-boundary value problem. Furthermore, these solutions satisfy the initial conditions

$$u_k(x,0) = a_k \sin(k\pi x) \quad \text{and} \quad (u_k)_t(x,0) = b_k k\pi \sin(k\pi x).$$

In order to obtain more solutions, we can add solutions of the form (5.11) and obtain

$$u(x,t) = \sum_{k=1}^{N} \sin(k\pi x)\big(a_k \cos(k\pi t) + b_k \sin(k\pi t)\big) \tag{5.12}$$

with initial conditions

$$u(x,0) = \sum_{k=1}^{N} a_k \sin(k\pi x) \quad \text{and} \quad u_t(x,0) = \sum_{k=1}^{N} b_k k\pi \sin(k\pi x). \tag{5.13}$$

EXAMPLE 5.1 Consider the problem (5.4) with $f(x) = 2\sin(\pi x)$ and $g(x) = -\sin(2\pi x)$. Hence, the initial data is of the form (5.13) with

$$a_1 = 2, \quad a_k = 0 \quad \text{for} \quad k > 1$$

and

$$b_2 = -\frac{1}{2\pi}, \quad b_k = 0 \quad \text{for} \quad k \neq 2.$$

The solution $u(x,t)$ is therefore given by

$$u(x,t) = 2\sin(\pi x)\cos(\pi t) - \frac{1}{2\pi}\sin(2\pi x)\sin(2\pi t).$$

This solution is plotted in Fig. 5.1. ■

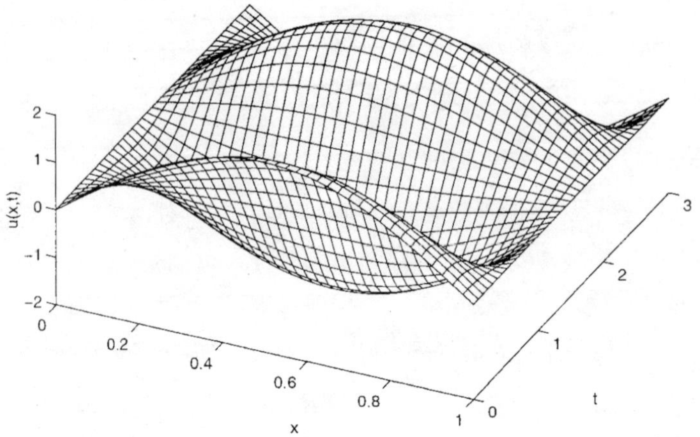

FIGURE 5.1. *The solution $u(x,t)$ derived in Example 5.1 for $(x,t) \in ([0,1]\times[0,3])$.*

In order to cover a larger class of initial functions, we allow general Fourier sine series as initial functions, i.e. we let N tend to infinity in (5.12) and (5.13). Hence, if

$$f(x) = \sum_{k=1}^{\infty} a_k \sin(k\pi x) \quad \text{and} \quad g(x) = \sum_{k=1}^{\infty} b_k \sin(k\pi x), \tag{5.14}$$

then we obtain a formal solution of the initial-boundary value problem (5.4) given by

$$u(x,t) = \sum_{k=1}^{\infty} \sin(k\pi x)\left(a_k \cos(k\pi t) + \frac{b_k}{k\pi}\sin(k\pi t)\right). \tag{5.15}$$

EXAMPLE 5.2 Consider the initial-boundary value problem (5.4) with

$$f(x) = x(1-x) \quad \text{and} \quad g(x) = 0.$$

We recall from Exercise 3.1(c) on page 107 that the Fourier sine series of f is given by

$$f(x) = \sum_{k=1}^{\infty} \frac{8}{\pi^3(2k-1)^3}\sin\big((2k-1)\pi x\big).$$

Hence, by (5.15) the formal solution is given by

$$u(x,t) = \sum_{k=1}^{\infty} \frac{8}{\pi^3(2k-1)^3}\sin\big((2k-1)\pi x\big)\cos\big((2k-1)\pi t\big).$$

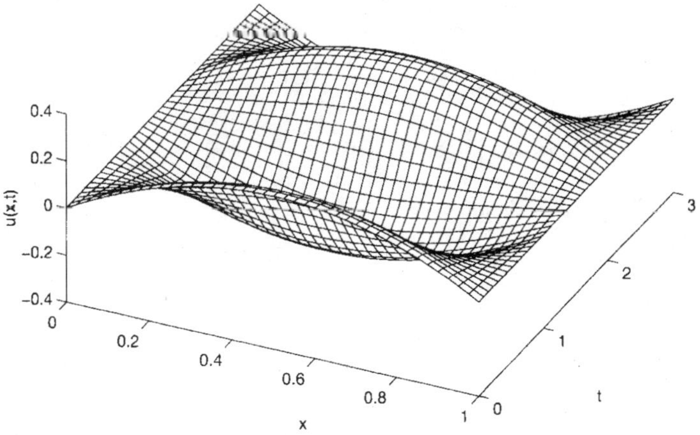

FIGURE 5.2. *The solution $u(x,t)$ derived in Example 5.2 for $(x,t) \in ([0,1]\times[0,3])$ using the first 20 terms of the series.*

In Fig. 5.2 we have plotted this formal solution by using the first 20 terms of this infinite series.

■

5.2 Uniqueness and Energy Arguments

Above we derived formal solutions of the initial-boundary value problem (5.4) by separation of variables. We will continue the study of the wave equation by applying energy arguments in this case. One of the consequences of this analysis is that the problem (5.4) has at most one smooth solution.

Assume that $u(x,t)$ is a solution of (5.4) such that $u \in C^2([0,1] \times [0,\infty))$.[1] For each $t \geq 0$ we define the "energy," $E(t)$, by

$$E(t) = \int_0^1 \left(u_x^2(x,t) + u_t^2(x,t)\right)dx.$$

Note that for $t = 0$ the energy is known from the initial functions f and g. The idea is to consider how this nonnegative scalar variable evolves with

[1] Here $u \in C^2([0,1] \times [0,\infty))$ means that all partial derivatives of order less than or equal 2 are continuous on $[0,1] \times [0,\infty)$, i.e., $u, u_x, u_t, u_{xx}, u_{xt}, u_{tt} \in C([0,1] \times [0,\infty))$.

time. By differentiating $E(t)$ with respect to time, we obtain

$$\begin{aligned} E'(t) &= \frac{d}{dt}\int_0^1 \big(u_x^2(x,t)+u_t^2(x,t)\big)dx \\ &= 2\int_0^1 \big(u_x(x,t)u_{xt}(x,t)+u_t(x,t)u_{tt}(x,t)\big)dx. \end{aligned} \tag{5.16}$$

Here we have assumed that the energy can be differentiated by differentiating under the integral sign. However, if $u \in C^2([0,1]\times[0,\infty))$, this can be justified by applying Proposition 3.1 on page 107. The term u_{xt} appearing on the right-hand side of (5.16) should be interpreted as

$$u_{xt} = \frac{\partial}{\partial t}\left(\frac{\partial}{\partial x}u\right).$$

However, it is a well-known result from calculus that if u is a C^2-function, then we can change the order of differentiation, i.e.

$$u_{xt} = \frac{\partial}{\partial t}\left(\frac{\partial}{\partial x}u\right) = \frac{\partial}{\partial x}\left(\frac{\partial}{\partial t}u\right) = u_{tx}.$$

Hence, using integration by parts, we obtain

$$\int_0^1 u_x u_{xt}\,dx = \int_0^1 u_x u_{tx}\,dx = u_x(x,t)u_t(x,t)\Big|_{x=0}^{x=1} - \int_0^1 u_{xx}u_t\,dx.$$

If u solves (5.4), it now follows that $u_t(x,t)=0$ for $x=0,1$, and therefore we have

$$\int_0^1 u_x u_{xt}\,dx = -\int_0^1 u_{xx}u_t\,dx = -\int_0^1 u_{tt}u_t\,dx,$$

where last equality follows from the differential equation. By inserting this into (5.16) we simply obtain

$$E'(t) = 0,$$

or

$$E(t) = E(0) \quad \text{for} \quad t \geq 0. \tag{5.17}$$

Hence, for the wave equation the energy $E(t)$ is preserved for all time. In the same way as for the heat equation in Section 3.7, we can use the equality (5.17) to obtain a stability estimate for the problem (5.4).

Let u_1 and u_2 be two solutions of (5.4) with initial functions (f_1, g_1) and (f_2, g_2), respectively, and let $w = u_1 - u_2$. It is a straightforward consequence of the linearity of the problem that w is a solution of (5.4) with initial functions $f = f_1 - f_2$ and $g = g_1 - g_2$. Hence, it follows from the equality (5.17), applied to the solution w, that

$$\begin{aligned}&\int_0^1 \left[\left((u_1 - u_2)_x(x,t)\right)^2 + \left((u_1 - u_2)_t(x,t)\right)^2\right] dx \\ &= \int_0^1 \left[\left((f_1 - f_2)_x(x)\right)^2 + \left((g_1 - g_2)(x)\right)^2\right] dx \qquad (5.18)\end{aligned}$$

This equality tells us that if the initial data of the two solutions are close, then the solutions will stay close for all time. In particular, we have the following uniqueness result:

Theorem 5.1 *If $u_1, u_2 \in C^2\big([0,1]\times[0,\infty)\big)$ are two solutions of the initial-boundary value problem (5.4) with the same initial data, then $u_1 \equiv u_2$.*

Proof: If the initial data are the same, then the right-hand side of (5.18) is zero. Hence, the left-hand side is zero, and as a consequence $(u_1)_x = (u_2)_x$ and $(u_1)_t = (u_2)_t$. Hence, the two solutions can only differ by a constant. However, since they have the same initial and boundary data, this implies that $u_1 \equiv u_2$. ■

5.3 A Finite Difference Approximation

In this section we shall study an explicit finite difference approximation of the initial value problem (5.4). An alternative implicit method will be studied in Exercise 5.9.

In order to derive the difference method, let us first recall that the problem (5.4) takes the form

$$\begin{aligned} u_{tt} &= u_{xx} \quad \text{for} \quad x \in (0,1),\ t > 0, \\ u(0,t) &= u(1,t) = 0, \quad t > 0, \qquad (5.19) \\ u(x,0) &= f(x), \quad u_t(x,0) = g(x), \quad x \in (0,1). \end{aligned}$$

Let us also repeat some of the notation introduced in Chapter 4. The grid spacing in the x-direction is $\Delta x = 1/(n+1)$, where $n \geq 1$ is an integer, and the associated grid points are $x_j = j\Delta x$ for $j = 0, 1, 2, \ldots, n+1$. The discrete time levels are given by $t_m = m\Delta t$ for integers $m \geq 0$, where $\Delta t > 0$ is the time step. Furthermore, the grid function v, with $v_j^m = v(x_j, t_m)$, approximates u.

The difference schemes for the heat equation studied in Chapter 4 were based on the approximation

$$u_{xx}(x,t) = \frac{u(x-\Delta x, t) - 2u(x,t) + u(x+\Delta x, t)}{(\Delta x)^2} + O\big((\Delta x)^2\big)$$

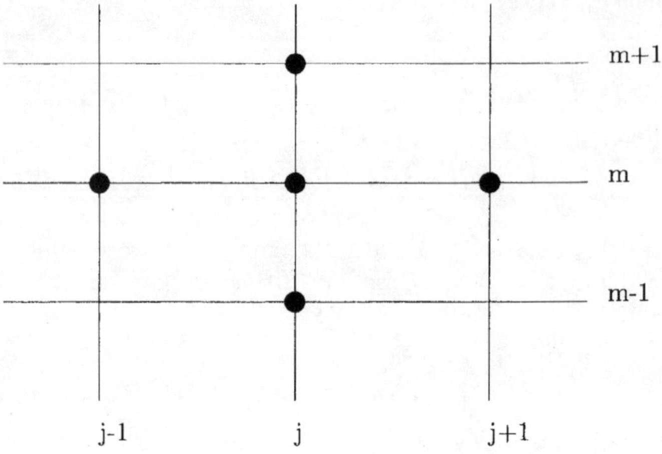

FIGURE 5.3. *The computational molecule of the scheme (5.20).*

for the spatial derivative $\partial^2/\partial x^2$. In the present case it seems rather natural to use a similar approximation also for the second-order derivative with respect to time. Hence, we have motivated the difference scheme

$$\frac{v_j^{m-1} - 2v_j^m + v_j^{m+1}}{(\Delta t)^2} = \frac{v_{j-1}^m - 2v_j^m + v_{j+1}^m}{(\Delta x)^2}. \tag{5.20}$$

The computational molecule for this scheme is illustrated in Fig. 5.3.

The difference scheme (5.20) will be assumed to hold for all interior grid points in x-direction, i.e. for $j = 1, 2, \ldots, n$, and for $m \geq 1$. Of course, we also require the discrete solution to satisfy the boundary conditions in (5.19), i.e.

$$v_0^m = v_{n+1}^m = 0 \quad \text{for} \quad m \geq 0.$$

It is easy to see that if $\{v_j^m\}_{j=1}^n$ and $\{v_j^{m-1}\}_{j=1}^n$ are known, then the solutions $\{v_j^{m+1}\}_{j=1}^n$ can be computed directly from (5.20). Therefore, the scheme is explicit, i.e. we do not need to solve linear systems. However, we note that in order to start the process, we need to know v at the first two time levels. We obviously choose

$$v_j^0 = f(x_j) \quad \text{for} \quad j = 1, 2, \ldots, n. \tag{5.21}$$

In order to obtain approximations v_j^1 for $u(x, \Delta t)$ we use a Taylor's expansion with respect to time to obtain

$$\begin{aligned} u(x, \Delta t) &= u(x, 0) + (\Delta t)u_t(x, 0) + \frac{(\Delta t)^2}{2} u_{tt}(x, 0) + O((\Delta t)^3) \\ &= f(x) + (\Delta t)g(x) + \frac{(\Delta t)^2}{2} f''(x) + O((\Delta t)^3). \end{aligned}$$

Here the last equality follows from (5.19), since

$$u_{tt}(x,0) = u_{xx}(x,0) = f''(x).$$

Hence, we have motivated the following approximation v_j^1 for $u(x_j, \Delta t)$:

$$v_j^1 = v_j^0 + (\Delta t)g(x_j) + \frac{(\Delta t)^2}{2(\Delta x)^2}\left(v_{j-1}^0 - 2v_j^0 + v_{j+1}^0\right). \tag{5.22}$$

In order to write the finite difference scheme in a more compact form, we let $v^m \in \mathbb{R}^n$ be the vector $v^m = (v_1^m, v_2^m, \ldots, v_n^m)^T$ and $A \in \mathbb{R}^{n,n}$ the tridiagonal matrix

$$A = \frac{1}{(\Delta x)^2}\begin{pmatrix} 2 & -1 & 0 & \ldots & 0 \\ -1 & 2 & -1 & \ddots & \vdots \\ 0 & \ddots & \ddots & \ddots & 0 \\ \vdots & \ddots & -1 & 2 & -1 \\ 0 & \ldots & 0 & -1 & 2 \end{pmatrix}. \tag{5.23}$$

Then the difference scheme above can be written

$$v^{m+1} = \left(2I - (\Delta t)^2 A\right)v^m - v^{m-1} \quad \text{for} \quad m \geq 1, \tag{5.24}$$

where the initial approximations v^0 and v^1 are determined by (5.21) and (5.22).

EXAMPLE 5.3 Consider the initial-boundary value problem studied in Example 5.2, i.e.

$$f(x) = x(1-x) \quad \text{and} \quad g(x) = 0.$$

We will compare the exact solution derived in Example 5.2 with solutions obtained by the finite difference scheme above. First we choose $\Delta x = \Delta t = 1/20$. The numerical solution for $t = 1.00$ is compared to the analytical solution, obtained in Example 5.2, in the left part of Fig. 5.4. As we observe, the two solutions are so close that we cannot see the difference between them. Next we change the parameters in the numerical scheme to $\Delta x = 1/21$ and $\Delta t = 1/20$. This has the effect of modifying the mesh ratio $\Delta t/\Delta x$ from 1 to 1.05. The result is given in the right part of Fig. 5.4. We observe that the numerical solution now contains undesired oscillations not present in the analytical solution.

Hence, it appears that there is a stability condition for the scheme (5.20) which has been violated in the second computation. We will investigate the stability properties of the difference scheme further in the next section. ■

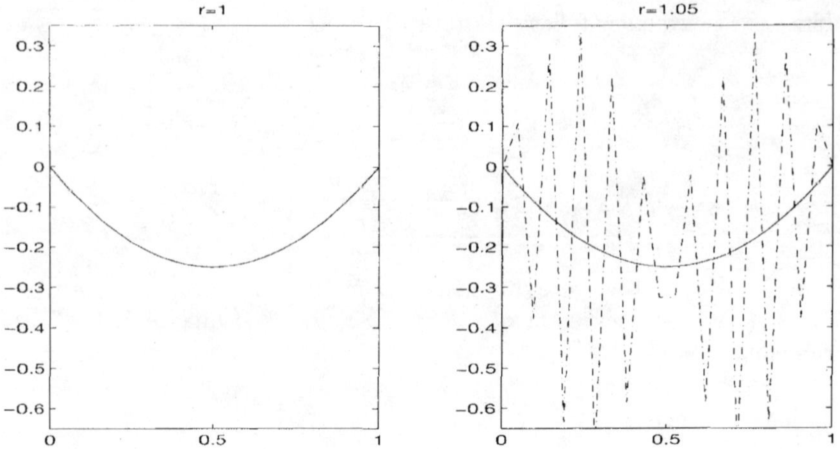

FIGURE 5.4. *The function* $u(x,t)$ *for* $0 \le x \le 1$ *and* $t = 1$ *using* $r = 1$ *(left) and* $r = 1.05$ *(right). The numerical solutions are dashed, while the analytic solution is solid.*

5.3.1 Stability Analysis

In order to explain the instability phenomenon observed in Example 5.3 above, we will perform a stability analysis of the finite difference scheme (5.20). In order to motivate our approach, let us recall the particular solutions (5.11) for the continuous problem (5.19). If we use the complex form (5.10) for the functions $T_k(t)$, then these solutions take the form

$$u_k(x,t) = \sin(k\pi x)e^{\pm ik\pi t}. \tag{5.25}$$

As before we let $X_k \in \mathbb{R}^n$ be the vector with components $X_{k,j} = \sin(k\pi x_j)$. For the finite difference scheme (5.24) we will consider possible solutions of the form

$$v^m = X_k a^m \quad \text{or} \quad v_j^m = X_{k,j} a^m, \tag{5.26}$$

where a is a complex number.

In order to see that this will define particular solutions of the difference scheme, we simply need to recall that if $1 \le k \le n$, X_k is an eigenvector of the matrix A given by (5.23). Furthermore, from Lemma 2.9 it follows that the corresponding eigenvalues $\mu_k = \frac{4}{(\Delta x)^2}\sin^2(k\pi\Delta x/2)$. Therefore, if we insert the ansatz (5.26) into (5.24), we obtain

$$a^2 - (2-s)a + 1 = 0, \tag{5.27}$$

where $s = (\Delta t)^2\mu_k = 4\frac{(\Delta t)^2}{(\Delta x)^2}\sin(k\pi\Delta x/2)$. Hence, if we let r be the mesh ratio, $r = \Delta t/\Delta x$, then $s \in (0, 4r^2)$.

The particular solutions given by (5.25) will always have the property that

$$|u_k(x,t)| \leq 1.$$

It is therefore reasonable to demand that the particular solutions (5.26) of the difference scheme have a corresponding property. We shall therefore require that

$$|a| \leq 1. \tag{5.28}$$

To be more precise, let us note that the roots a of (5.27) will depend on s, i.e. $a = a(s)$. Since $s \in (0, 4r^2)$, we will therefore define the scheme to be *stable* as long as the roots $a(s)$ satisfy (5.28) for all $s \in (0, 4r^2)$.

Lemma 5.1 *Let $s \geq 0$ be given. The roots of (5.27) satisfy (5.28) if and only if $s \leq 4$.*

Proof: The roots of (5.27) are given by

$$a = \frac{2 - s \pm \sqrt{s(s-4)}}{2}. \tag{5.29}$$

If $s = 0$, there is a double root for $a = 1$, and if $s = 4$, the only root is -1. If $s \in (0,4)$, there are two complex roots a_1 and a_2. Written in polar coordinates, these are of the form

$$a_1 = \rho e^{i\theta} \quad \text{and} \quad a_2 = \rho e^{-i\theta}$$

for $\rho > 0$ and $\theta \in (0, \pi)$. Furthermore, from (5.27) it follows that the product of the roots is 1, i.e.

$$a_1 a_2 = \rho^2 = 1.$$

Hence, the roots are of the form $e^{\pm i\theta}$, and therefore the bound (5.28) holds. On the other hand, if $s > 4$, there are two distinct real roots a_1 and a_2, with $a_1 a_2 = 1$. Hence, one of them must have absolute value greater than 1 in this case. ■

As a consequence of this Lemma, the roots will satisfy (5.28) for all $s \in (0, 4r^2)$ if and only if

$$\boxed{r = \Delta t / \Delta x \leq 1.} \tag{5.30}$$

We recall that this stability bound is consistent with the observation done in Example 5.3. If the mesh parameters satisfy this bound, the numerical solution behaves qualitatively as the exact solution. However, if the bound is violated, we observe oscillations in the numerical solution which are not present in the exact solution.

5.4 Exercises

EXERCISE 5.1 Find the formal solutions of the problem

$$\begin{aligned} u_{tt} &= u_{xx} \quad \text{for} \quad x \in (0,1),\ t > 0, \\ u(0,t) &= u(1,t) = 0, \\ u(x,0) &= f(x), \quad u_t(x,0) = g(x), \end{aligned}$$

for the initial functions

(a) $f(x) = 3\sin(2\pi x)$, $g(x) = \sin(\pi x)$,

(b) $f(x) = 3\sin(2\pi x)$, $g(x) = x(1-x)$,

(c) $f(x) = 3\sin(2\pi x)$, $g(x) = \sin(x)\cos(4x)$.

EXERCISE 5.2 (a) Assume that $u = u(x,t)$ solves a wave equation of the form

$$u_{tt} = c^2 u_{xx},$$

where c is a constant. Let $v(x,t) = u(x,\alpha t)$. Determine $\alpha > 0$ such that v satisfies the corresponding equation with $c = 1$, i.e.

$$v_{tt} = v_{xx}.$$

(b) Find the formal solution of the problem:

$$\begin{aligned} u_{tt} &= c^2 u_{xx} \quad \text{for} \quad x \in (0,1),\ t > 0, \\ u(0,t) &= u(1,t) = 0, \\ u(x,0) &= f(x), \qquad u_t(x,0) = g(x), \end{aligned}$$

when

$$f(x) = \sum_{k=1}^{\infty} a_k \sin(k\pi x), \quad g(x) = \sum_{k=1}^{\infty} b_k \sin(k\pi x).$$

EXERCISE 5.3 (a) Find the formal solution of the problem:

$$\begin{aligned} u_{tt} &= u_{xx} \quad \text{for} \quad x \in (0,1),\ t > 0, \\ u_x(0,t) &= u_x(1,t) = 0, \\ u(x,0) &= f(x), \qquad u_t(x,0) = g(x), \end{aligned}$$

when

$$f(x) = \sum_{k=0}^{\infty} a_k \cos(k\pi x); \quad g(x) = \sum_{k=0}^{\infty} b_k \cos(k\pi x).$$

(Note that the boundary conditions are of Neumann-type.)

(b) Show that the energy

$$E(t) = \int_0^1 \left(u_x^2(x,t) + u_t^2(x,t)\right)dx$$

is constant in time if u is a smooth solution of the problem above.

EXERCISE 5.4 Find the formal solution of the following problem:

$$\begin{aligned} u_{tt} &= u_{xx} \quad \text{for} \quad x \in (0,1),\ t > 0, \\ u(0,t) &= a, \qquad u(1,t) = b, \\ u(x,0) &= f(x), \qquad u_t(x,0) = g(x), \end{aligned}$$

for given constants a and b.

EXERCISE 5.5 Find the formal solution of the following problem:

$$\begin{aligned} u_{tt} &= u_{xx} + 2x \quad \text{for} \quad x \in (0,1),\ t > 0, \\ u(0,t) &= u(1,t) = 0, \\ u(x,0) &= f(x), \qquad u_t(x,0) = g(x). \end{aligned}$$

EXERCISE 5.6 Implement the scheme (5.20) for the initial-boundary value problem (5.19). Investigate the performance of the method by comparing the numerical results with the analytical solutions given in

(a) Example 5.1,

(b) Example 5.2.

EXERCISE 5.7 In this problem we study the wave equation as a two-point boundary value problem with respect to time. For $T > 0$ consider the problem

$$\begin{aligned} u_{tt} &= u_{xx} \quad \text{for} \quad x \in (0,1),\ 0 < t < T, \\ u(0,t) &= u(1,t) = 0, \\ u(x,0) &= f(x), \quad u(x,T) = g(x). \end{aligned}$$

(a) Assume that

$$f(x) = \sum_{k=1}^{\infty} a_k \sin(k\pi x); \quad g(x) = \sum_{k=1}^{\infty} b_k \sin(k\pi x).$$

Find a formal solution of the problem when $T = \sqrt{2}$.

(b) Assume $T = 1$. Does the problem have a unique solution in this case?

EXERCISE 5.8 Consider a damped wave equation of the form

$$\begin{aligned} u_{tt} + u_t &= u_{xx} \quad \text{for} \quad x \in (0,1),\ t > 0, \\ u(0,t) &= u(1,t) = 0, \\ u(x,0) &= f(x), \quad u_t(x,0) = g(x). \end{aligned}$$

(a) Find a formal solution of the problem.

(b) Assume that $u = u(x,t)$ is a smooth solution of the problem above and let

$$E(t) = \int_0^1 \left(u_x^2(x,t) + u_t^2(x,t)\right) dx.$$

Show that

$$E(t) \le E(0) \quad \text{for} \quad t \ge 0.$$

EXERCISE 5.9 In this problem we shall study an implicit finite difference scheme for the problem (5.19). Instead of the scheme (5.24) we consider a finite difference scheme of the form

$$v^{m+1} - 2v^m + v^{m-1} = -\frac{(\Delta t)^2}{4} A\left(v^{m+1} + 2v^m + v^{m-1}\right), \tag{5.31}$$

where the matrix A is given by (5.23).

(a) Write the difference scheme in component form (i.e. similar to (5.20)) and sketch the computational molecule for the scheme.

(b) Assume that $v^{m-1}, v^m \in \mathbb{R}^n$ are known. Explain how we compute v^{m+1} from (5.31) and show that v^{m+1} is uniquely determined.

(c) Perform a stability analysis for the scheme (5.31). Show that the scheme is stable independent of the mesh ratio $\Delta t/\Delta x$.

EXERCISE 5.10 Implement the implicit scheme (5.31) for the initial-boundary value problem (5.19). By doing experiments similar to those in Example 5.3 for the explicit method (5.20), try to verify that the method is stable independent of the mesh ratio $\Delta t/\Delta x$.

Exercise 5.11 In this problem we shall study first-order initial-boundary value problems of the form

$$\begin{aligned} u_t + cu_x &= 0, \qquad x \in (0,1),\ t > 0, \\ u(0,t) &= g(t), \\ u(x,0) &= f(x), \end{aligned} \tag{5.32}$$

where c is a constant. If $c > 0$, the unique solution of this problem is given by

$$u(x,t) = \begin{cases} f(x-ct) & \text{for} \quad x > ct, \\ g(t - x/c) & \text{for} \quad x < ct; \end{cases}$$

cf. the discussion in Example 1.1 on page 12. From this formula we easily derive

$$|u(x,t)| \leq \max\{|f(x)|, |g(\tau)| : x \in [0,1], \tau \in [0,t]\}.$$

A finite difference method will be called stable if its solution satisfies a bound similar to this.

(a) Consider the explicit finite difference method

$$\frac{v_j^{m+1} - v_j^m}{\Delta t} + c\frac{v_j^m - v_{j-1}^m}{\Delta x} = 0 \tag{5.33}$$

for $m \geq 0$ and $j = 1, 2, \ldots, n+1$, where, as usual, $\Delta x = 1/(n+1)$. Sketch the computational molecule for the scheme.

(b) Assume that $c > 0$. Explain how we can use the difference scheme (5.33), together with the initial and boundary values, to compute v_j^m for $m \geq 0$ and $j = 1, 2, \ldots, n+1$. Show that the scheme is stable if

$$c\frac{\Delta t}{\Delta x} \leq 1.$$

(c) Assume that $c < 0$. Show that the scheme is never stable in the sense defined above. How does this correspond to properties of the continuous problem (5.32)?

(d) Assume that $c > 0$ and consider the implicit scheme

$$\frac{v_j^{m+1} - v_j^m}{\Delta t} + c\frac{v_j^{m+1} - v_{j-1}^{m+1}}{\Delta x} = 0.$$

Sketch the computational molecule, explain how v_j^m for $m \geq 0$ and $j = 1, 2, \ldots, n+1$ can be computed from data. Show that the scheme is always stable.

6

Maximum Principles

The purpose of this chapter is to study maximum principles. Such principles state something about the solution of an equation without having to solve it.

We will start by studying two-point boundary value problems. For a class of such problems, we will prove that the solution does not have any interior maxima or minima; thus the extreme values are attained at the boundaries. The method for proving this fact can readily be generalized to the case of time-dependent problems, and we will study the heat equation. Finally, we will consider Poisson's equation in the case of two space dimensions.

6.1 A Two-Point Boundary Value Problem

Before we start studying the maximum principle for the heat equation, let us take one step back and consider a similar problem in a simpler framework. We consider a two-point boundary value problem of the form

$$u''(x) + a(x)u'(x) = 0, \qquad x \in (0,1),$$

where a is a given function and u is known at the endpoints $x = 0$ and $x = 1$. For this problem, we will prove that the solution cannot exceed the boundary values.

The basic idea in deriving maximum principles is usually the following elementary property of functions well known from calculus; in a local maximum x_0 of a smooth function $v(x)$, we have $v'(x_0) = 0$ and $v''(x_0) \leq 0$; see Fig. 6.1.

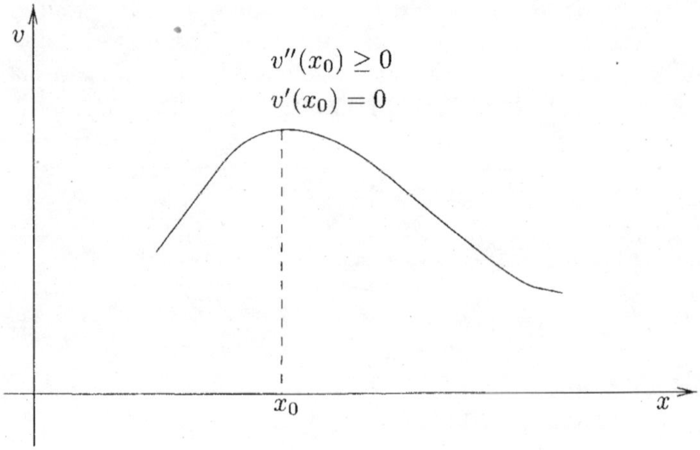

FIGURE 6.1. *A smooth function $v = v(x)$ close to a local maximum.*

We will use this property of a smooth function to prove a maximum principle for the solution of a two-point boundary value problem. In order to do this, we start by considering a differential inequality. Let $v \in C^2\big((0,1)\big) \cap C\big([0,1]\big)$ be a function satisfying the following inequality:

$$v''(x) + a(x)v'(x) > 0, \qquad x \in (0,1), \tag{6.1}$$

where a is continuous on $[0,1]$. Suppose now that v has a local maximum in an interior point x_0, i.e. $x_0 \in (0,1)$. Then, as explained above, we have

(a) $v'(x_0) = 0$ and

(b) $v''(x_0) \leq 0$.

But clearly (a) and (b) imply that $v''(x_0) + a(x_0)v'(x_0) \leq 0$ which is a contradiction of (6.1). Consequently, a smooth function v satisfying the strict inequality (6.1) cannot have a local maximum in the interval $(0,1)$. We have the following result:

Lemma 6.1 *A function $v \in C^2\big((0,1)\big) \cap C\big([0,1]\big)$ satisfying (6.1), where $a \in C\big([0,1]\big)$, satisfies the following maximum principle:*

$$v(x) \leq V \quad \textit{for all} \quad x \in [0,1],$$

where $V = \max(v(0), v(1))$.

This is a nice result, but not exactly what we are looking for. Our aim is to replace the inequality of (6.1) with an equality and still get the same conclusion. The argument given above almost covers this case, but not completely. However, by introducing an auxiliary function, we can use the result of Lemma 6.1 to prove the result we are aiming at.

Consider the two-point boundary value problem

$$u''(x) + a(x)u'(x) = 0, \qquad x \in (0,1), \tag{6.2}$$

with boundary conditions

$$u(0) = u_0 \quad \text{and} \quad u(1) = u_1. \tag{6.3}$$

Here u_0 and u_1 are given constants and $a = a(x)$ is a given continuous function on $[0,1]$.

We want to prove that if $u \in C^2\big((0,1)\big) \cap C\big([0,1]\big)$ is a solution of (6.2),(6.3), then u cannot exceed $\max(u_0, u_1)$ in the interval $(0,1)$. We do this by constructing a sequence of functions $v_\epsilon = v_\epsilon(x)$ satisfying the inequality (6.1) and converging towards u as ϵ tends to zero. For this purpose, let $c = \sup_{x\in[0,1]} |a(x)|$, and define

$$v_\epsilon(x) = u(x) + \epsilon e^{(1+c)x} \tag{6.4}$$

for $\epsilon \geq 0$. Observe that

$$v_\epsilon''(x) + a(x)v_\epsilon'(x) = \epsilon(1+c)(1+c+a(x))e^{(1+c)x},$$

thus,

$$v_\epsilon''(x) + a(x)v_\epsilon'(x) > 0$$

for all $\epsilon > 0$. Hence, it follows from Lemma 6.1 that

$$v_\epsilon(x) \leq \max(v_\epsilon(0), v_\epsilon(1)). \tag{6.5}$$

Going back to (6.4), we observe that

$$\begin{aligned} u(x) &= v_\epsilon(x) - \epsilon e^{(1+c)x} \\ &\leq v_\epsilon(x) \\ &\leq \max(v_\epsilon(0), v_\epsilon(1)) \\ &= \max(u_0 + \epsilon, u_1 + \epsilon e^{1+c}). \end{aligned}$$

Now, by letting $\epsilon \to 0$ from above, we get[1]

$$u(x) \leq \max(u(0), u(1)) = \max(u_0, u_1). \tag{6.6}$$

[1] Here you may wonder how on earth we found such a smart auxiliary function v_ϵ. The answer is that we found it in the book of Protter and Weinberger [21]. But how can such a trick be invented if you do not know the answer? The basic idea here is of course to exploit the fact that we already have a maximum principle for functions satisfying $v'' + av' > 0$, and we want to use this in order to derive a similar maximum principle for u satisfying $u'' + au' = 0$. Thus we want to change u slightly such that the perturbed function satisfies an inequality rather than an equality. If we put $v_\epsilon(x) = u(x) + \epsilon y(x)$, we get $v_\epsilon'' + av_\epsilon' = \epsilon(y'' + ay')$. Hence, any function y satisfying $y''(x) + a(x)y'(x) > 0$ for all $x \in [0,1]$ will do the job. Now, it is not too hard to see that it is reasonable to try some kind of exponential function for y.

So far we have only been concerned with upper bounds for u. Of course, lower bounds are equally important. In order to derive a similar lower bound, we could go through the same steps once more. However, a slick trick enables us to use the result we have already obtained.

Define

$$w(x) = -u(x),$$

and observe that

$$w''(x) + a(x)w'(x) = 0.$$

Then, by the argument given above, we get

$$w(x) \leq \max(w(0), w(1)).$$

Hence

$$-u(x) \leq \max(-u_0, -u_1) = -\min(u_0, u_1),$$

and consequently

$$u(x) \geq \min(u_0, u_1).$$

By summarizing our observations, we have the following result:

Theorem 6.1 *Suppose* $u \in C^2((0,1)) \cap C([0,1])$ *is a solution of (6.2)–(6.3). Then* $u(x)$ *satisfies*

$$\min(u_0, u_1) \leq u(x) \leq \max(u_0, u_1)$$

for all $x \in [0,1]$.

We observe that the derivation of the maximum principle given above is done without finding a formula for the solution u. However, for the rather simple problem above, it is easy to find an explicit formula for u, and this formula can be used to give a direct proof of Theorem 6.1 (see Exercise 6.1). The main reason for presenting the argument above is that this proof may serve as a guidline for how to construct similar proofs for more complex problems, where a simple explicit formula is not available.

6.2 The Linear Heat Equation

In the section above, we saw that a maximum principle can be derived for the solution of a two-point boundary value problem by applying only elementary properties of smooth functions. In this section we will use exactly the same technique in order to derive a maximum principle for the linear heat equation. We will prove that the maximum value of the solution cannot be attained in the interior of the solution domain; a maximum value must be attained either initially or at one of the boundaries. In the next section, we go one step further and apply this technique to the nonlinear heat equation.

When reading this section, it may be useful to have a physical interpretation of the heat equation in mind. Consider a uniform rod of unit length with an initial temperature given by $f(x)$. The temperatures at the left and right boundaries are given by $u_\ell(t)$ and $u_r(t)$ respectively. Then the temperature $u = u(x,t)$ in the rod is governed[2] by the following model:

$$\begin{aligned} u_t &= u_{xx} \quad \text{for} \quad x \in (0,1), \quad t \in (0,T], \\ u(0,t) &= u_\ell(t), \quad u(1,t) = u_r(t), \quad t \in [0,T], \\ u(x,0) &= f(x), \quad x \in [0,1], \end{aligned} \tag{6.7}$$

for an appropriate choice of scales. Here $T > 0$ is a finite constant. We assume that at $t = 0$ the boundary conditions coincide with the initial data at the endpoints, i.e. we assume $u_\ell(0) = f(0)$ and $u_r(0) = f(1)$.

Let us start by considering the special case $u_\ell(t) = u_r(t) = 0$, i.e. the temperature is kept equal to zero at the endpoints. Furthermore, we assume that the initial temperature is positive throughout the rod. Then, just from experience, we would expect the temperature to decrease in the rod and eventually converge towards zero. Also, it would come as a bit of a surprise if we found a spot within the rod that is hotter than the highest initial temperature of the rod. Carrying this a bit further by allowing nonzero temperature on the boundaries, we would expect the highest temperature to appear either initially or at one of the boundaries. For instance, if we have a rod with the temperature initially equal to zero and then start heating the left endpoint but keep the temperature at the right endpoint equal to zero, we expect, for some fixed time greater than zero, to see a monotonically

[2] In the field of applied mathematics we often say that a physical phenomenon is "governed" by a certain mathematical model. Obviously, this should not be interpreted literally; what we mean is usually that the model gives a reasonable description of the phenomenon under consideration. Keep in mind that we are only capable of deriving models. In some fortunate situations, they may provide very accurate predictions, but they are still models.

On the other hand, you should also be aware of the fact that results from physical experiments and observations can never, with any rigor, be used as evidence for properties of the mathematical model. Thus, although we know that there exists a physical temperature in the rod we are considering, we cannot use this as an argument for the existence of a solution of the mathematical problem. If the modeling has been properly done, one may certainly hope that properties that are apparent in the real-world physical situation may be carried over to our model, but such similarities are not evident.

decreasing temperature profile. These rather obvious properties[3] will be proved for the heat equation in this section.

As for the boundary value problem, we will also consider a finite difference scheme and prove the proper maximum principle for the discrete solutions.

6.2.1 *The Continuous Case*

Our aim is to derive a maximum principle for the solution of (6.7). But as for the two-point boundary value problem, we find it convenient to start by considering a differential inequality. Then, through a regularization[4] of the problem (6.7), we prove the maximum principle by letting the regularization parameter go to zero.

Define R to be the rectangle in the (x, t) plane given by

$$R = \{(x, t) : x \in [0, 1], \quad t \in [0, T]\}. \tag{6.8}$$

Let $v = v(x, t)$ be a smooth function satisfying the inequality

$$v_t < v_{xx} \quad \text{for} \quad 0 < x < 1,\ 0 < t \leq T. \tag{6.9}$$

Here we refer to v as a smooth function if v is continuous on the closed rectangle R, with v_t, v_x and v_{xx} continuous for $x \in (0, 1)$ and $t > 0$. We will show that v cannot have any maximum in the interior of R. More precisely, a maximum of v has to be attained at the "lower" boundary of R. The "lower" boundary is defined by

$$\begin{aligned} B = \{(x, t) : x = 0,\ 0 \leq t \leq T\} \cup \{(x, t) : t = 0,\ 0 \leq x \leq 1\} \\ \cup \{(x, t) : x = 1,\ 0 \leq t \leq T\}; \end{aligned} \tag{6.10}$$

see Fig. 6.2.

We will prove the maximum principle by assuming that v has a maximum in the interior of R, and then derive a contradiction to (6.9).

Suppose that (x_0, t_0) is a local maximum of v in the interior of R, i.e. $x_0 \in (0, 1)$ and $t_0 \in (0, T)$. Then, by the properties discussed in the previous section, we have

(i) $v_t(x_0, t_0) = 0$ and

---

[3] Albeit obvious from a physical point of view, the maximum principle is not at all trivial from a mathematical point of view. One natural way to try to prove the principle is to consider the Fourier solution derived in Chapter 3. This attempt fails; it is very hard to prove a maximum principle based on such a series expansion.

[4] The term "regularization" is often used in mathematics. Usually, it means to change something slightly in a favorable direction. For instance, looking at the two-point boundary value problem above, we "regularized" the problem by adding a little term such that a differential equation was changed to a differential inequality which we know something about.

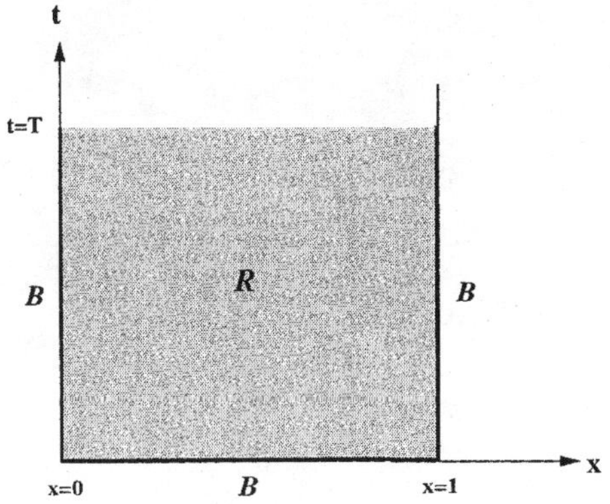

FIGURE 6.2. *Definition of the rectangle R and the lower boundary B.*

(ii) $v_{xx}(x_0, t_0) \leq 0$.

But now (i) and (ii) imply that

$$v_t(x_0, t_0) \geq v_{xx}(x_0, t_0),$$

which clearly contradicts (6.9). Consequently, we cannot have an interior maximum for a smooth function satisfying the inequality (6.9). But what about the upper boundary, i.e. $t = T$; can we have a local maximum for some $x_0 \in (0, 1)$ and $t_0 = T$? Suppose that is the case. Then it follows that

(iii) $v_t(x_0, t_0) \geq 0$ and

(iv) $v_{xx}(x_0, t_0) \leq 0$,

which again contradicts (6.9). Thus we have derived the following result:

Lemma 6.2 *A function $v \in C(R)$, with $v_t, v_x, v_{xx} \in C((0, 1) \times (0, T])$, satisfying the inequality (6.9) obeys the following maximum principle:*

$$v(x, t) \leq V \quad \text{for all} \quad x \in [0, 1], \quad t \in [0, T],$$

where

$$V = \sup_{(x,t) \in B} v(x, t).$$

This lemma is exactly the tool we need to prove the maximum principle for the heat equation (6.7). As for the boundary value problem, we utilize this result by introducing a regularization of the solution of the heat equation.

Let u be a smooth solution of (6.7). With a smooth solution we mean that u is continuous on the closed rectangle R with u_t, u_x, and u_{xx} continuous for $x \in (0, 1)$ and $t > 0$. Define

$$v^\epsilon(x, t) = u(x, t) + \epsilon x^2 \tag{6.11}$$

for $\epsilon > 0$. Then

$$v_t^\epsilon = v_{xx}^\epsilon - 2\epsilon.$$

Hence, for any $\epsilon > 0$, we have

$$v_t^\epsilon < v_{xx}^\epsilon,$$

and it follows from the lemma that

$$v^\epsilon(x,t) \leq V^\epsilon, \tag{6.12}$$

where V^ϵ denotes the maximum of v^ϵ on the boundary B. Now, it follows from (6.11) that

$$u(x,t) = v^\epsilon(x,t) - \epsilon x^2 \leq v^\epsilon(x,t).$$

Hence, for any $\epsilon > 0$, we have[5]

$$u(x,t) \leq V^\epsilon = \sup_{(x,t)\in B} (f(x) + \epsilon x^2, u_\ell(t), u_r(t) + \epsilon).$$

By letting ϵ tend to zero from above, we get

$$u(x,t) \leq \sup_{(x,t)\in B} (f(x), u_\ell(t), u_r(t)). \tag{6.13}$$

In order to derive a similar lower bound for u, we apply the same trick as for the two-point boundary value problem. Let $w(x,t) = -u(x,t)$. Then $w_t = w_{xx}$ and, using (6.13), we have

$$w(x,t) \leq \sup_{(x,t)\in B} (-f(x), -u_\ell(t), -u_r(t)) = -\inf_{(x,t)\in B} (f(x), u_\ell(t), u_r(t)),$$

and consequently

$$u(x,t) = -w(x,t) \geq \inf_{(x,t)\in B} (f(x), u_\ell(t), u_r(t)).$$

We can summarize these observations as follows:

Theorem 6.2 *Suppose* $u \in C(R)$, *with* u_t, u_x, $u_{xx} \in C((0,1) \times (0,T])$, *is a solution of (6.7). Then* u *satisfies the maximum principle*[6]

$$\inf_{(x,t)\in B} (f(x), u_\ell(t), u_r(t)) \leq u(x,t) \leq \sup_{(x,t)\in B} (f(x), u_\ell(t), u_r(t)).$$

for all $(x,t) \in R$.

---

[5] Note that $\sup_y (a(y), b(y), c(y))$ is shorthand for $\max(\sup_y a(y), \sup_y b(y), \sup_y c(y))$. A similar notation is used for inf.

[6] Recall that the domain R and its lower boundary B are defined in Fig. 6.2.

We should remark here that the proof above requires that u is smooth, i.e. u is continuous on the closed rectangle R. In particular, this implies that u is continuous at $(x,t) = (0,0)$ and $(x,t) = (1,0)$ or

$$u_\ell(0) = f(0) \quad \text{and} \quad u_r(0) = f(1).$$

Later, in Chapter 10, we shall refer to these relations as compatibility conditions for the data. Note that these conditions are not satisfied for the problem studied in Example 3.2. Hence, the maximum principle has not been established for this case.

6.2.2 *Uniqueness and Stability*

A maximum principle for a linear differential equation will frequently imply a uniqueness and stability result for the solution (see Exercise 6.2). This is also the case for the present model. In order to derive this result, we let u denote a solution of

$$\begin{aligned} u_t &= u_{xx} \quad \text{for} \quad x \in (0,1), \quad t \in [0,T], \\ u(0,t) &= u_\ell(t), \quad u(1,t) = u_r(t), \quad t \in [0,T], \\ u(x,0) &= f(x), \quad x \in [0,1], \end{aligned} \tag{6.14}$$

and similarly, $\bar{u}$ denotes a solution of

$$\begin{aligned} \bar{u}_t &= \bar{u}_{xx} \quad \text{for} \quad x \in (0,1), \quad t \in [0,T], \\ \bar{u}(0,t) &= \bar{u}_\ell(t), \quad \bar{u}(1,t) = \bar{u}_r(t), \quad t \in [0,T], \\ \bar{u}(x,0) &= \bar{f}(x), \quad x \in [0,1]. \end{aligned} \tag{6.15}$$

Furthermore, we let e denote the difference between these solutions, i.e. $e = u - \bar{u}$. Then e is a solution of the following initial-boundary value problem:

$$\begin{aligned} e_t &= e_{xx} \quad \text{for} \quad x \in (0,1), \quad t \in [0,T], \\ e(0,t) &= \Delta u_\ell(t), \quad e(1,t) = \Delta u_r(t), \quad t \in [0,T], \\ e(x,0) &= \Delta f(x), \quad x \in [0,1], \end{aligned} \tag{6.16}$$

where $\Delta u_\ell(t) = u_\ell(t) - \bar{u}_\ell(t)$, $\Delta u_r(t) = u_r(t) - \bar{u}_r(t)$ and $\Delta f(x) = f(x) - \bar{f}(x)$. By Theorem 6.2 above, we get

$$\inf_{(x,t)\in B} (\Delta f(x), \Delta u_\ell(t), \Delta u_r(t)) \le e(x,t) \le \sup_{(x,t)\in B} (\Delta f(x), \Delta u_\ell(t), \Delta u_r(t)),$$

and then we have the following result:

Corollary 6.1 *The problem (6.14) has at most one smooth solution. Furthermore, the solution is stable with respect to perturbations in the sense that*

$$\sup_{(x,t)\in R} |u(x,t) - \bar{u}(x,t)| \le \sup_{(x,t)\in B} (|f(x) - \bar{f}(x)|, |u_\ell(t) - \bar{u}_\ell(t)|, |u_r(t) - \bar{u}_r(t)|)$$

where u is the solution of (6.14) and $\bar{u}$ is the solution of (6.15). Here, the domain R and the lower boundary B are defined in Fig. 6.2.

6.2.3 The Explicit Finite Difference Scheme

Having derived the proper maximum principle for the heat equation, we proceed by analyzing a numerical method for this initial-boundary value problem. In the present section we will consider the explicit scheme introduced in Section 4.1. Using a discrete version of Fourier's method, we derived a certain stability condition for this scheme. The same type of stability condition was derived for several other problems in Section 4.3 using the method of von Neumann. Now we will prove that the stability condition derived earlier is sufficient in order for the numerical solutions to satisfy a discrete version of the maximum principle.

In the next section we will address the same question for the implicit finite difference scheme introduced in Section 4.4. It turns out that the numerical solutions generated by the implicit scheme satisfy the discrete maximum principle for any relevant choice of grid parameters.

We consider the explicit finite difference scheme for the initial boundary value problem (6.7). Let us start by briefly recapitulating the basic notation. Let v_j^m denote an approximation to the exact solution u at the grid point (x_j, t_m). As usual, $x_j = j\Delta x$, where $\Delta x = 1/(n+1)$ for a given integer $n \ge 1$, and $t_m = m\Delta t$, where $\Delta t > 0$ is referred to as the time step.

The explicit scheme derived in Section 4.1 applied to the initial-boundary value problem (6.7) can be written in the following form:

$$v_j^{m+1} = rv_{j-1}^m + (1-2r)v_j^m + rv_{j+1}^m, \qquad j = 1, \ldots, n, \quad m \ge 0, \tag{6.17}$$

where $r = \Delta t/\Delta x^2$. The boundary conditions of (6.7) give

$$v_0^m = u_\ell(t_m) \quad \text{and} \quad v_{n+1}^m = u_r(t_m) \tag{6.18}$$

for $m \ge 0$, and the initial condition leads to

$$v_j^0 = f(x_j) \quad \text{for} \quad j = 1, \ldots, n. \tag{6.19}$$

Our aim is now to prove that the discrete solution v_j^m defined by this explicit scheme satisfies a maximum principle similar to the result in the continuous case. In order to prove this, we will need some notation which is very closely related to the notation we used in the continuous case. We define B_Δ to be the grid points located on the lower boundary B, and R_Δ to be the collection of grid points in the rectangle R. Here B and R are sketched in Fig. 6.2 above. More specifically, we define the "discrete rectangle"

$$R_\Delta = \{(x_j, t_m) : x_j \in [0,1], \quad t_m \in [0,T]\} \tag{6.20}$$

and the associated "lower boundary"

$$\begin{aligned} B_\Delta = &\{(x_j, t_m) : x_j = 0,\ 0 \le t_m \le T\} \cup \{(x_j, t_m) : t_m = 0,\ 0 \le x_j \le 1\} \\ &\cup \{(x_j, t_m) : x_j = 1,\ 0 \le t_m \le T\}; \end{aligned} \tag{6.21}$$

see Fig. 6.3. For brevity, we define[7]

$$V^- = \min_{(x_i, t_k) \in B_\Delta} (f(x_i), u_\ell(t_k), u_r(t_k))$$

and

$$V^+ = \max_{(x_i, t_k) \in B_\Delta} (f(x_i), u_\ell(t_k), u_r(t_k)).$$

We want to show that a numerical solution generated by the scheme (6.17)–(6.19) is bounded below by V^- and above by V^+. This will be done under the assumption that the grid parameters satisfy the following condition:

$$r = \frac{\Delta t}{(\Delta x)^2} \le 1/2. \tag{6.22}$$

This is exactly the condition we derived in Section 4.1 using discrete Fourier analysis; see (4.25) on page 130.

Theorem 6.3 *Suppose that the grid sizes Δx and Δt satisfy the condition (6.22), and let v_j^m be the numerical approximation of (6.7) generated by the scheme (6.17)–(6.19). Then*

$$V^- \le v_j^m \le V^+$$

for all grid points $(x_j, t_m) \in R_\Delta$.

Proof: The proof of the lower and the upper bound, are similar, so we concentrate on the upper bound, which is verified by induction on the time level. Consider one fixed time level t_m, and assume that

$$v_j^m \le V^+ \quad \text{for} \quad j = 1, \ldots, n.$$

[7] Here $\min_i(a_i, b_i, c_i)$ is shorthand for $\min(\min_i a_i, \min_i b_i, \min_i c_i)$. We use a similar notation for max.

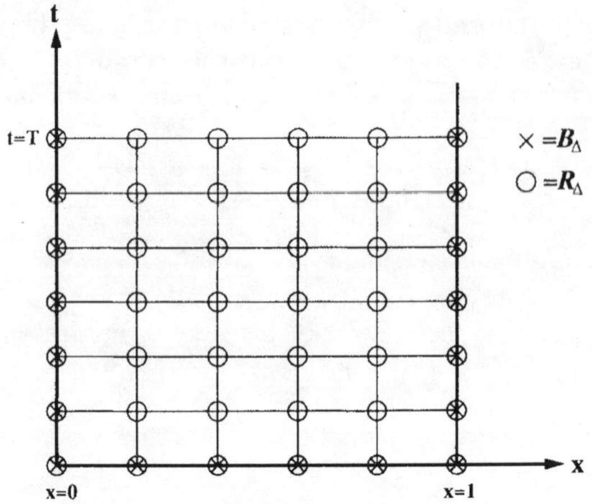

FIGURE 6.3. *Definition of the rectangle R_Δ and the lower boundary B_Δ.*

By (6.17) we have

$$v_j^{m+1} = rv_{j-1}^m + (1-2r)v_j^m + rv_{j+1}^m, \quad \text{for} \quad j = 1, \dots, n,$$

and by (6.22), we have $1 - 2r \geq 0$. These facts imply that

$$v_j^{m+1} \leq rV^+ + (1-2r)V^+ + rV^+ = V^+.$$

Since this holds for any $j = 0, \dots, n+1$, the result follows by induction on m. ■

As in the continuous case, we can use this result to prove stability of the numerical solutions with respect to perturbations in the initial or boundary data. You are asked to formulate and prove such a result in Exercise 6.10.

6.2.4 The Implicit Finite Difference Scheme

We recall from the discussion in Section 4.4 that explicit schemes tend to become very CPU-time demanding as Δx is reduced. This is due to the stability condition (6.22) which forces the number of time steps to be of order $O(n^2)$, where n is the number of grid points in the x direction. This fact motivated the development of an implicit scheme in Section 4.4. According to the von Neumann method, the implicit scheme is stable for any positive values of the grid parameters. In this section we will look at this problem once more, and prove that the scheme indeed satisfies the discrete maximum principle for any relevant choice of grid parameters. You should note that this is not a consequence of the von Neumann method, which merely guarantees that each of the discrete particular solutions are well behaved.

Using the same notation as for the explicit scheme, we recall that the implicit scheme has the following form:

$$\frac{v_j^{m+1} - v_j^m}{\Delta t} = \frac{v_{j-1}^{m+1} - 2v_j^{m+1} + v_{j+1}^{m+1}}{\Delta x^2} \quad \text{for} \quad j = 1, \dots, n, \quad m \geq 0. \tag{6.23}$$

The boundary conditions and initial data lead to

$$v_0^m = u_\ell(t_m) \quad \text{and} \quad v_{n+1}^m = u_r(t_m), \quad m \geq 0, \tag{6.24}$$

and

$$v_j^0 = f(x_j) \quad \text{for} \quad j = 1, \dots, n \tag{6.25}$$

respectively. In Section 4.4 we proved that this scheme is well defined; see in particular Exercise 4.5 on page 149. Now we want to show that the numerical solution generated by this scheme satisfies the maximum principle. As above, it is sufficient to consider the problem of deriving an upper bound, since the lower bound can be derived in exactly the same manner. Thus we want to show that

$$v_i^k \leq V^+ = \max_{(x_i, t_k) \in B_\Delta} (f(x_i), u_\ell(t_k), u_r(t_k))$$

for all grid points $(x_i, t_k) \in R_\Delta$.

Note that the scheme can be rewritten in the form

$$(1 + 2r)v_j^{m+1} = v_j^m + r(v_{j-1}^{m+1} + v_{j+1}^{m+1}), \qquad j = 1, \dots, n, \quad m \geq 0, \tag{6.26}$$

where we recall that $r = \Delta t/(\Delta x)^2$. Consider a fixed time level t_m and assume that $v_j^m \leq V^+$ for $j = 0, \dots, n+1$. Then

$$(1 + 2r)v_j^{m+1} \leq V^+ + 2r \max_{i=0,\dots,n+1} v_i^{m+1}, \tag{6.27}$$

for all $j = 1, \dots, n$. Since both v_0^{m+1} and v_{n+1}^{m+1} are bounded by V^+, it follows that the inequality (6.27) holds for all $j = 0, \dots, n+1$. Consequently,

$$(1 + 2r) \max_{i=0,\dots,n+1} v_i^{m+1} \leq V^+ + 2r \max_{i=0,\dots,n+1} v_i^{m+1},$$

and thus

$$\max_{i=0,\dots,n+1} v_i^{m+1} \leq V^+.$$

Now the upper bound is proved by induction on m. A similar argument leads to a lower bound, and we have the following result:

Theorem 6.4 *Let v_j^m be the numerical approximation of (6.7) generated by the implicit scheme (6.23)–(6.25). Then*

$$V^- \leq v_j^m \leq V^+$$

for all grid points $(x_j, t_m) \in R_\Delta$.

It is important to note that this result holds for any positive values of Δx and Δt.

6.3 The Nonlinear Heat Equation

In the previous section we studied the linear heat equation. In the derivation of this equation, one major simplification has been made: the parameters describing the physical properties of the rod are assumed to be constants. Thus, the parameters do not change as the temperature varies along the rod. For small variations in the temperature, such approximations can be justified, but for large variations it is a dubious assumption. For large variations it is desirable, from a modeling point of view, to allow physical quantities like the thermal conductivity to be a function of the temperature. This refinement of the model leads to a nonlinear heat equation, and it motivates the analysis of problems of the following form:

$$\begin{aligned} u_t &= (k(u)u_x)_x \quad \text{for} \quad x \in (0,1), \quad t \in (0,T], && (6.28)\\ u(0,t) &= u_\ell(t), \quad u(1,t) = u_r(t), \quad t \in [0,T], && (6.29)\\ u(x,0) &= f(x), \quad x \in [0,1]. && (6.30) \end{aligned}$$

From physical considerations it is reasonable to assume that the function $k = k(u)$ is smooth and strictly positive. Specifically, we assume that there exist constants k_0 and K_0 such that

$$0 < k_0 \leq k(u) \leq K_0 \qquad (6.31)$$

for all u. Problems of this form are usually referred to as nonlinear heat equations. In this section we prove that solutions of the problem (6.28)–(6.30) satisfy a maximum principle of the same type as the one we derived in the linear case.

It is important to note that in this section we leave the space of exactly solvable problems. The Fourier technique derived in Chapter 3 no longer applies, and there is, in general, no technique available for solving nonlinear heat equations explicitly. Luckily, the finite difference schemes still work fine, and we will show that a numerical solution generated by an explicit finite difference scheme satisfies a discrete version of the maximum principle. Certainly, a stability condition must be satisfied in the discrete case, and as in the linear case, this implies very short time steps. Thus, we want to consider implicit schemes. However, implicit schemes in the nonlinear case lead to tridiagonal systems of nonlinear algebraic equations. We consider the analysis of such equations to be slightly beyond our scope and thus we shall confine ourselves to the analysis of explicit schemes. Some computations showing typical features of implicit schemes will be given, but no analysis will be presented.

6.3.1 The Continuous Case

We start by considering the continuous case, using the same technique as above.

Suppose that $u = u(x,t)$ is a smooth solution of (6.28)–(6.30) and define v^ϵ by

$$v^\epsilon(x,t) = u(x,t) - \epsilon t \tag{6.32}$$

for any $\epsilon > 0$. Since, by (6.28),

$$u_t = k(u)u_{xx} + k'(u)(u_x)^2,$$

we have

$$v_t^\epsilon = k(v^\epsilon + \epsilon t)v_{xx}^\epsilon + k'(v^\epsilon + \epsilon t)(v_x^\epsilon)^2 - \epsilon,$$

and thus

$$v_t^\epsilon < k(v^\epsilon + \epsilon t)v_{xx}^\epsilon + k'(v^\epsilon + \epsilon t)(v_x^\epsilon)^2 \tag{6.33}$$

because $\epsilon > 0$.

Let the rectangle R and the lower boundary B be as above (see page 180). Furthermore, we assume that v^ϵ has a local maximum in the interior of R, say in $(x_0, t_0) \in R \backslash B$ with $t_0 \leq T$. Then

$$0 = v_t^\epsilon(x_0, t_0) = v_x^\epsilon(x_0, t_0) \geq v_{xx}^\epsilon(x_0, t_0). \tag{6.34}$$

But since $k(v_\epsilon + \epsilon t) \geq k_0 > 0$, it follows that (6.34) contradicts (6.33), and consequently there is no local maximum in the interior of R. The upper boundary ($t = T$) can be excluded as in the linear case, and thus we have

$$v^\epsilon(x,t) \leq \sup_{(x,t)\in B} v^\epsilon(x,t)$$

for all $(x,t) \in R$. Since $v = u - \epsilon t$, it follows that

$$u(x,t) - \epsilon t \leq \sup_{(x,t)\in B} \left(f(x), u_\ell(t) - \epsilon t, u_r(t) - \epsilon t\right)$$

for all $(x,t) \in R$. By letting ϵ tend to zero from above, we get the desired upper bound for u. As usual, a corresponding lower bound is derived by considering $w = -u$ and using the upper bound for w. We leave the details of this to the reader, and state the maximum principle for the nonlinear heat equation:

Theorem 6.5 *Suppose $u \in C(R)$, with u_t, u_x, $u_{xx} \in C((0,1) \times (0,T])$, is a solution of (6.28)–(6.30). Then u satisfies the maximum principle*[8]

$$\inf_{(x,t)\in B} (f(x), u_\ell(t), u_r(t)) \leq u(x,t) \leq \sup_{(x,t)\in B} (f(x), u_\ell(t), u_r(t))$$

for all $(x,t) \in R$.

Note that, in contrast to the linear case, this theorem cannot be used to derive a stability result for the initial-boundary value problem. This indicates a property that is true quite generally; it is much harder to prove properties of the nonlinear problems than of their linear counterparts.

6.3.2 An Explicit Finite Difference Scheme

Obviously, we would like to be able to solve the nonlinear problem (6.28)–(6.30) numerically. Furthermore, we would like to compute numerical solutions that satisfy a discrete version of the maximum principle given in Theorem 6.5. We have briefly touched upon this problem earlier. In Example 4.5 on page 138 we studied a numerical method for a nonlinear heat equation. We derived there, somewhat heuristically, a stability condition by freezing the coefficients in the scheme and then applying von Neumann's method. Here we will show that the condition we arrived at using this technique is sufficient to imply a maximum principle for the discrete solutions.

We consider the following explicit finite difference scheme:

$$v_j^{m+1} = rk_{j-1/2}^m v_{j-1}^m + (1 - r(k_{j-1/2}^m + k_{j+1/2}^m))v_j^m + rk_{j+1/2}^m v_{j+1}^m \quad (6.35)$$

for $j = 1, \ldots, n, \quad m \geq 0$. The initial values and the boundary conditions are handled as in the linear case (see (6.18)–(6.19) on page 184). As usual we have $r = \Delta t/\Delta x^2$, and in addition we have defined

$$k_{j+1/2}^m = \frac{1}{2}(k(v_j^m) + k(v_{j+1}^m)).$$

The derivation of this scheme is discussed in Exercise 4.20 on page 155.

In order to state the maximum principle for a discrete solution generated by this finite difference scheme, we recall the definition of V^- and V^+,

$$V^- = \min_{(x_i,t_k)\in B_\Delta} (f(x_i), u_\ell(t_k), u_r(t_k))$$

and

$$V^+ = \max_{(x_i,t_k)\in B_\Delta} (f(x_i), u_\ell(t_k), u_r(t_k)),$$

---

[8]The domain R and its lower boundary B are sketched in Fig. 6.2 on page 181; see also (6.8) and (6.10).

where B_Δ is defined in (6.21). As in the linear case, a certain stability condition has to be satisfied. We assume that Δt and Δx satisfy

$$K_0 \frac{\Delta t}{(\Delta x)^2} \leq 1/2, \tag{6.36}$$

where K_0 is a upper bound for $k(u)$; see (6.31).

As in the linear case, the discrete maximum principle is derived by induction on the time level. Consider a given time level t_m and assume that

$$v_j^m \leq V^+ \quad \text{for} \quad j = 1, \dots, n.$$

Then by using the scheme (6.35) we get

$$v_j^{m+1} \leq rk_{j-1/2}^m V^+ + (1 - r(k_{j-1/2}^m + k_{j+1/2}^m))V^+ + rk_{j+1/2}^m V^+ = V^+,$$

where we have utilized the fact that, by (6.36),

$$(1 - r(k_{j-1/2}^m + k_{j+1/2}^m)) \geq 0.$$

Since this holds for $j = 0, \dots, n+1$, it follows by induction that the numerical solution is bounded above by V^+. In a similar way we can prove that the discrete solution is bounded below by V^-. We summarize these observations in the following theorem:

Theorem 6.6 *Suppose that the grid sizes Δx and Δt satisfy the condition (6.36) and that the function $k = k(u)$ satisfies the requirement (6.31). Furthermore, we let v_j^m be the numerical approximation of (6.28)–(6.30) generated by the scheme (6.35) with boundary conditions and initial data given by (6.18) and (6.19). Then*

$$V^- \leq v_j^m \leq V^+$$

for all grid points $(x_j, t_m) \in R_\Delta$.

In this theorem we only consider functions $k = k(u)$ which satisfy the requirement (6.31) for all values of u. Thus, the theorem covers the case of e.g. $k(u) = 2 + \sin(u)$ but not $k(u) = e^u$. This requirement is too strong, and we will discuss how to weaken it in Exercise 6.12.

6.4 Harmonic Functions

Recall that in Chapter 2 we studied two-point boundary value problems for the differential equation

$$-u_{xx} = f,$$

defined on a bounded interval. In particular, if $f \equiv 0$, we obtain the homogeneous equation

$$-u_{xx} = 0. \tag{6.37}$$

Of course, the solutions of this equation are all linear functions of x. In this section we shall study an analog of the equation (6.37) in two space dimensions.

The differential equation will be studied on a bounded, connected, and open domain, Ω, in $\mathbb{R}^2$. The boundary of Ω will be denoted $\partial\Omega$, and $\bar{\Omega}$ will be the corresponding closed set given by

$$\bar{\Omega} = \Omega \cup \partial\Omega.$$

For the one-dimensional problems studied in Chapter 2, the open interval $(0,1)$ corresponds to the domain Ω, with the two endpoints as its boundary. Hence, the closed interval $[0,1]$ corresponds to $\bar{\Omega}$ in this case.

EXAMPLE 6.1 Assume that $\Omega = \{(x,y) \mid x^2 + y^2 < 1\}$. Then

$$\partial\Omega = \{(x,y) \mid x^2 + y^2 = 1\} \quad \text{and} \quad \bar{\Omega} = \{(x,y) \mid x^2 + y^2 \le 1\}.$$

■

The Laplace operator $\boldsymbol{\Delta}$, in two space dimensions, is defined by

$$\boldsymbol{\Delta} u = u_{xx} + u_{yy} \equiv \frac{\partial^2 u}{\partial x^2} + \frac{\partial^2 u}{\partial y^2}.$$

Definition 6.1 *A function $u \in C^2(\Omega) \cap C(\bar{\Omega})$ is said to be harmonic in Ω if*

$$\boldsymbol{\Delta} u = 0 \quad \text{for all} \quad (x,y) \in \Omega. \tag{6.38}$$

Here the statement $u \in C^2(\Omega)$ means that all partial derivatives of total order ≤ 2 are continuous, i.e. the functions $u, u_x, u_y, u_{xx}, u_{xy}$, and u_{yy} are all continuous in Ω. The equation (6.38) is frequently referred to as the *Laplace equation.* Hence, a function $u \in C^2(\Omega)$ is harmonic in Ω if it satisfies the Laplace equation in Ω and if it is continuous in the closed domain $\bar{\Omega}$. The corresponding inhomogeneous equation

$$-\boldsymbol{\Delta} u = f,$$

where $f = f(x,y)$ is a given function, is usually called *Poisson's equation.*

EXAMPLE 6.2 Let $u(x,y)$ be a polynomial function of the form

$$u(x,y) = a + bx + cy + dxy,$$

where a, b, c, d are real coefficients. Then it is straightforward to check that $\boldsymbol{\Delta} u = 0$. Hence, u is harmonic in any domain Ω. ■

We recall that in one dimension the set of harmonic functions is exactly all linear functions. From the example above we might think that also in two dimensions it will be the case that any harmonic function is necessarily a polynomial function. However, the next example shows that this is not true.

EXAMPLE 6.3 Let $r = r(x, y) = \sqrt{x^2 + y^2}$ and define

$$u(x, y) = \ln(r(x, y)).$$

This function is continuous in all of $\mathbb{R}^2$ except for the origin where it is not defined. A direct calculation shows that

$$u_{xx} = \frac{1}{r^2}(1 - \frac{2x^2}{r^2}) \quad \text{and} \quad u_{yy} = \frac{1}{r^2}(1 - \frac{2y^2}{r^2}),$$

and this implies that $\Delta u = 0$. Hence, the function u is harmonic in any domain which is bounded away from the origin. ■

This example indicates that the set of harmonic functions in two space dimensions is a more complicated and richer set of functions than the corresponding set in one dimension. In fact, as will be clearer below, the set of harmonic functions in a two-dimensional domain Ω can be identified with (smooth) functions defined on its boundary $\partial\Omega$.

6.4.1 Maximum Principles for Harmonic Functions

We will now focus our attention on the maximum principle for harmonic functions. Recall first that if $u = u(x)$ is a linear (or harmonic) function of one variable, then it clearly satisfies the inequality

$$\min(u(a), u(b)) \leq u(x) \leq \max(u(a), u(b)) \quad \text{for all} \quad x \in (a, b).$$

This inequality is in fact also a special case of the more general result given in Theorem 6.1. The maximum principle for harmonic functions in two space variables states that a similar inequality holds for such functions.

Theorem 6.7 *Assume that u is harmonic in Ω. Then u satisfies the inequality*

$$M_0 \leq u(x, y) \leq M_1 \quad \text{for all} \quad (x, y) \in \Omega,$$

where

$$M_0 = \min_{(x,y)\in\partial\Omega} u(x, y) \quad \text{and} \quad M_1 = \max_{(x,y)\in\partial\Omega} u(x, y).$$

Proof: The proof is rather similar to the proof of Theorem 6.2 in the sense that the argument requires a similar regularization of the function u. For any $\epsilon > 0$ define

$$v^\epsilon(x, y) = u(x, y) + \epsilon(x^2 + y^2).$$

Then, since $\Delta u = 0$, it follows that

$$\Delta v^\epsilon = 4\epsilon > 0 \quad \text{for all} \quad (x, y) \in \Omega. \tag{6.39}$$

However, if v^ϵ has a maximum at an interior point (x_0, y_0) of Ω, then by the second derivative test of calculus, it follows that

$$\Delta v^\epsilon = v^\epsilon_{xx} + v^\epsilon_{yy} \leq 0$$

at the point (x_0, y_0). Since this contradicts (6.39), we conclude that v^ϵ has no interior maximum point. Therefore,

$$v^\epsilon(x, y) \leq M_1^\epsilon \quad \text{for all} \quad (x, y) \in \Omega,$$

where $M_1^\epsilon = \max_{(x,y)\in\partial\Omega} v^\epsilon(x, y)$. By letting $\epsilon \to 0$ we obtain

$$u(x, y) \leq M_1.$$

The desired lower bound can be demonstrated by similar arguments. ■

If we inspect the proof above, we will discover that the upper bound, $u \leq M_1$, will follow as long as u satisfies $\Delta u \geq 0$ in Ω. Functions with this property are referred to as *subharmonic functions*.

Corollary 6.2 *Assume that u is subharmonic, i.e. $(\Delta u)(x, y) \geq 0$ for all $(x, y) \in \Omega$. Then*

$$u(x, y) \leq M_1 \quad \text{for all} \quad (x, y) \in \Omega.$$

Proof: Since $\Delta u \geq 0$, we still have (6.39), i.e. $\Delta v^\epsilon > 0$. The desired inequality is therefore derived exactly as in the proof above. ■

Corollary 6.3 *If u is harmonic in Ω, then*

$$|u(x, y)| \leq M \quad \text{for all} \quad (x, y) \in \Omega,$$

where $M = \max_{(x,y)\in\partial\Omega} |u(x, y)|$.

Proof: From Theorem 6.7 we have

$$|u(x, y)| = \max(u(x, y), -u(x, y)) \leq \max(M_1, -M_0) = M.$$

■

As an application of the maximum principle, we will consider the Dirichlet problem for Poisson's equation. For a given function f, defined on Ω, and a given function g, defined on the boundary $\partial\Omega$, this problem takes the form

$$\begin{aligned} -\Delta u &= f \quad \text{in} \quad \Omega, \\ u &= g \quad \text{on} \quad \partial\Omega. \end{aligned} \tag{6.40}$$

The function f will be referred to as the right-hand side, and g is called the Dirichlet data. Under proper conditions on the domain Ω and on the functions f and g, it can be established that there always exists a solution u of this problem. Furthermore, the solution is unique. These facts explain our claim above that the set of harmonic functions on Ω can, in some sense, be identified with the set of functions on the boundary $\partial\Omega$. The solution of the problem (6.40), with $f = 0$, will exactly be a harmonic function with its restriction to $\partial\Omega$ prescribed to be g.

We will return to the construction of solutions of problems of the form (6.40) in later chapters. However, here we shall use the maximum principle for harmonic functions to show that this problem can have at most one solution.

Theorem 6.8 *Assume that $u^1, u^2 \in C^2(\Omega) \cap C(\bar{\Omega})$ are two solutions of the problem (6.40) with the same right-hand side f and the same Dirichlet data g. Then $u^1 \equiv u^2$.*

Proof: Let $v = u^1 - u^2$. Then

$$\Delta v = 0,$$

i.e. v is harmonic. Furthermore, since $u^1 = u^2 = g$ on $\partial\Omega$, it follows that $v \equiv 0$ on $\partial\Omega$. Hence, we derive from Corollary 6.3 that $v \equiv 0$ in Ω. ∎

6.5 Discrete Harmonic Functions

The purpose of this section is to study a finite difference approximation of Poisson's equation (6.40). In particular, we shall establish a maximum principle for the numerical solution defined by this difference scheme. The finite difference approximation will be the obvious generalization of the scheme introduced for the corresponding one-dimensional problem in Section 2.2.

Even if one of the main advantages of numerical methods is that they can be adopted to rather general domains, in this section we shall, for notational simplicity, restrict ourselves to a rectangular domain. More precisely, we let Ω be the unit square, i.e.

$$\Omega = \{(x, y) \mid 0 < x, y < 1\}.$$

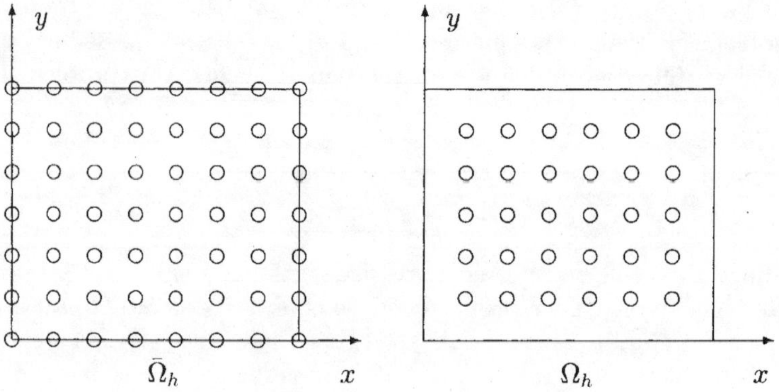

FIGURE 6.4. *Definition of* $\bar{\Omega}_h$ *and* Ω_h.

If $n \geq 1$ is an integer, then the spacing is given by $h = 1/(n+1)$, and the grid points are $(x_j, y_k) = (jh, kh)$ for $0 \leq j, k \leq n+1$. The set of all the grid points will be denoted by $\bar{\Omega}_h$, i.e.

$$\bar{\Omega}_h = \{(x_j, y_k) \mid 0 \leq j, k \leq n+1\},$$

while the set of interior grid points, Ω_h, is given by

$$\Omega_h = \{(x_j, y_k) \mid 1 \leq j, k \leq n\};$$

see Fig. 6.4.

The grid points on the boundary, $\partial\Omega_h$, are then given by

$$\partial\Omega_h = \bar{\Omega}_h \backslash \Omega_h.$$

For a grid function v, defined on $\bar{\Omega}_h$, we frequently write $v_{j,k}$ instead of $v(x_j, y_k)$. For such functions the finite difference operator L_h, approximating the negative Laplace operator $-\Delta$, is now defined by

$$\begin{aligned}(L_h v)(x_j, y_k) &= \frac{1}{h^2}\big[(-v_{j+1,k} + 2v_{j,k} - v_{j-1,k}) + (-v_{j,k+1} + 2v_{j,k} - v_{j,k-1})\big] \\ &= \frac{1}{h^2}\big[4v_{j,k} - v_{j+1,k} - v_{j-1,k} - v_{j,k+1} - v_{j,k-1}\big]\end{aligned}$$

for all interior grid points (x_j, y_k); see Fig. 6.5. This operator is usually referred to as the *five-point operator*, since the value of $(L_h v)$ at an interior grid point is defined from the value of v at the same point and at the four neighboring grid points. A finite difference approximation of Poisson's equation (6.40) is now defined by

$$\begin{aligned}(L_h v)(x, y) &= f(x, y) \quad \text{for all} \quad (x, y) \in \Omega_h, \\ v(x, y) &= g(x, y) \quad \text{for all} \quad (x, y) \in \partial\Omega_h.\end{aligned} \tag{6.41}$$

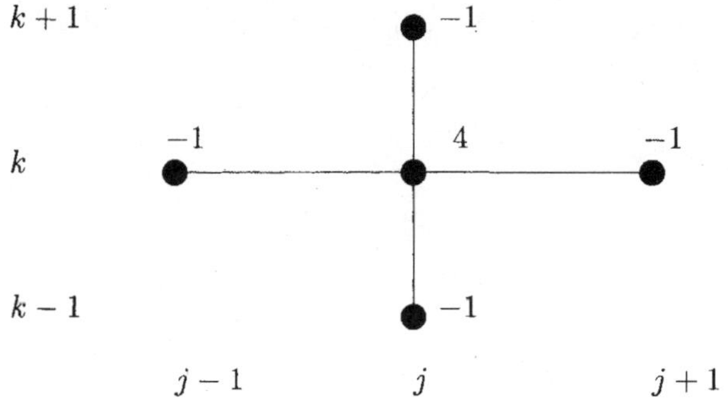

FIGURE 6.5. *The computational molecule of the five-point operator.*

Since the values of v at the grid points on the boundary are given explicitly, the system (6.41) is a linear system of n^2 equations with n^2 unknowns given by $\{v_{j,k}\}_{j,k=1}^{n}$. We will return to the study of the system (6.41) and its relations to Poisson's problem (6.40) in the next chapter. Here, we will focus the attention on maximum principles for this system which will be discrete analogs of the properties derived above for harmonic functions and for the problem (6.40).

Definition 6.2 *A grid function v is called a discrete harmonic function if*

$$L_h v = 0 \quad \text{for all} \quad (x,y) \in \Omega_h.$$

It turns out that discrete harmonic functions satisfy a maximum principle similar to their analytical counterparts.

Theorem 6.9 *If v is a discrete harmonic function, then*

$$M_0 \le v(x,y) \le M_1 \quad \text{for all} \quad (x,y) \in \bar{\Omega}_h,$$

where

$$M_0 = \min_{(x,y)\in\partial\Omega_h} v(x,y) \quad \text{and} \quad M_1 = \max_{(x,y)\in\partial\Omega_h} v(x,y).$$

Proof: We will only show the upper bound, since the lower bound follows by a completely analogous argument. Assume the contrary: that there exists an interior grid point $(\bar{x},\bar{y}) \in \Omega_h$ such that

$$v(\bar{x},\bar{y}) = \max_{(x,y)\in\bar{\Omega}_h} v(x,y) > M_1. \tag{6.42}$$

Since v is a discrete harmonic function, we derive from (6.42) that

$$\begin{aligned} v(\bar{x},\bar{y}) &= \frac{1}{4}\big[v(\bar{x}+h,\bar{y}) + v(\bar{x}-h,\bar{y}) + v(\bar{x},\bar{y}+h) + v(\bar{x},\bar{y}-h)\big] \\ &\le \max_{(x,y)\in\bar{\Omega}_h} v(x,y) = v(\bar{x},\bar{y}). \end{aligned} \tag{6.43}$$

We conclude that the inequality has to be an equality, and therefore the value of v at the four grid points which are neighbors to $(\bar{x}, \bar{y})$ must also be $v(\bar{x}, \bar{y})$. By repeating this argument until we reach a boundary point, we conclude that there must be a grid point $(\tilde{x}, \tilde{y}) \in \partial\Omega_h$ such that $v(\tilde{x}, \tilde{y}) = v(\bar{x}, \bar{y})$. However, this contradicts (6.42). ■

In the corollary below, the upper bound in Theorem 6.9 is proved under a weaker hypothesis on v. This result will be useful below.

Corollary 6.4 *If $L_h v \leq 0$ for all $(x, y) \in \Omega_h$, then*

$$v(x, y) \leq M_1 = \max_{(x,y)\in\partial\Omega_h} v(x, y) \quad \text{for all} \quad (x, y) \in \bar{\Omega}_h.$$

Proof: We can use the same proof as above. From the assumption $L_h v \leq 0$ we conclude that (6.43) still holds if the first equality is replaced by $\leq$. The rest of the argument can be used unchanged. ■

Corollary 6.5 *If v is a discrete harmonic function, then*

$$|v(x, y)| \leq M \quad \text{for all} \quad (x, y) \in \bar{\Omega}_h,$$

where $M = \max_{(x,y)\in\partial\Omega_h} |v(x, y)|$.

Proof: Argue exactly as in the proof of Corollary 6.3. ■

A consequence of these results is that the discrete system (6.41) will always have a unique solution.

Corollary 6.6 *There is a unique grid function v which solves the discrete Poisson's problem (6.41).*

Proof: Recall that the system (6.41) can be viewed as a linear system with n^2 equations and n^2 unknowns $\{v_{j,k}\}_{j,k=1}^n$. However, for a square linear system existence and uniqueness of the solution will follow if we can show that the corresponding homogeneous system only has the solution $v \equiv 0$ (see Project 1.2).

Hence, assume that the grid function v satisfies

$$\begin{aligned} (L_h v)(x, y) &= 0 \quad \text{for all} \quad (x, y) \in \Omega_h, \\ v(x, y) &= 0 \quad \text{for all} \quad (x, y) \in \partial\Omega_h. \end{aligned}$$

We have to show that $v \equiv 0$ is the only solution of this system. However, this is now a direct consequence of Corollary 6.5. ■

Recall that in Chapter 2 we were able to prove precise results about the error between the exact solution of a two point boundary value problem and the corresponding solution of a finite difference scheme (see Theorem

2.2). Our aim is to establish a similar bound for the difference between the solution of Poisson's equation (6.40) and the corresponding discrete solution of (6.41). This is achieved in the next chapter.

In order to derive such an error bound, we shall use a stability property for the difference scheme (6.41) which is a generalization to two space dimensions of Proposition 2.6. This result will be established below. The following bound represents a preliminary step in the derivation of the desired stability property.

Lemma 6.3 *Assume that v is a grid function such that $v \equiv 0$ on $\partial\Omega_h$ and*

$$L_h v = 1 \quad \text{for all} \quad (x, y) \in \Omega_h.$$

Then

$$0 \le v(x, y) \le 1/8 \quad \text{for all} \quad (x, y) \in \bar{\Omega}_h.$$

Proof: We first observe that the grid function $-v$ satisfies $L_h(-v) \le 0$ on Ω_h. Therefore, it follows from Corollary 6.4 that $-v(x, y) \le 0$, or

$$v(x, y) \ge 0 \quad \text{for all} \quad (x, y) \in \bar{\Omega}_h.$$

In order to obtain the upper bound, we will compare v with the grid function w defined by

$$w(x, y) = \frac{1}{2}x(1 - x) \quad \text{for all} \quad (x, y) \in \bar{\Omega}_h.$$

Since w is independent of y, it follows directly from Exercise 2.16 that

$$L_h w = 1 \quad \text{for all} \quad (x, y) \in \Omega_h.$$

Hence, $w - v$ is a discrete harmonic function. Since $w - v \ge 0$ on $\partial\Omega_h$, we therefore obtain from Theorem 6.9 that

$$w - v \ge 0 \quad \text{for all} \quad (x, y) \in \bar{\Omega}_h.$$

However, since

$$\max_{(x,y) \in \bar{\Omega}_h} w(x, y) \le 1/8,$$

we then have

$$v(x, y) \le w(x, y) \le 1/8 \quad \text{for all} \quad (x, y) \in \bar{\Omega}_h.$$

■

This result is easily generalized to any constant right-hand side.

Lemma 6.4 *Assume that w is a grid function such that $w = 0$ on $\partial\Omega_h$ and*

$$(L_h w)(x, y) = q \quad \text{for all} \quad (x, y) \in \Omega_h,$$

where $q \in \mathbb{R}$ is constant. Then

$$min(0, q/8) \leq w(x, y) \leq max(0, q/8) \quad \text{for all} \quad (x, y) \in \bar{\Omega}_h.$$

Proof: This follows directly from Lemma 6.3. In fact the linearity of L_h implies that $w = qv$, where v is specified in Lemma 6.3. The bounds therefore follow from the bound given in Lemma 6.3. ∎

We now have the following generalization of Proposition 2.6:

Proposition 6.1 *Assume that v is a grid function which solves the system (6.41) with $g \equiv 0$. Then*

$$\|v\|_{h,\infty} \leq 1/8 \, \|f\|_{h,\infty},$$

where $\|v\|_{h,\infty} = \max_{(x,y)\in\bar{\Omega}_h} |v(x, y)|$.

Proof: Let w be a grid function such that $w = 0$ on $\partial\Omega_h$ and

$$L_h w = \|f\|_{h,\infty} \quad \text{for all} \quad (x, y) \in \Omega_h.$$

Hence, $v - w = 0$ on $\partial\Omega_h$ and

$$L_h(v - w) = f - \|f\|_{h,\infty} \leq 0 \quad \text{on } \Omega_h.$$

From Corollary 6.4 we therefore obtain that

$$v - w \leq 0 \quad \text{on } \bar{\Omega}_h$$

and, as a consequence of Lemma 6.4,

$$v(x, y) \leq w(x, y) \leq \frac{1}{8} \, \|f\|_{h,\infty} \quad \text{for all} \quad (x, y) \in \bar{\Omega}_h.$$

A similar argument, using a grid function w satisfying $w = 0$ on $\partial\Omega_h$ and

$$L_h w = -\|f\|_{h,\infty} \quad \text{on } \Omega_h,$$

now implies

$$v(x, y) \geq -\frac{1}{8} \, \|f\|_{h,\infty}.$$

∎

6.6 Exercises

In these exercises we use the notation defined in this chapter. In particular, we use R, B, R_Δ, and B_Δ, which are defined in Fig. 6.2 on page 181 and Fig. 6.3 on page 186.

Exercise 6.1

(a) Find the solution of problem (6.2)–(6.3) by multiplying the equation (6.2) by a proper integrating factor.

(b) Use this formula for u to establish Theorem 6.1.

Exercise 6.2 Consider a two-point boundary value problem of the form

$$\begin{aligned} u''(x) + a(x)u'(x) &= f(x), \\ u(0) = u_0, \quad u(1) &= u_1. \end{aligned}$$

Here $a \in C([0,1])$ and $f \in C((0,1))$ are given functions.

Assume that $u, \bar{u} \in C^2((0,1)) \cap C([0,1])$ are two solutions of this problem, with the same functions a and f, but with boundary values u_0, u_1 and $\bar{u}_0, \bar{u}_1$ respectively. Show that

$$\|u - \bar{u}\|_\infty \leq \max(|u_0 - \bar{u}_0|, |u_1 - \bar{u}_1|),$$

where

$$\|g\|_\infty = \sup_{x\in[0,1]} |g(x)|.$$

Exercise 6.3 Let $a = a(u)$ be a uniformly bounded function and suppose that u is a solution of the following two-point boundary value problem:

$$u'' + a(u)u' = 0, \qquad x \in (0,1),$$

with boundary conditions

$$u(0) = u_0 \quad \text{and} \quad u(1) = u_1.$$

Show that u satisfies the following maximum principle:

$$\min(u_0, u_1) \leq u(x) \leq \max(u_0, u_1)$$

for all $x \in [0,1]$.

EXERCISE 6.4 Suppose that u is a solution of the following two-point boundary value problem:

$$u'' + \sin(u)u' = 1, \qquad x \in (0,1), \tag{6.44}$$

with boundary conditions

$$u(0) = u(1) = 0.$$

Show that $u(x) \leq 0$ for all $x \in [0,1]$.

EXERCISE 6.5 Consider the equation

$$\begin{aligned} u_t &= \alpha u_{xx} \quad \text{for} \quad x \in (0,1), \quad t \in (0,T], \\ u(0,t) &= u(1,t) = 0, \quad t \in [0,T], \\ u(x,0) &= f(x), \quad x \in [0,1], \end{aligned}$$

where $\alpha > 0$ is a given constant. We define an explicit scheme

$$\frac{v_j^{m+1} - v_j^m}{\Delta t} = \alpha \frac{v_{j-1}^m - 2v_j^m + v_{j+1}^m}{\Delta x^2} \quad \text{for} \quad j = 1, \dots, n, \ m \geq 0,$$

and an implicit scheme

$$\frac{v_j^{m+1} - v_j^m}{\Delta t} = \alpha \frac{v_{j-1}^{m+1} - 2v_j^{m+1} + v_{j+1}^{m+1}}{\Delta x^2} \quad \text{for} \quad j = 1, \dots, n, \ m \geq 0.$$

(a) State and prove a maximum principle for the continuous problem.

(b) Derive a stability condition for the explicit scheme such that the numerical solutions satisfy a discrete version of the maximum principle derived in (a).

(c) Show that the implicit scheme is unconditionally stable in the sense that the numerical solutions generated by this scheme satisfy the discrete version of the maximum principle for any relevant mesh sizes.

(d) Compare the results derived in (b) and (c) with the results obtained by the method of von Neumann in Exercise 4.12 on page 151.

EXERCISE 6.6 Consider the following initial-boundary value problem:

$$\begin{aligned} u_t &= a(x,t)u_{xx} + b(x,t)u_x \quad \text{for} \quad x \in (0,1), \quad t \in (0,T], \\ u(0,t) &= u_\ell(t), \quad u(1,t) = u_r(t), \quad t \in [0,T], \\ u(x,0) &= f(x), \quad x \in [0,1], \end{aligned}$$

where $a(x,t) \geq a_0 > 0$ for all $(x,t) \in R$ and where $b = b(x,t)$ is a bounded function.

(a) State and prove a maximum principle for this initial-boundary value problem.

(b) Derive an explicit finite difference scheme for this problem and prove a maximum principle for the discrete solutions.

(c) Derive an implicit scheme and investigate whether numerical solutions generated by this scheme satisfy a discrete maximum principle for all positive mesh parameters.

EXERCISE 6.7 Consider the following initial-boundary value problem:

$$\begin{aligned} u_t + cu_x &= u_{xx} \quad \text{for} \quad x \in (0,1), \quad t \in (0,T], \\ u(0,t) &= u(1,t) = 0, \quad t \in [0,T], \\ u(x,0) &= f(x), \quad x \in [0,1], \end{aligned}$$

where $c \geq 0$ is a given constant and where $f(0) = f(1) = 0$.

(a) Show that a solution of this problem satisfies the following maximum principle:

$$\inf_{x\in[0,1]} f(x) \leq u(x,t) \leq \sup_{x\in[0,1]} f(x)$$

for all $(x,t) \in R$.

Derive stability conditions for the following numerical methods such that the corresponding discrete solutions satisfy a discrete version of the maximum principle stated in (a).

(b)

$$\frac{v_j^{m+1} - v_j^m}{\Delta t} + c\frac{v_{j+1}^m - v_{j-1}^m}{2\Delta x} = \frac{v_{j-1}^m - 2v_j^m + v_{j+1}^m}{\Delta x^2}$$

(c)

$$\frac{v_j^{m+1} - v_j^m}{\Delta t} + c\frac{v_j^m - v_{j-1}^m}{\Delta x} = \frac{v_{j-1}^m - 2v_j^m + v_{j+1}^m}{\Delta x^2}$$

(d)

$$\frac{v_j^{m+1} - v_j^m}{\Delta t} + c\frac{v_{j+1}^m - v_j^m}{\Delta x} = \frac{v_{j-1}^m - 2v_j^m + v_{j+1}^m}{\Delta x^2}$$

(e)

$$\frac{v_j^{m+1} - v_j^m}{\Delta t} + c\frac{v_j^m - v_{j-1}^m}{\Delta x} = \frac{v_{j-1}^{m+1} - 2v_j^{m+1} + v_{j+1}^{m+1}}{\Delta x^2}$$

(f)

$$\frac{v_j^{m+1} - v_j^m}{\Delta t} + c\frac{v_j^{m+1} - v_{j-1}^{m+1}}{\Delta x} = \frac{v_{j-1}^{m+1} - 2v_j^{m+1} + v_{j+1}^{m+1}}{\Delta x^2}.$$

(g) Compare the results derived in (b)–(f) with the results obtained in Exercise 4.18 on page 153, using the method of von Neumann.

EXERCISE 6.8 Consider the following initial-boundary value problem:

$$\begin{aligned} u_t &= a(x,t)u_{xx} + \alpha u(x,t) \quad \text{for} \quad x \in (0,1), \quad t \in (0,T], \\ u(0,t) &= u(1,t) = 0, \quad t \in [0,T], \\ u(x,0) &= f(x), \quad x \in [0,1], \end{aligned}$$

where $a(x,t) \geq a_0 > 0$ for all $(x,t) \in R$ and where α is a given constant.

(a) Show that

$$||u(\cdot,t)||_\infty \leq e^{\alpha t}||f||_\infty \quad \text{for} \quad 0 \leq t \leq T, \tag{6.45}$$

where we recall

$$||u(\cdot,t)||_\infty = \sup_{x\in[0,1]} |u(x,t)|.$$

Here you may find it useful to consider $w = e^{-\alpha t}u$.

(b) Discuss how the stability with respect to perturbations in the initial data depends on the parameter α.

(c) Derive an explicit finite difference scheme and prove that the numerical solutions generated by this scheme satisfy a discrete version of (6.45) provided that the proper condition on the mesh sizes is satisfied.

(d) Consider an implicit scheme and show that a discrete version of (6.45) is satisfied for any relevant values of the mesh parameters.

EXERCISE 6.9 Consider the following nonlinear initial-boundary value problem:

$$\begin{aligned} u_t + f(u)_x &= u_{xx} \quad \text{for} \quad x \in (0,1), \quad t \in (0,T], \\ u(0,t) &= u_\ell(t), \quad u(1,t) = u_r(t), \quad t \in [0,T], \\ u(x,0) &= h(x), \quad x \in [0,1], \end{aligned}$$

where $f = f(u)$ is a smooth given function.

(a) Show that a solution of this problem equation satisfies the following maximum principle:

$$\inf_{(x,t)\in B} (h(x), u_\ell(t), u_r(t)) \leq u(x,t) \leq \sup_{(x,t)\in B} (h(x), u_\ell(t), u_r(t)).$$

(b) Consider the following numerical scheme:

$$\frac{v_j^{m+1} - v_j^m}{\Delta t} + \frac{f(v_{j+1}^m) - f(v_{j-1}^m)}{2\Delta x} = \frac{v_{j+1}^m - 2v_j^m + v_{j-1}^m}{\Delta x^2}.$$

The boundary conditions give

$$v_0^m = u_\ell(t_m) \quad \text{and} \quad v_{n+1}^m = u_r(t_m)$$

for $m \geq 0$, and the initial condition leads to

$$v_j^0 = h(x_j) \quad \text{for} \quad j = 1, \dots, n.$$

Suppose that the grid parameters are chosen such that

$$r = \Delta t/\Delta x^2 \leq 1/2, \tag{6.46}$$

and

$$\frac{\Delta x}{2} \max_{U^- \leq u \leq U^+} |f'(u)| \leq 1, \tag{6.47}$$

where

$$U^- = \inf_{(x,t)\in B} (h(x), u_\ell(t), u_r(t)) \quad \text{and} \quad U^+ = \sup_{(x,t)\in B} (h(x), u_\ell(t), u_r(t)).$$

Show that if these conditions are satisfied, the numerical solutions generated by this scheme satisfy a discrete version of the maximum principle in (a).

EXERCISE 6.10 Consider the finite difference scheme (6.17)–(6.19). Formulate and prove a discrete version of Corollary 6.1 for this difference scheme.

EXERCISE 6.11 Consider the finite difference scheme (6.23)–(6.25). Formulate and prove a discrete version of Corollary 6.1 for this difference scheme.

EXERCISE 6.12 In Theorem 6.6 we assume that the function $k = k(u)$ satisfies the bound $k_0 \leq k(u) \leq K_0$ for all values of u. Obviously, this is a rather strict requirement, and it is the purpose of this exercise to weaken this assumption considerably.

As usual we define

$$V^- = \min_{(x_i,t_k)\in B_\Delta} (f(x_i), u_\ell(t_k), u_r(t_k))$$

and

$$V^+ = \max_{(x_i,t_k)\in B_\Delta} (f(x_i), u_\ell(t_k), u_r(t_k)).$$

Furthermore, we let

$$k_0 = \inf_{u\in[V^-,V^+]} k(u) \quad \text{and} \quad K_0 = \sup_{u\in[V^-,V^+]} k(u).$$

Suppose that $0 < k_0 \leq K_0 < \infty$, and that Δx and Δt satisfy the inequality (6.36) on page 191.

(a) Prove that the numerical solution generated by the scheme (6.35) with boundary conditions and initial data given by (6.18) and (6.19) satisfies

$$V^- \leq v_j^m \leq V^+$$

for all grid points $(x_j, t_m) \in R_\Delta$.

(b) Let $k(u) = e^u$ and $u_\ell(t) = u_r(t) =$ for $t \geq 0$. Furthermore, we define the initial condition $f(x) = x(1-x)$. State a precise maximum principle for the discrete solution of this problem.

EXERCISE 6.13 Assume that u is a solution of Poisson's equation (6.40) with $f \geq 0$ and $g \geq 0$. Show that

$$u(x, y) \geq 0 \quad \text{for all} \quad (x, y) \in \bar{\Omega}.$$

EXERCISE 6.14 Let Ω be the unit square, i.e. $\Omega = \{(x,y) \mid 0 < x, y < 1\}$. Assume that u solves (6.40) with $f \equiv 1$ and $g \equiv 0$. Show that $0 \le u \le 1/8$ in $\bar{\Omega}$. (Hint: Compare u with $w(x,y) = \frac{1}{2}x(1-x)$ as in the proof of Lemma 6.3 above.)

EXERCISE 6.15 Let Ω be as in Exercise 6.14 and assume that $u \in C^2(\Omega) \cap C(\bar{\Omega})$ solves (6.40) with $g \equiv 0$. Show that

$$\|u\|_\infty \le \frac{1}{8}\|f\|_\infty,$$

where $\|u\|_\infty = \sup_{(x,y)\in\bar{\Omega}} |u(x,y)|$.

EXERCISE 6.16 In this problem we shall study Poisson's equation (6.40) in a general domain Ω. We assume that $\Omega \subseteq \{(x,y) \mid x^2 + y^2 < r^2\}$ for a suitable $r > 0$.

(a) Assume that $v = 0$ on $\partial\Omega$ and satisfies $-\Delta v = 1$ in Ω.

Show that

$$0 \le v(x,y) \le r^2/4 \quad \text{for all} \quad (x,y) \in \bar{\Omega}.$$

(Hint: Compare v with $w(x,y) = \frac{1}{4}(r^2 - x^2 - y^2)$.)

(b) Assume that $u \in C^2(\Omega) \cap C(\bar{\Omega})$ is a solution of (6.40) with $g \equiv 0$.

Show that

$$\|u\|_\infty \le \frac{r^2}{4}\|f\|_\infty.$$

7

Poisson's Equation in Two Space Dimensions

Poisson's equation is a fundamental partial differential equation which arises in many areas of mathematical physics, for example in fluid flow, flow in porous media, and electrostatics. We have already encountered this equation in Section 6.4 above, where we studied the maximum principle for harmonic functions. As a corollary of the maximum principle we have in fact already established that the Dirichlet problem for Poisson's equation has at most one solution (see Theorem 6.8).

The main focus in this chapter will therefore be on how to construct solutions of this problem. We shall also derive a new qualitative property of harmonic functions, the mean value property, which in fact will lead to an alternative proof of the maximum principle. We shall start by utilizing the separation of variables technique for Poisson's equation. We will see that if the geometry of the domain has certain simple structures, then this method leads to exact (formal) solutions. In the final section of this chapter we shall also discuss properties of corresponding finite difference solutions.

7.1 Rectangular Domains

Recall that Poisson's problem, with Dirichlet boundary conditions, takes the form

$$\begin{aligned} -\Delta u &= f \quad \text{in} \quad \Omega, \\ u &= g \quad \text{in} \quad \partial\Omega. \end{aligned} \tag{7.1}$$

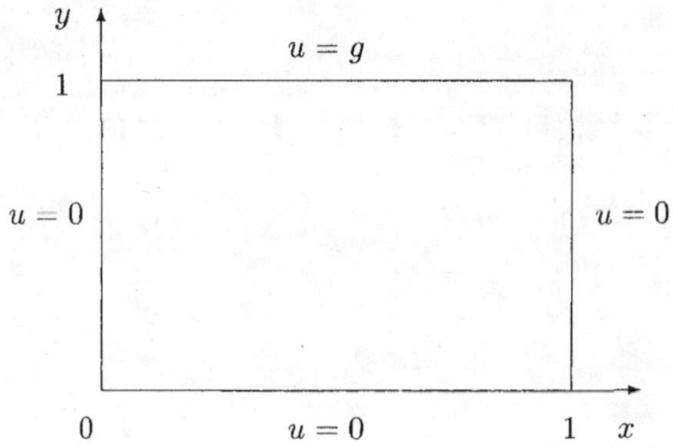

FIGURE 7.1. *Dirichlet boundary conditions on the unit square.*

Here the domain $\Omega \subset \mathbb{R}^2$ is assumed to be bounded, connected, and open. Furthermore, $\partial\Omega$ denotes the boundary of Ω. The purpose of this section is to show how we can use separation of variables to solve this problem when the domain Ω is a rectangle. In fact, we will only carry out the analysis when Ω is the unit square,

$$\Omega = \{(x,y) \mid 0 < x, y < 1\},$$

but the techniques can be adapted to any rectangle. Furthermore, we shall only consider the homogeneous problem, i.e. $f \equiv 0$. Other examples will be treated in the exercises (see Exercises 7.2 and 7.18). In order to simplify the analysis below, we shall consider boundary conditions of the form (see Fig. 7.1)

$$u(0,y) = u(1,y) = 0, \qquad 0 \le y \le 1, \tag{7.2}$$
$$u(x,0) = 0, \qquad 0 \le x \le 1, \tag{7.3}$$
$$u(x,1) = g(x), \qquad 0 < x < 1. \tag{7.4}$$

We now make the ansatz that the solution has the form

$$u(x,y) = X(x)Y(y).$$

Substituting this in the equation $\Delta u = 0$, we obtain

$$X''(x)Y(y) + X(x)Y''(y) = 0,$$

or, by dividing with XY,

$$-\frac{X''(x)}{X(x)} = \frac{Y''(y)}{Y(y)}.$$

Since the left-hand side only depends on x, while the right-hand side depends on y, we conclude as before that both sides have to be equal to a constant λ, independent of x and y. For the function $X(x)$ we therefore obtain the eigenvalue problem

$$\begin{aligned} -X''(x) &= \lambda X(x), \qquad 0 < x < 1, \\ X(0) &= X(1) = 0, \end{aligned} \tag{7.5}$$

where the boundary conditions are derived from (7.2). The eigenvalue problem (7.5) is by now familiar to us, and we conclude immediately from Lemma 2.7 that the eigenvalues are

$$\lambda_k = (k\pi)^2, \qquad k = 1, 2, \dots ,$$

with corresponding eigenfunctions given by

$$X_k(x) = \sin(k\pi x), \qquad k = 1, 2, \dots .$$

In particular this means that $\lambda = \beta^2 > 0$ for a suitable $\beta > 0$.

The function $Y(y)$ has to satisfy

$$\begin{aligned} Y''(y) &= \lambda Y(y), \qquad 0 < y < 1, \\ Y(0) &= 0. \end{aligned} \tag{7.6}$$

Here the boundary condition at $y = 0$ is derived from (7.3), while the nonhomogeneous conditions (7.4) will be incorporated later.

The general solution of (7.6), with $\lambda = \beta^2$, (observe that there is no minus sign in front of Y'' in this case) is linear combinations of $e^{\beta y}$ and $e^{-\beta y}$. Furthermore, from the boundary condition at $y = 0$ we conclude that Y has to be a multiple of the function[1]

$$Y(y) = \sinh(\beta y).$$

Hence, with $\beta = k\pi$, we obtain particular solutions $u_k(x, y)$ of the form

$$u_k(x, y) = \sin(k\pi x)\sinh(k\pi y) \quad \text{for} \quad k = 1, 2, \dots .$$

All these solutions will be harmonic and satisfy the boundary conditions (7.2) and (7.3). Furthermore, the same properties will carry over to linear combinations; i.e. we obtain formal solutions

$$u(x, y) = \sum_{k=1}^{\infty} c_k \sin(k\pi x)\sinh(k\pi y), \tag{7.7}$$

where c_k denotes arbitrary real coefficients.

---

[1] We recall that $\sinh(z) = (e^z - e^{-z})/2$.

Consider now the final boundary condition (7.4). Assume that the function $g(x)$ admits a Fourier sine series

$$g(x) = \sum_{k=1}^{\infty} g_k \sin(k\pi x), \tag{7.8}$$

where, as before, the Fourier coefficients g_k are given by

$$g_k = 2\int_0^1 g(x)\sin(k\pi x)\,dx.$$

By comparing the series (7.7), with $y = 1$, and the series (7.8), we obtain from (7.4) that

$$c_k = g_k/\sinh(k\pi) \quad \text{for} \quad k = 1, 2, \ldots. \tag{7.9}$$

The formulas (7.7) and (7.9) give the complete formal solution of the problem given by the homogeneous equation $\mathbf{\Delta} u = 0$ and the boundary conditions (7.2)–(7.4). More general Dirichlet boundary conditions are considered in Exercise 7.3 below.

7.2 Polar Coordinates

If the geometry of the domain Ω is naturally described in polar coordinates, then it is convenient to consider the unknown function u as a function of the polar coordinates r and ϕ, where

$$x = r\cos\phi \quad \text{and} \quad y = r\sin\phi,$$

or equivalently

$$r = \sqrt{x^2 + y^2} \quad \text{and} \quad \phi = \arctan\left(\frac{y}{x}\right).$$

Here $r \geq 0$ and $\phi \in (-\pi, \pi)$.

The Jacobian matrix of this transformation is given by

$$\begin{pmatrix} \partial r/\partial x & \partial r/\partial y \\ \partial\phi/\partial x & \partial\phi/\partial y \end{pmatrix} = \begin{pmatrix} \cos\phi & \sin\phi \\ -(\sin\phi)/r & (\cos\phi)/r \end{pmatrix}. \tag{7.10}$$

Hence we obtain[2]

$$u_x = u_r\frac{\partial r}{\partial x} + u_\phi\frac{\partial\phi}{\partial x} = (\cos\phi)\,u_r - \left(\frac{\sin\phi}{r}\right)u_\phi,$$

---

[2]It would have been more precise to distinguish the function of r and ϕ from the original function $u = u(x, y)$. For example, we could have used $U(r, \phi) = u(x, y)$. However, as is more or less standard, we use u to denote both functions.

and, with some effort (see Exercise 7.6),

$$
\begin{aligned}
u_{xx} &= \frac{\partial}{\partial x}\left[(\cos\phi)\, u_r - \frac{\sin\phi}{r} u_\phi\right] \\
&= \left(\cos^2\phi\right) u_{rr} + \frac{\sin^2\phi}{r^2} u_{\phi\phi} - 2\frac{\sin\phi\cos\phi}{r} u_{r\phi} \qquad (7.11)\\
&+ \frac{\sin^2\phi}{r} u_r + 2\frac{\sin\phi\cos\phi}{r^2} u_\phi.
\end{aligned}
$$

A similar calculation gives

$$
\begin{aligned}
u_{yy} &= \left(\sin^2\phi\right) u_{rr} + \frac{\cos^2\phi}{r^2} u_{\phi\phi} + 2\frac{\sin\phi\cos\phi}{r} u_{r\phi} \qquad (7.12)\\
&+ \frac{\cos^2\phi}{r} u_r - 2\frac{\sin\phi\cos\phi}{r^2} u_\phi.
\end{aligned}
$$

By adding the identities (7.11) and (7.12), we therefore obtain that

$$\Delta u = u_{xx} + u_{yy} = u_{rr} + \frac{1}{r^2} u_{\phi\phi} + \frac{1}{r} u_r. \qquad (7.13)$$

EXAMPLE 7.1 Assume we want to find a harmonic function which is rotation invariant, i.e. u is independent of ϕ. In polar coordinates $u = u(r)$, and from (7.13) we obtain that $u(r)$ has to satisfy the ordinary differential equation

$$u_{rr} + \frac{1}{r} u_r = 0.$$

After multiplication by r this can be written

$$(r u_r)_r = 0.$$

Hence, we obtain that any harmonic function which only depends on r has to be of the form

$$u(r) = c_1 \ln(r) + c_2,$$

where c_1 and c_2 are arbitrary real coefficients. We recall that we have already encountered the harmonic function $\ln(r)$ in Example 6.3 on page 193. ■

7.2.1 *The Disc*

In order to illustrate the application of the representation (7.13) of the Laplace operator in polar coordinates, we shall consider a problem of the form

$$
\begin{aligned}
\Delta u &= 0 \quad \text{in} \quad \Omega, \\
u &= g \quad \text{on} \quad \partial\Omega,
\end{aligned}
\qquad (7.14)
$$

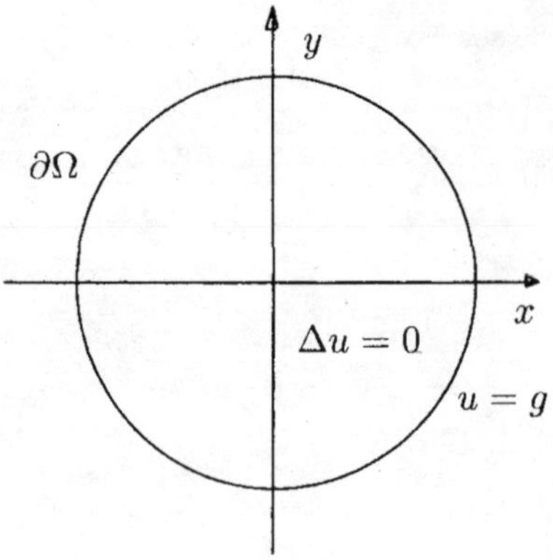

FIGURE 7.2. *The Dirichlet problem on a disc.*

where Ω is the disc of radius $\rho > 0$ with center at the origin, i.e.

$$\Omega = \{(x, y) \mid x^2 + y^2 < \rho^2\},$$

(see Fig. 7.2).

Our aim is to show that the problem (7.14) on this domain Ω can be solved by separation of variables with respect to r and ϕ.

Let us first observe that it is reasonable to assume that the function g is a 2π-periodic function with respect to ϕ. We therefore assume that g is written in a Fourier series of the form

$$g(\phi) = \frac{a_0}{2} + \sum_{k=1}^{\infty} \left[a_k \cos(k\phi) + b_k \sin(k\phi)\right], \tag{7.15}$$

where[3]

$$a_k = \frac{1}{\pi}\int_{-\pi}^{\pi} g(\phi)\cos(k\phi)\, d\phi \quad \text{and} \quad b_k = \frac{1}{\pi}\int_{-\pi}^{\pi} g(\phi)\sin(k\phi)\, d\phi.$$

We now make the ansatz that $u(r, \phi)$ has the form

$$u(r, \phi) = R(r)\Phi(\phi).$$

Substituting this into the equation $u_{rr} + \frac{1}{r}u_r + \frac{1}{r^2}u_{\phi,\phi} = 0$, we obtain

$$0 = \Delta u = R''\Phi + \frac{1}{r}R'\Phi + \frac{1}{r^2}R\Phi'',$$

which implies

$$r^2\frac{R''}{R} + r\frac{R'}{R} = -\frac{\Phi''}{\Phi}.$$

[3]See section 8.1.4.

Since the left-hand side only depends on r and the right-hand side only depends on ϕ, we must have

$$-\Phi'' = \lambda\Phi \tag{7.16}$$

and

$$r^2R'' + rR' - \lambda R = 0, \tag{7.17}$$

where λ is independent of r and ϕ. Since u should be smooth around the negative x-axis, we must require that Φ is 2π-periodic with respect to ϕ. We therefore impose the periodic boundary conditions

$$\Phi(-\pi) = \Phi(\pi) \quad \text{and} \quad \Phi'(-\pi) = \Phi'(\pi)$$

to the differential equation (7.16). Hence we obtain the eigenvalues

$$\lambda_k = k^2, \qquad k = 0, 1, 2, \dots,$$

with possible eigenfunctions of the form

$$\Phi_k(\phi) = c_1 \cos(k\phi) + c_2 \sin(k\phi), \qquad k = 0, 1, \dots.$$

Here c_1 and c_2 are arbitrary constants. The equation (7.17) is an ordinary differential equation with respect to r. The equation is linear, but with variable coefficients. An equations of the form (7.17) is usually said to be of Euler type. These equations can be solved analytically; see Exercise 7.7. The solutions are of the form

$$R(r) = r^\beta.$$

Substituting this ansatz into (7.17), together with the fact that $\lambda = k^2$ we immediately obtain

$$\beta(\beta-1)r^\beta + \beta r^\beta - k^2 r^\beta = 0,$$

which implies that

$$\beta = \pm k.$$

Hence, for $k \geq 1$ we obtain the two linearly independent solutions r^{-k} and r^k. For $k = \lambda = 0$, the equation (7.17) has already been studied in Example 7.1 above. There we found that any solution in this case is of the form

$$R(r) = c_1 \ln(r) + c_2.$$

Observe however that if $R(r)$ is of the form r^{-k}, for $k > 0$, or $\ln(r)$, then $u(r,\phi) \to \infty$ as $r \to 0$. Since the origin is in the interior of the domain Ω, this is not acceptable. We therefore adopt the boundary condition

$$\lim_{r \to 0} R(r) \quad \text{is finite}$$

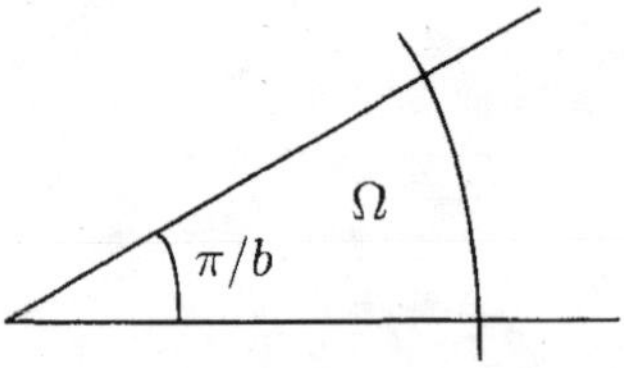

FIGURE 7.3. *Wedge with angle* π/b.

for the equation (7.17). Hence, the solutions $\ln(r)$ and r^{-k} are rejected, and we are left with the solutions

$$R_k(r) = r^k \quad \text{for} \quad k = 0, 1, 2, \dots .$$

By taking linear combinations of the particular solutions of the form $R_k\Phi_k$, we obtain

$$u(r,\phi) = \frac{a_0'}{2} + \sum_{k=1}^{\infty} r^k \big(a_k' \cos(k\phi) + b_k' \sin(k\phi)\big), \tag{7.18}$$

where a_k' and b_k' are constants to be determined. Comparing this with the representation (7.15) for $g(\phi)$ we derive, from the boundary condition $u(\rho,\phi) = g(\phi)$, that

$$a_k' = \rho^{-k} a_k, \qquad b_k' = \rho^{-k} b_k,$$

and hence the solution (7.18) is determined.

7.2.2 A Wedge

Another interesting application of polar coordinates arises when the domain Ω is a wedge. Let Ω be of the form

$$\Omega = \big\{(r,\phi) \mid 0 < r < \rho,\ 0 < \phi < \pi/b\big\},$$

where $b > \frac{1}{2}$ (see Fig. 7.3).

Assume that we want to find a function u which is harmonic in Ω and satisfies the boundary conditions

$$u(r,0) = u(r,\pi/b) = 0, \qquad 0 < r < \rho, \tag{7.19}$$

and

$$u(\rho,\phi) = g(\phi), \qquad 0 < \phi < \pi/b. \tag{7.20}$$

Again we are looking for solutions of the form

$$u(r,\phi) = R(r)\Phi(\phi).$$

As above, we derive the equations (7.16) and (7.17) for Φ and R. Furthermore, equation (7.16) should be associated with the boundary conditions

$$\Phi(0) = \Phi(\pi/b) = 0$$

obtained from (7.19). Hence, we obtain the eigenfunctions

$$\Phi_k(\phi) = \sin(kb\phi) \quad \text{for} \quad k = 1, 2, \ldots$$

and the associated eigenvalues

$$\lambda_k = (bk)^2 \quad \text{for} \quad k = 1, 2, \ldots .$$

The corresponding solutions of (7.17), satisfying $|R_k(0)| < \infty$, are given by

$$R_k(r) = r^{bk}, \qquad k = 1, 2, \ldots .$$

Therefore, by taking linear combinations of the particular solutions, we obtain solutions of the form

$$u(r, \phi) = \sum_{k=1}^{\infty} a_k r^{bk} \sin(kb\phi). \tag{7.21}$$

These solutions are harmonic and satisfy the two boundary conditions (7.19). From the boundary condition (7.20) we derive that the coefficients a_k should be determined by g such that

$$g(\phi) = \sum_{k=1}^{\infty} a_k \rho^{kb} \sin(kb\phi).$$

7.2.3 A Corner Singularity

There is one special property of the solution (7.21) we should be aware of. Consider the first particular solution $u_1(r, \phi)$ given by

$$u_1(r, \phi) = r^b \sin(b\phi).$$

If we differentiate this function with respect to r, we obtain

$$\frac{\partial u_1}{\partial r}(r, \phi) = br^{b-1} \sin(b\phi).$$

Hence, if $b < 1$, the first derivative with respect to r is unbounded as $r \to 0$. For example, if $b = 1/3$, then $\frac{\partial u_1}{\partial r}$ behaves like $r^{-2/3}$ as $r \to 0$. The function r^b, for $b = 1/3$ and $b = 2$, is graphed in Fig. 7.4. Observe that $b < 1$ corresponds to a wedge with angle greater than π (see Fig. 7.5). It is well known that corners which form angles greater than π will generally create such singularities in the derivatives of the solution, and that special care should be taken when such problems are solved numerically.

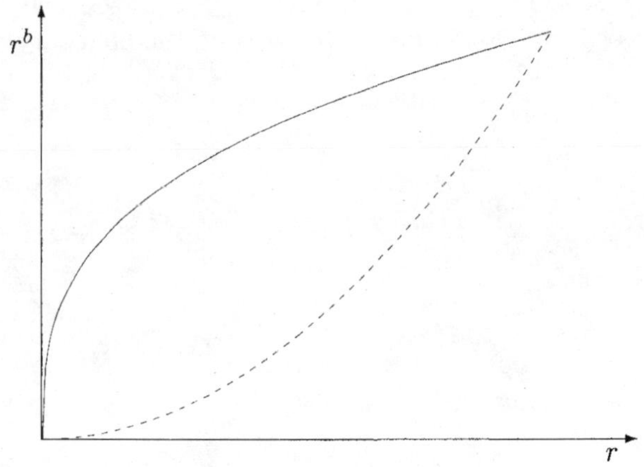

FIGURE 7.4. *The function r^b for $b = 1/3$ (solid) and $b = 2$ (dashed).*

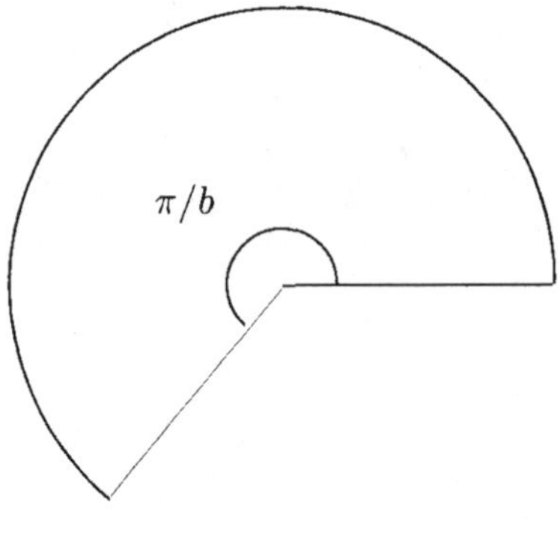

FIGURE 7.5. *Wedge with angle greater than π.*

7.3 Applications of the Divergence Theorem

In the two first sections of this chapter we used separation of variables to find formal solutions of Poisson's problem when the geometry of the domain Ω is simple. However, for more complex domains this approach does not work. The purpose of the present section is to establish that even if an analytical expression of the solution is not available, Poisson's problem in two space dimensions still has qualitative properties which resemble the one dimensional case studied in Chapter 2. These properties are for example useful for the understanding of numerical approximations of Poisson's problem. Such approximations will be studied later in this chapter.

Recall that integration by parts is a very useful tool in the study of Poisson's equation in one space dimension. In particular, in Section 2.3 we use

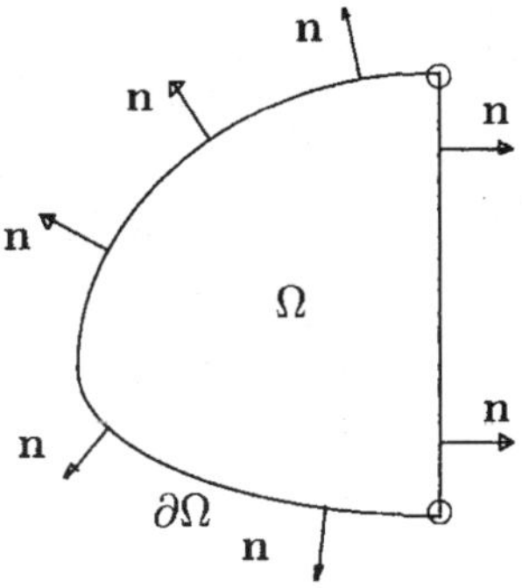

FIGURE 7.6. *The vector* **n** *is not defined at the two points where* $\partial\Omega$ *is nonsmooth.*

integration by parts to establish the symmetry and the positive definiteness of the operator $L = -\frac{d^2}{dx^2}$ defined on a suitable space of functions. In two space dimensions applications of the divergence theorem, well known from any calculus course, will replace the use of integration by parts.

We will use the divergence theorem with respect to the domain Ω. Hence, in order for the divergence theorem to hold, we need to assume that the boundary of Ω, $\partial\Omega$, satisfies some smoothness assumptions. For example, it will be sufficient to assume that $\partial\Omega$ is a piecewise C^1-curve. Hence, we will allow domains with a smooth boundary, for example a disc, and also domains with piecewise smooth boundaries like a triangle or a rectangle, or in fact any polygonal domain.[4] We will not state these requirements on $\partial\Omega$ explicitly below. Throughout this section we will simply implicitly assume that $\partial\Omega$ allows the divergence theorem to hold.

Assume first that

$$\mathbf{F} = \mathbf{F}(x,y) = \begin{bmatrix} F_1(x,y) \\ F_2(x,y) \end{bmatrix}$$

is a differentiable vector-valued function defined on Ω. At each point $(x,y) \in \partial\Omega$, where $\partial\Omega$ is C^1, let $\mathbf{n} = \mathbf{n}(x,y)$ be the unit outer normal vector (see Fig. 7.6).

Recall that

$$\operatorname{div}\mathbf{F} = F_{1,x} + F_{2,y} \equiv \frac{\partial}{\partial x}F_1 + \frac{\partial}{\partial y}F_2.$$

The divergence theorem now states that

$$\iint_\Omega \operatorname{div}\mathbf{F}\; dx\, dy = \int_{\partial\Omega} \mathbf{F}\cdot\mathbf{n}\; ds, \tag{7.22}$$

[4] A polygonal domain is a domain where the boundary consists of a collection of straight lines.

where s denotes the arc length along $\partial\Omega$. Now let $\mathbf{F}(x,y)$ be a vector-valued function of the form

$$\mathbf{F} = v\nabla u,$$

where u and v are scalar functions defined on Ω. Here ∇u denotes the gradient of u given by

$$\nabla u = \begin{pmatrix} u_x \\ u_y \end{pmatrix}.$$

Then

$$\begin{aligned} \operatorname{div} \mathbf{F} &= v\Delta u + \nabla v \cdot \nabla u \\ &= v(u_{xx} + u_{yy}) + (v_x u_x + v_y u_y). \end{aligned}$$

Therefore, it follows from (7.22) that

$$\iint_\Omega (v\Delta u + \nabla v \cdot \nabla u)\, dx\, dy = \int_{\partial\Omega} v\frac{\partial u}{\partial \mathbf{n}}\, ds. \tag{7.23}$$

Here $\frac{\partial u}{\partial \mathbf{n}}$ denotes the normal derivative of u on $\partial\Omega$ given by

$$\frac{\partial u}{\partial \mathbf{n}} = \nabla u \cdot \mathbf{n}.$$

The formula (7.23) is frequently referred to as *Green's first identity*.
The following observation is a simple consequence of (7.23).

Lemma 7.1 *If u is harmonic in Ω, then*

$$\int_{\partial\Omega} \frac{\partial u}{\partial \mathbf{n}}\, ds = 0.$$

Proof: Apply the identity (7.23) with $v \equiv 1$. Since $\Delta u = 0$ and $\nabla v = 0$, we obtain the desired result. ■

The identity (7.23) can be written in the form

$$\iint_\Omega \nabla u \cdot \nabla v\, dx\, dy = -\iint_\Omega v\Delta u\, dx\, dy + \int_{\partial\Omega} v\frac{\partial u}{\partial \mathbf{n}}\, ds.$$

Furthermore, by changing the role of u and v we also obtain

$$\iint_\Omega \nabla u \cdot \nabla v\, dx\, dy = -\iint_\Omega u\Delta v\, dx\, dy + \int_{\partial\Omega} u\frac{\partial v}{\partial \mathbf{n}}\, ds.$$

Hence, since the two expressions for $\iint_\Omega \nabla u \cdot \nabla v\, dx\, dy$ must be equal, we have

$$\iint_\Omega (u\Delta v - v\Delta u)\, dx\, dy = \int_{\partial\Omega} \left(u\frac{\partial v}{\partial \mathbf{n}} - v\frac{\partial u}{\partial \mathbf{n}}\right) ds. \tag{7.24}$$

This formula is usually referred to as *Green's second identity.*

Consider Poisson's equation with homogeneous Dirichlet boundary conditions, i.e.

$$\begin{aligned} -\mathbf{\Delta} u &= f \quad \text{in} \quad \Omega, \\ u &= 0 \quad \text{on} \quad \partial\Omega. \end{aligned} \tag{7.25}$$

We shall let L denote the negative Laplace operator, i.e.

$$L = -\mathbf{\Delta}.$$

Furthermore, let

$$C_0^2(\Omega) = \{v \in C^2(\Omega) \cap C(\bar{\Omega}) \mid u|_{\partial\Omega} = 0\}.$$

Hence, roughly speaking, $C_0^2(\Omega)$ consists of all functions in $C^2(\Omega)$ which are zero on the boundary of Ω. With this notation the homogeneous Poisson's equation (7.25) can be written as follows: Find $u \in C_0^2(\Omega)$ such that

$$Lu = f, \tag{7.26}$$

where the right-hand side $f \in C(\bar{\Omega})$. For functions u and v defined on Ω, we let $\langle \cdot, \cdot \rangle$ denote the inner product

$$\langle u, v \rangle = \iint_\Omega u(x,y)v(x,y)\, dx\, dy.$$

Recall that the operator L is a generalization of the operator $-\frac{d^2}{dx^2}$ studied in Section 2.3. The following result generalizes corresponding results for the one-dimensional operator given in the Lemmas 2.2 and 2.4.

Lemma 7.2 (i) *The operator L is symmetric in the sense that*

$$\langle Lu, v \rangle = \langle u, Lv \rangle \quad \text{for all} \quad u, v \in C_0^2(\Omega).$$

(ii) *Furthermore, the operator L is positive definite in the sense that*

$$\langle Lu, u \rangle \geq 0 \quad \text{for all} \quad u \in C_0^2(\Omega),$$

with equality only if $u \equiv 0$.

Proof: If $u, v \in C_0^2(\Omega)$, then it follows from Green's second identity (7.24) that

$$\iint_\Omega (u\mathbf{\Delta} v - v\mathbf{\Delta} u)\, dx\, dy = 0,$$

or

$$\langle Lu, v \rangle = \langle v, Lu \rangle.$$

On the other hand, Green's first identity (7.23) implies that if $u \in C_0^2(\Omega)$, then

$$-\iint_{\Omega} u\Delta u \, dx\, dy = \iint_{\Omega} \nabla u \cdot \nabla u \, dx\, dy = \iint_{\Omega} (u_x^2 + u_y^2)\, dx\, dy$$

or

$$\langle Lu, u\rangle = \iint_{\Omega} (u_x^2 + u_y^2)\, dx\, dy \geq 0.$$

Furthermore, if $\langle Lu, u\rangle = 0$, then u_x and u_y are zero throughout Ω. Therefore u is constant in $\bar{\Omega}$, and since $u = 0$ on $\partial\Omega$, we conclude that $u \equiv 0$. ■

A consequence of this result is that Poisson's problem (7.25) has at most one solution. Hence, we obtain an alternative proof of Theorem 6.8.

Lemma 7.3 *Assume that* $u^1, u^2 \in C^2(\Omega) \cap C(\bar{\Omega})$ *are two solutions of Poisson's problem (7.1) with the same data* f *and* g. *Then* $u^1 \equiv u^2$.

Proof: Let $v = u^1 - u^2$. Then $Lv = 0$ in Ω and $v \in C_0^2(\Omega)$. Since $\langle Lv, v\rangle = 0$, it therefore follows from Lemma 7.2 that $v \equiv 0$. ■

7.4 The Mean Value Property for Harmonic Functions

The purpose of this section is to derive the mean value property, or Poisson's formula, for harmonic functions defined on a domain in $\mathbb{R}^2$. In order to motivate the mean value property, let us first explain the corresponding property for functions u of one variable. Obviously, the requirement $u'' = 0$ implies that u is a linear function. Therefore, u is harmonic if and only if u is of the form

$$u(x) = c_1 + c_2 x$$

for $c_1, c_2 \in \mathbb{R}$. For any x and $a > 0$, we compute the mean value

$$\frac{1}{2}\big(u(x-a) + u(x+a)\big) = c_1 + c_2\left(\frac{1}{2}(x-a) + \frac{1}{2}(x+a)\right) = u(x).$$

Hence, $u(x)$ is always equal to the mean value $\frac{1}{2}\big(u(x-a) + u(x+a)\big)$; see Fig. 7.7. The mean value property for harmonic functions is a generalization of this identity to higher dimensions.

In this section it will be more convenient to refer to a point in $\mathbb{R}^2$ as a vector $\mathbf{x} = (x_1, x_2)$ instead of the coordinates (x, y) as we have used above.

Let $\mathbf{x} \in \mathbb{R}^2$ be fixed. For any real $a > 0$ let

$$B_a(\mathbf{x}) = \{\mathbf{y} \in \mathbb{R}^2 \mid |\mathbf{x} - \mathbf{y}| < a\}.$$

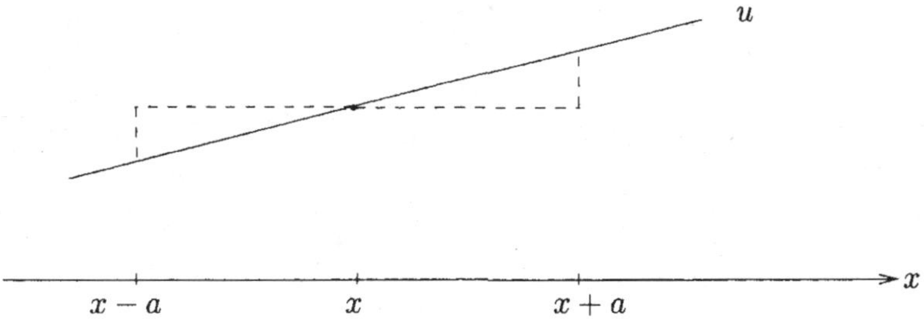

FIGURE 7.7. *The mean value property.*

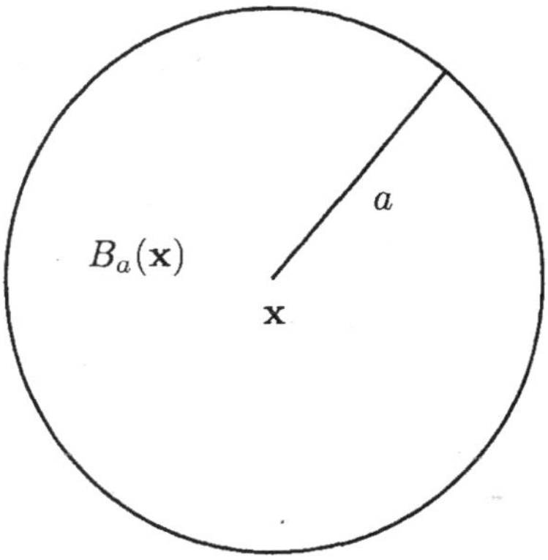

FIGURE 7.8. *The disc* $B_a(\mathbf{x})$

Here $|\mathbf{x}|$ denotes the Euclidean distance given by

$$|\mathbf{x}| = \sqrt{x_1^2 + x_2^2}.$$

Hence, the set $B_a(\mathbf{x})$ is the disc with center at $\mathbf{x}$ and with radius a; see Fig. 7.8.

Assume now that u is a harmonic function in $B_a(\mathbf{x})$, i.e. $\Delta u = 0$ in Ω. From the discussion in Section 7.2 above, we can conclude that u must be determined by its values on the boundary of $B_a(\mathbf{x})$. This follows since the solution of problem (7.14) appears to be uniquely determined by the Dirichlet data g. In fact, there is a simple formula for u at the center $\mathbf{x}$ expressed with respect to the boundary values of u. The mean value property for harmonic functions states that

$$u(\mathbf{x}) = \frac{1}{2\pi a} \int_{|\mathbf{x}-\mathbf{y}|=a} u(\mathbf{y})\, ds. \tag{7.27}$$

Hence, the value of the harmonic function u at the center of the disc $B_a(\mathbf{x})$ is equal to the average of u on its circumference. We state this beautiful relation as a theorem.

Theorem 7.1 *If u is harmonic in the disc $B_a(\mathbf{x})$, then u satisfies the identity (7.27).*

Proof: In order to establish (7.27), it is sufficient to consider the case when $\mathbf{x} = \mathbf{0} = (0,0)$, since a translated harmonic function is harmonic; see Exercise 7.4. Hence, it is sufficient to show that if u is harmonic in $B_a(\mathbf{0})$, then

$$u(\mathbf{0}) = \frac{1}{2\pi a} \int_{|\mathbf{x}|=a} u(\mathbf{x})\, ds, \tag{7.28}$$

or by introducing polar coordinates

$$u(\mathbf{0}) = \frac{1}{2\pi} \int_{-\pi}^{\pi} u(a,\phi)d\phi. \tag{7.29}$$

For each $r \in (0,a]$ define

$$U(r) = \frac{1}{2\pi} \int_{-\pi}^{\pi} u(r,\phi)d\phi.$$

Hence, the desired formula (7.29) will follow if we can show that

$$u(\mathbf{0}) = U(r) \quad \text{for} \quad 0 < r \le a. \tag{7.30}$$

Since u is continuous at the origin, we obviously have

$$\lim_{r\to 0} U(r) = u(\mathbf{0}).$$

Therefore, (7.30) will follow if $U'(r) = 0$ for $0 < r < a$. In order to see this, note first that Lemma 7.1 implies that

$$\int_{-\pi}^{\pi} \frac{\partial u}{\partial r}(r,\phi)d\phi = 0.$$

Furthermore, Proposition 3.1 on page 107 implies that

$$U'(r) = \frac{1}{2\pi} \int_{-\pi}^{\pi} \frac{\partial u}{\partial r}(r,\phi)d\phi = 0.$$

Hence (7.30) follows. ∎

The mean value property can be used to prove the maximum principle for harmonic functions, which we have already established in Section 6.4 (see Theorem 6.7 on page 193). In fact, we can prove a stronger form of the maximum principle. Assume that u is harmonic in Ω and that there is a point $\mathbf{z} \in \Omega$ (i.e. in the interior) such that

$$u(\mathbf{z}) \geq u(\mathbf{x}) \quad \text{for all } \mathbf{x} \in \bar{\Omega}.$$

Then the mean value property implies that u is constant in $\bar{\Omega}$. The details in the derivation of this strong form of the maximum principle from the mean value property are discussed in Exercise 7.11.

7.5 A Finite Difference Approximation

On a general domain Ω it is not possible to find an analytical expression of the solution of Poisson's equation. Therefore, such problems are frequently replaced by a corresponding discrete problem.

The purpose of this section is to discuss a finite difference approximation of Poisson's equation. In fact, the difference scheme we shall study was already introduced in Section 6.5.

7.5.1 The Five-Point Stencil

For notational simplicity we shall again consider the case where the domain Ω is the unit square. We first recall some of our notation introduced in Section 6.5. The domain Ω is given by

$$\Omega = \{(x,y) \mid 0 < x, y < 1\},$$

while the set of grid points, $\bar{\Omega}_h$, is of the form

$$\bar{\Omega}_h = \{(x_j, y_k) \mid 0 \leq j, k \leq n+1\}.$$

Here $x_j = jh, y_k = kh$ for a suitable spacing $h = 1/(n+1)$; see Fig. 7.9.

The set of interior grid points is

$$\Omega_h = \{(x_j, y_k) \mid 1 \leq j, k \leq n\},$$

while $\partial\Omega_h = \bar{\Omega}_h \setminus \Omega_h$ is the set of grid points on $\partial\Omega$.

To be consistent with the notation used in Section 2.3, we let D_h denote the set of all grid functions defined on $\bar{\Omega}_h$, i.e.

$$D_h = \{v \mid v : \bar{\Omega}_h \to \mathbb{R}\},$$

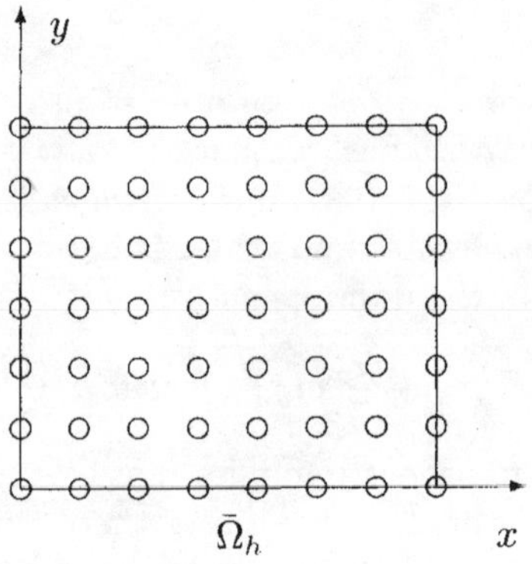

FIGURE 7.9. *The grid points* $\bar{\Omega}_h$.

while $D_{h,0}$ is the subset

$$D_{h,0} = \{v \in D_h \mid v|_{\partial\Omega_h} = 0\}.$$

Hence a function in $D_{h,0}$ is specified by its values on the interior grid points Ω_h.

We consider a finite difference approximation of Poisson's equation on Ω with homogeneous Dirichlet boundary conditions, i.e.

$$\begin{aligned} Lu = -\Delta u = f \quad &\text{in} \quad \Omega, \\ u = 0 \quad &\text{on} \quad \partial\Omega. \end{aligned} \tag{7.31}$$

The finite difference operator L_h, approximating the differential operator L, is of the form (see Fig. 7.10)

$$(L_h v)(x_j, y_k) = \frac{1}{h^2}\left[4v_{j,k} - v_{j+1,k} - v_{j-1,k} - v_{j,k+1} - v_{j,k-1}\right] \tag{7.32}$$

where, as usual, $v_{j,k} = v(x_j, y_k)$.

When we approximate a problem with homogeneous boundary conditions as in (7.31), it is natural to consider $(L_h v)$ for functions $v \in D_{h,0}$ (i.e. with zero boundary conditions). The values $(L_h v)(x_j, y_k)$ are then defined for all interior grid points $(x_j, y_k) \in \Omega_h$.

We recall that the finite difference approximation of the problem (7.31) can be formulated as follows:

Find $v \in D_{h,0}$ such that

$$(L_h v)(x_j, y_k) = f(x_j, y_k) \quad \text{for all} \quad (x_j, y_k) \in \Omega_h. \tag{7.33}$$

This is a system of n^2 linear equations in the n^2 unknowns $\{v_{j,k}\}_{j,k=1}^n$. Furthermore, the existence of a unique solution of this problem was already established in Corollary 6.6 as a consequence of the maximum principle for

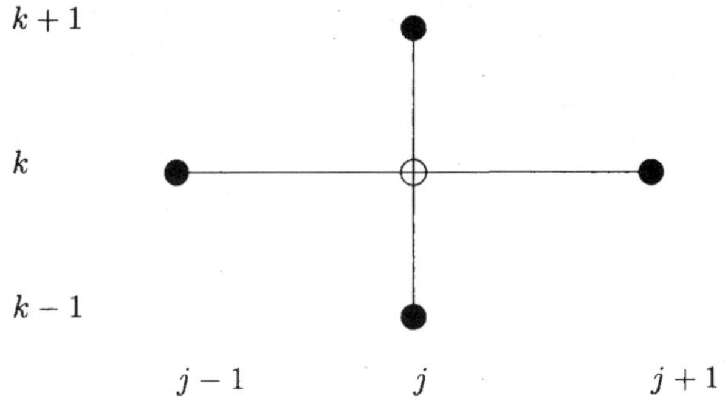

FIGURE 7.10. *The computational molecule for the operator L_h.*

discrete harmonic functions. Here we shall establish that the operator L_h has symmetry and positive definite properties which are discrete analogs of the properties for the continuous operator L given in Lemma 7.2 above.

Define the discrete inner product $\langle \cdot, \cdot \rangle_h$ by

$$\langle u, v \rangle_h = h^2 \sum_{j,k=1}^{n} u_{j,k} v_{j,k}$$

for $u, v \in D_{h,0}$. Then we have:

Lemma 7.4 i) *The operator L_h is symmetric in the sense that*

$$\langle L_h u, v \rangle_h = \langle u, L_h v \rangle_h \quad \textit{for all } u, v \in D_{h,0}.$$

ii) *Furthermore, the operator L_h is positive definite in the sense that*

$$\langle L_h v, v \rangle_h \geq 0 \quad \textit{for all } v \in D_{h,0},$$

with equality only if $v \equiv 0$.

Proof: Using the summation by parts formula (2.31), it is straightforward to show that (see Exercise 7.12)

$$\begin{aligned} \langle L_h u, v \rangle_h &= \sum_{j=0}^{n} \sum_{k=0}^{n} \big[(u_{j+1,k} - u_{j,k})(v_{j+1,k} - v_{j,k}) \\ &\qquad + (u_{j,k+1} - u_{j,k})(v_{j,k+1} - v_{j,k}) \big] \qquad (7.34) \\ &= \langle u, L_h v \rangle_h \end{aligned}$$

for $u, v \in D_{h,0}$. This establishes part i).

Furthermore, (7.34) implies that

$$\langle L_h v, v\rangle_h = \sum_{j=0}^{n}\sum_{k=0}^{n}\left[(v_{j+1,k} - v_{j,k})^2 + (v_{j,k+1} - v_{j,k})^2\right] \geq 0.$$

Also, if $\langle L_h v, v\rangle_h = 0$, then $v_{j+1,k} = v_{j,k}$ and $v_{j,k+1} = v_{j,k}$ for $0 \leq j,k \leq n$. Since $v_{0,k} = 0$ and $v_{j,0} = 0$, we conclude that $v \equiv 0$. ■

As in the continuous case, the positive definite property of the operator L_h immediately implies that the discrete system has at most one solution. Hence, we obtain an alternative proof of Corollary 6.6. You are asked to complete this proof in Exercise 7.13.

7.5.2 An Error Estimate

Finally, in this section we shall establish an error estimate for the finite difference scheme (7.33). More precisely, we shall give a bound for the error between the solution u of the continuous Poisson problem (7.31) and the solution v of the discrete Poisson problem (7.33). In order to motivate this result, we will first consider a numerical example.

EXAMPLE 7.2 Let us consider Poisson's problem (7.31) with $\Omega = (0,1) \times (0,1)$ and

$$f(x,y) = \left[(3x + x^2)y(1-y) + (3y + y^2)x(1-x)\right]e^{x+y}.$$

The function f is chosen such that the solution u of (7.31) is known. It is straightforward to check that u is given by

$$u(x,y) = x(1-x)y(1-y)e^{x+y}.$$

In the same way as we did in Example 2.5 on page 48, we compare u and the corresponding solution v of (7.33). For different values of h we compute the error

$$E_h = \max_{(x,y)\in\bar{\Omega}_h} |u(x,y) - v(x,y)|.$$

Furthermore, these values are used to estimate the rate of convergence; see Project 1.1. The results are given in Table 7.1.

From the table we observe that the error seems to satisfy a bound of the form

$$E_h = O(h^2).$$

This will in fact be established theoretically in Theorem 7.2 below. ■

n	h	E_h	Rate of convergence
5	1/6	0.00244618	
10	1/11	0.00076068	1.926
20	1/21	0.00021080	1.9846
40	1/41	0.00005533	1.9992
80	1/81	0.00001418	1.9996

TABLE 7.1. *Maximum error and estimated rate of convergence*

For the discussion in this section we will assume that the solution u of Poisson's problem (7.31) is four-times differentiable, i.e. $u \in C^4(\bar{\Omega})$. Let α be the finite constant given by

$$\alpha = \max_{0 \leq j+k \leq 4} \left\| \frac{\partial^{j+k} u}{\partial x^j \partial y^k} \right\|_\infty, \tag{7.35}$$

where, as usual, $\|u\|_\infty = \sup_{(x,y)\in\Omega} |u(x,y)|$. Hence, α bounds all partial derivatives of u of total order less than or equal to 4. In correspondence with the notation used in Chapter 2, we also let

$$\|v\|_{h,\infty} = \max_{(x,y)\in\bar{\Omega}_h} |v(x,y)|$$

for any grid function v. As in Section 2.3, we introduce the truncation error

$$\tau_h(x_j, y_k) = \big(L_h u - f\big)(x_j, y_k)$$

for all $(x_j, y_k) \in \Omega_h$. The following result is a generalization of Lemma 2.6 on page 64.

Lemma 7.5 *Assume that $u \in C^4(\bar{\Omega})$. The truncation error τ_h satisfies*

$$\|\tau_h\|_{h,\infty} \leq \frac{\alpha h^2}{6},$$

where α is given by (7.35).

Proof: This can be proved exactly the same way as we proved Lemma 2.6 and follows essentially from the error bound (2.12) on page 46. You are asked to carry this out in Exercise 7.14. ∎

The following error estimate for the finite difference method (7.33) is a generalization of Theorem 2.2 on page 65.

Theorem 7.2 *Let u and v be corresponding solutions of (7.31) and (7.33), respectively. If $u \in C^4(\bar{\Omega})$, then*

$$\|u - v\|_{h,\infty} \leq \frac{\alpha h^2}{48},$$

where α is given by (7.35).

Proof: The proof follows the same pattern as the proof of Theorem 2.2 on page 65. Define the error function $e \in D_{h,0}$ by

$$e(x_j, y_k) = (u - v)(x_j, y_k)$$

for all $(x_j, y_k) \in \bar{\Omega}_h$. Then

$$L_h e = L_h(u - v) = f + \tau - f = \tau$$

for all $(x, y) \in \Omega_h$. From the stability property for the operator L_h, established in Proposition 6.1 on page 200, it therefore follows that

$$||e||_{h,\infty} \leq \frac{1}{8}||\tau_h||_{h,\infty}.$$

The proof is completed by applying the bound for τ_h presented in Lemma 7.5 above. ■

7.6 Gaussian Elimination for General Systems

In Chapter 2.2 we studied Gaussian elimination for tridiagonal systems. However, the system (7.33) will not be tridiagonal. In order to solve such systems on a computer, we therefore need a more general algorithm than Algorithm 2.1 on page 53. You have probably already encountered Gaussian elimination in a linear algebra course. We will give a brief reminder of the algorithm here, focusing particularly on computational issues.

7.6.1 *Upper Triangular Systems*

Consider an $n \times n$ system of linear equations of the form

$$Av = b, \tag{7.36}$$

where the matrix A is of the form

$$A = \begin{pmatrix} a_{1,1} & a_{1,2} & \cdots & a_{1,n} \\ a_{2,1} & a_{2,2} & \cdots & a_{2,n} \\ \vdots & \vdots & \ddots & \vdots \\ a_{n,1} & a_{n,2} & \cdots & a_{n,n} \end{pmatrix}. \tag{7.37}$$

Alternatively, this system can be written in component form, i.e.

$$\begin{array}{ccccccccc} a_{1,1}v_1 & + & a_{1,2}v_2 & + & \cdots & + & a_{1,n}v_n & = & b_1 \\ a_{2,1}v_1 & + & a_{2,2}v_2 & + & \cdots & + & a_{2,n}v_n & = & b_2 \\ \vdots & & \vdots & & & & \vdots & & \vdots \\ a_{n,1}v_1 & + & a_{n,2}v_2 & + & \cdots & + & a_{n,n}v_n & = & b_n. \end{array} \tag{7.38}$$

We recall that this system is referred to as a tridiagonal system if $a_{i,j} = 0$ for $|i-j| > 1$. We refer to the system as a general system, or a full system, if all the elements are allowed to be nonzero.

In order to derive an algorithm for a general system, let us start with a special example. The system (7.38) is called upper triangular if $a_{i,j} = 0$ for $i > j$, i.e. all elements below the diagonal are zero. Hence, an upper triangular system is of the form

$$\begin{array}{ccccccc} a_{1,1}v_1 & + & a_{1,2}v_2 & + & \cdots & + & a_{1,n}v_n & = & b_1 \\ & & a_{2,2}v_2 & + & \cdots & + & a_{2,n}v_n & = & b_2 \\ & & & \ddots & & & \vdots & & \vdots \\ & & & & a_{n-1,n-1}v_{n-1} & + & a_{n-1,n}v_n & = & b_{n-1} \\ & & & & & & a_{n,n}v_n & = & b_n. \end{array} \qquad (7.39)$$

Furthermore, this system is nonsingular if and only if $a_{i,i} \neq 0$ for $i = 1, 2, \ldots, n$; see Exercise 7.22.

A nonsingular upper triangular system can easily be solved by so-called *back substitution*, given in Algorithm 7.1 below. We simply compute v_n from the last equation in (7.39), then v_{n-1} from the previous equation and so on.

Algorithm 7.1

```
for  i = n, n − 1, ..., 1
     v_i = b_i
     for  j = i + 1, i + 2, ..., n
          v_i = v_i − a_{i,j} v_j
     v_i = v_i / a_{i,i}.
```

7.6.2 *General Systems*

The main strategy in Gaussian elimination is to transform the general system (7.38) into upper triangular form, and then use the algorithm above on this transformed system.

A system $Av = b$ is said to be *upper k-triangular* if $a_{i,j} = 0$ for $i > j$ and $j < k$, i.e. all the elements below the diagonal in the first $(k-1)$ columns are zero. Hence, an upper k-triangular system is of the form

$$\begin{array}{ccccccccc}
a_{1,1}v_1 & + & a_{1,2}v_2 & + & & \cdots & + & a_{1,n}v_n & = & b_1 \\
& \ddots & & & & & & & & \\
& & \ddots & & & & & & & \\
& & & & a_{k,k}v_k & + & \cdots & + & a_{k,n}v_n & = & b_k \\
& & & & a_{k+1,k}v_k & + & \cdots & + & a_{k+1,n}v_n & = & b_{k+1} \\
& & & & \vdots & & & & \vdots & & \vdots \\
& & & & a_{n,k}v_k & + & \cdots & + & a_{n,n}v_n & = & b_n
\end{array} \tag{7.40}$$

If $a_{k,k} \neq 0$, we can transform this system to an equivalent system, i.e. with the same solutions, which is upper $(k+1)$-triangular. We just keep the first k equations, while for $i > k$ we replace equation i by

$$(\text{equation } i) - m_{i,k}\,(\text{equation } k),$$

where the multiplier is given by $m_{i,k} = a_{i,k}/a_{k,k}$. The new coefficients $a'_{i,j}$ and b'_i are given by

$$a'_{i,j} = a_{i,j} - m_{i,k}a_{k,j} \quad \text{and} \quad b'_i = b_i - m_{i,k}b_k \tag{7.41}$$

for $i > k$. In particular, $a'_{i,k} = 0$ for $i > k$ and hence the new system is upper $(k+1)$-triangular.

Observe that the original system (7.38) is upper 1-triangular, while an upper n-triangular system is upper triangular, i.e. of the form (7.39). Hence, if we perform the transformation above $(n-1)$ times, we obtain a sequence of equivalent systems

$$A^{(1)}v = b^{(1)},\ A^{(2)}v = b^{(2)},\ \ldots,\ A^{(n)}v = b^{(n)}$$

where the first system is the original one and the final system is upper triangular. Thus the final system can be solved by Algorithm 7.1.

Let $a_{i,j}^{(k)}$ denote the elements of the matrix $A^{(k)}$. The formulas (7.41) lead to the following recurrence relations for those elements:

$$\begin{aligned}
m_{i,k} &= a_{i,k}^{(k)}/a_{k,k}^{(k)} \\
a_{i,j}^{(k+1)} &= a_{i,j}^{(k)} - m_{i,k}a_{k,j}^{(k)} \\
b_i^{(k+1)} &= b_i^{(k)} - m_{i,k}b_k^{(k)}
\end{aligned} \tag{7.42}$$

for $i, j > k$. These algebraic formulas are the basic identities defining the Gaussian elimination algorithm.

Algorithm 7.2

```
for k = 1, 2, ..., n − 1
    for i = k + 1, k + 2, ..., n
        m_{i,k} = a_{i,k}/a_{k,k}
        for j = k + 1, k + 2, ..., n
            a_{i,j} = a_{i,j} − m_{i,k} a_{k,j}
```

This algorithm carries out part of the transformation (7.42) by storing all the elements $a_{i,j}^{(k)}$ in the original matrix A. In order to save storage we can also use the positions $a_{i,k}$ to store the multipliers $m_{i,k}$. These multipliers are needed in order to perform the transformation (7.42) on the right-hand side b. This part of the transformation (7.42) is usually carried out by a separate algorithm referred to as *forward substitution*. The reason for this is that in many practical applications one needs to solve many systems with the same coefficient matrix A, but with different right-hand sides b. By separating the calculations for A and b, Algorithm 7.2 is then only needed once. As we can see below, the algorithm for forward substitution is similar to Algorithm 7.1 for backward substitution.

Algorithm 7.3

```
for k = 1, 2, ..., n − 1
    for i = k + 1, k + 2, ..., n
        b_i = b_i − m_{i,k} b_k
```

Here we assume that the multipliers $m_{i,k}$ are obtained from Algorithm 7.2. When the complete triangular system is computed by the Algorithms 7.2 and 7.3, we finally apply the back substitution algorithm, Algorithm 7.1, to find the solution v.

The Gaussian elimination process will succeed in reducing the general system (7.38) to upper triangular form as long as all the elements $a_{k,k}^{(k)} \neq 0$ (see (7.42)). Below we shall show that if the original matrix A is symmetric and positive definite, then this will in fact be the case.

If the number of unknowns n in a linear system is very large, then the Gaussian elimination process above will be impractical, due to either the required computer time or lack of sufficient storage. The amount of computer time is roughly proportional to the number of arithmetic operations which are required, i.e. the sum of multiplications, divisions, additions, and subtractions. By this measure, the cost of carrying out Algorithm 7.2 is approximately $2n^3/3$ (see Exercise 7.24). Hence, if we need to solve a system with 10^6 unknowns on a computer which can perform $\frac{1}{3} \cdot 10^9$ operations per second,[5] the computing time will be approximately $2 \cdot 10^9$ seconds, or about 63 years.

7.6.3 Banded Systems

For many linear systems occuring in practice, one can obtain more effective algorithms by exploiting structures of the system. A simple property which can be utilized in Gaussian elimination is a so-called *banded* structure. An $n \times n$ system is called banded, with bandwidth d, if

$$a_{i,j} = 0 \quad \text{if} \quad |i-j| > d.$$

Hence, the coefficient matrix A of a banded system has the form

$$A = \begin{pmatrix} a_{1,1} & \cdots & a_{1,d+1} & 0 & \cdots & \cdots & \cdots & 0 \\ \vdots & \ddots & & \ddots & \ddots & & & \vdots \\ a_{d+1,1} & & \ddots & & \ddots & \ddots & & \vdots \\ 0 & \ddots & & \ddots & & \ddots & \ddots & \vdots \\ \vdots & \ddots & \ddots & & \ddots & & \ddots & 0 \\ \vdots & & \ddots & \ddots & & \ddots & & a_{n-d,n} \\ \vdots & & & \ddots & \ddots & & \ddots & \vdots \\ 0 & \cdots & \cdots & \cdots & 0 & a_{n,n-d} & \cdots & a_{n,n} \end{pmatrix}.$$

You can convince yourself (see (7.42)) that when Gaussian elimination is applied to a banded system, then all the elements $a_{i,j}^{(k)}$ and $m_{i,j}$ are zero for $|i-j| > d$. Hence, the computation can be restricted to the data structure required to represent the nonzero elements of the original coefficient matrix. Therefore, we obtain the following banded version of Algorithm 7.2.

Algorithm 7.4: Banded matrix

$$\begin{array}{l} \text{for } k = 1, 2, \ldots, n-1 \\ \qquad \text{for } i = k+1, k+2, \ldots, \min(k+d, n) \\ \qquad\qquad m_{i,k} = a_{i,k}/a_{k,k} \\ \qquad\qquad \text{for } j = k+1, k+2, \ldots, \min(k+d, n) \\ \qquad\qquad\qquad a_{i,j} = a_{i,j} - m_{i,k}a_{k,j} \end{array}$$

The number of multiplications required by this algorithm is approximately $2nd^2$; see Exercise 7.24. In order to compare the algorithm for a full matrix with the banded version above, we consider an example.

---

[5] This is, approximately, the performance of the *R*8000 processor from Silicon Graphics. The CPU time indicated here does not reflect the memory limitations for such a large, full system.

EXAMPLE 7.3 Consider the system (7.33), i.e. the discrete Poisson's equation. We let $h = 1/(\bar{n}+1)$ be the spacing, such that the number of unknowns is given by $n = (\bar{n})^2$. This system can be written as a linear system in the form $Aw = b$, where the vector $w \in \mathbb{R}^n$ is related to the unknowns $\{v_{j,k}\}$ by

$$v_{j,k} = w_{k+(j-1)\bar{n}}.$$

This corresponds to storing the unknowns $\{v_{j,k}\}$ row by row in the vector w. By multiplying each equation in (7.34) by h^2, we obtain the linear system $Aw = b$, with a banded coefficient matrix given by

$$A = \begin{pmatrix} 4 & \beta_1 & & & -1 & & & \\ \beta_1 & 4 & \beta_2 & & & \ddots & & \\ & \beta_2 & \ddots & \ddots & & & \ddots & \\ & & \ddots & \ddots & \ddots & & & \ddots \\ -1 & & & \ddots & \ddots & \ddots & & \\ & \ddots & & & \ddots & \ddots & \ddots & -1 \\ & & \ddots & & & \ddots & \ddots & \ddots \\ & & & \ddots & & & \ddots & \ddots \\ & & & & \ddots & & \ddots & 4 & \beta_{n-1} \\ & & & & -1 & & & \beta_{n-1} & 4 \end{pmatrix}.$$

Here we have only indicated the nonzero elements, and the solid lines indicate the boundary of the band. The coefficients β_j are given by

$$\beta_j = \begin{cases} 0 & \text{if } \frac{j}{\bar{n}} = \text{integer}, \\ -1 & \text{otherwise.} \end{cases}$$

The symmetric matrix A is a banded $n \times n$ matrix with bandwidth[6] $\bar{n} = \sqrt{n}$. If this system is solved as a full system, the number of operations in Algorithm 7.2 is approximately $\frac{2}{3}n^3$. However, if we use the banded version of the algorithm, this is reduced to $2n^2$. Thus, in the case of $n = 10^6$ (or $\bar{n} = 10^3$), with the computer considered above, the computing time is reduced from about 63 years to about 1 hour and 40 minutes.

[6]Note that the matrix A has only five nonzero diagonals, while the band contains $2\bar{n} + 1$ diagonals. However, only the elements outside the band will remain zero during the Gaussian elimination process. Therefore, we need to update all the elements inside the band.

Assume that we solve the system above with Algorithm 7.4 for a fixed $h > 0$, and that the computing time is approximately 1 second. If we instead use the spacing $h/4$, then the number of unknowns $n = \bar{n}^2$ is roughly multiplied by 16, and hence the estimated CPU time is increased with a factor of 256. This predicts that the CPU time for $h/4$ is more than 4 minutes. If the spacing is further decreased to $h/8$, the CPU time will be more than an hour. ■

7.6.4 Positive Definite Systems

We shall show that if the coefficient matrix A of the original system (7.38) is symmetric and positive definite, then all the elements $a_{k,k}^{(k)}$, defined by (7.42), are nonzero. Hence, the elimination process will not break down and the solution v can be computed.

We recall here that a symmetric $n \times n$ matrix A is positive definite if

$$v^T A v \geq 0 \quad \text{for all} \quad v \in \mathbb{R}^n,$$

with equality only if $v = 0$. The proof of the proposition below is a generalization of the proof of Proposition 2.4 on page 56.

Proposition 7.1 *If A is symmetric and positive definite, then the elements $a_{k,k}^{(k)}$, defined by (7.42), are strictly positive for $k = 1, 2, \ldots, n$.*

Proof: Assume that $a_{1,1}^{(1)}, a_{2,2}^{(2)} \ldots, a_{k-1,k-1}^{(k-1)} > 0$, but that $a_{k,k}^{(k)} \leq 0$. We shall show that this assumption leads to a contradiction. For any vector $b \in \mathbb{R}^n$ the linear system $Av = b$ is transformed by (7.42) to an equivalent upper k-triangular system $A^{(k)}v = b^{(k)}$ or (see (7.40))

$$\begin{array}{ccccccccc}
a_{1,1}^{(k)}v_1 & + \cdots + & a_{1,k-1}^{(k)}v_{k-1} & + & a_{1,k}^{(k)}v_k & + \cdots + & a_{1,n}^{(k)}v_n & = & b_1^{(k)} \\
 & \ddots & & & & & & & \\
 & & a_{k-1,k-1}^{(k)}v_{k-1} & + & a_{k-1,k}^{(k)}v_k & + \cdots + & a_{k-1,n}^{(k)}v_n & = & b_{k-1}^{(k)} \\
 & & & & a_{k,k}^{(k)}v_k & + \cdots + & a_{k,n}^{(k)}v_n & = & b_k^{(k)} \\
 & & & & \vdots & & & & \\
 & & & & a_{n,k}^{(k)}v_k & + \cdots + & a_{n,n}^{(k)}v_n & = & b_n^{(k)}.
\end{array}$$

It follows from (7.42) (see Exercise 7.25) that the vectors b and $b^{(k)}$ are related by

$$b_i = b_i^{(k)} + \sum_{j=1}^{\min(i-1,k-1)} m_{i,j} b_j^{(k)}. \tag{7.43}$$

The proof will be completed by constructing a nonzero vector $v \in \mathbb{R}^n$ such that

$$v^T A v \leq 0,$$

which will contradict the assumption that A is positive definite. Choose $v_k = 1$ and $v_{k+1} = \cdots = v_n = 0$. Furthermore, since $a_{j,j}^{(k)} = a_{j,j}^{(j)} > 0$ for $j < k$, it follows that we can find unique values $v_{k-1}, \ldots, v_1$ such that $b_1^{(k)} = b_2^{(k)} = \cdots = b_{k-1}^{(k)} = 0$. Hence, from (7.43) we obtain that $b = Av$ satisfies

$$b_1 = b_2 = \cdots = b_{k-1} = 0 \quad \text{and} \quad b_k = b_k^{(k)} = a_{k,k}^{(k)}.$$

The vector v is nonzero, since $v_k = 1$ and

$$v^T A v = \sum_{i=1}^{n} v_i b_i = v_k b_k = a_{k,k}^{(k)} \leq 0.$$

This is the desired contradiction, and consequently we conclude that

$$a_{k,k}^{(k)} > 0 \quad \text{for} \quad k = 1, 2, \ldots, n.$$

■

We end this section with a warning. Although Gaussian elimination, in theory, can be used to solve linear systems arising from discretizations of partial differential equations, they are not commonly applied to large systems of practical interest. They are simply too slow. There are much faster iterative methods available for these problems, which also require less storage. An introduction to such methods can be found in Hackbusch [13].

7.7 Exercises

EXERCISE 7.1 Let Ω be the unit square. Find a function $u(x, y)$ which is harmonic in Ω and satisfies the boundary conditions (7.2)–(7.4) with $g(x) = x(1 - x)$ (see Exercise 3.1).

EXERCISE 7.2 Let Ω be the unit square. Consider the problem

$$-\Delta u = 0 \quad \text{in} \quad \Omega,$$

together with the boundary conditions (7.2)–(7.3) and

$$u_y(x, 1) = g(x), \qquad 0 \leq x \leq 1.$$

Explain how this problem can be solved by separation of variables.

EXERCISE 7.3 Consider the Laplace equation

$$\Delta u = 0$$

with inhomogeneous boundary conditions

$$\begin{aligned} u(0,y) &= g_1(y), & 0 \le y \le 1, \\ u(1,y) &= g_2(y), & 0 \le y \le 1, \\ u(x,0) &= g_3(x), & 0 \le x \le 1, \\ u(x,1) &= g_4(x), & 0 \le x \le 1. \end{aligned}$$

Explain how we can obtain a formal solution of this problem from formal solutions of simpler problems with boundary conditions similar to (7.2)–(7.4), i.e. with an inhomogeneous boundary condition at only one edge.

EXERCISE 7.4 (Invariance under rigid motions.)

(a) (Translation) Let

$$x' = x + a, \qquad y' = y + b$$

for fixed real numbers a and b and let u and v be functions such that $v(x', y') = u(x, y)$. Explain why $\Delta u = \Delta v$, i.e.

$$\big(u_{xx} + u_{yy}\big)(x, y) = \big(v_{x'x'} + v_{y'y'}\big)(x', y').$$

(b) (Rotation) Let

$$\begin{aligned} x' &= x\cos(\phi) + y\sin(\phi), \\ y' &= -x\sin(\phi) + y\cos(\phi). \end{aligned}$$

Explain why this corresponds to a rotation of the coordinate system, and show that if $v(x', y') = u(x, y)$, then $\Delta v = \Delta u$.

EXERCISE 7.5 Let Ω be the rectangle

$$\Omega = \big\{(x, y) \bigm| 0 < x < L,\ 0 < y < M\big\},$$

where L and M are positive constants. Assume that u is harmonic in Ω and let

$$v(x, y) = u(xL, yM)$$

for (x, y) in the unit square. Show that v satisfies the equation

$$v_{xx} + \frac{L^2}{M^2} v_{yy} = 0.$$

EXERCISE 7.6 Give a detailed derivation of the identities (7.10), (7.11), and (7.12).

EXERCISE 7.7 Consider the Euler equation (7.17), i.e.

$$r^2R''(r) + rR'(r) - k^2R(r) = 0, \tag{7.44}$$

where $r > 0$.

Define a new function $v(z) = R(e^z)$ for $z \in \mathbb{R}$.

(a) Show that

$$v''(z) = k^2v(z).$$

(b) Find the general solution of equation (7.44).

EXERCISE 7.8 Let Ω be a wedge with angle π/b and radius $\rho > 0$, i.e.

$$\Omega = \{(r, \phi) \mid 0 < r < \rho,\ 0 < \phi < \pi/b\}.$$

(a) Explain how we can use separation of variables to find a harmonic function $u(r, \phi)$ which satisfies boundary conditions of the form

$$u_\phi(r, 0) = u_\phi(r, \pi/b) = 0$$

and

$$u(\rho, \phi) = g(\phi).$$

(b) Does this type of boundary condition imply a singularity in the solution of the form discussed for the Dirichlet condition in Section 7.2.3?

EXERCISE 7.9 Consider a Poisson problem with Neumann boundary conditions of the form

$$\begin{aligned} -\Delta u &= f \quad \text{in} \quad \Omega, \\ \frac{\partial u}{\partial \mathbf{n}} &= g \quad \text{on} \quad \partial\Omega. \end{aligned} \tag{7.45}$$

(a) Show that a necessary condition for this problem to have a solution is that

$$\iint_\Omega f\ dx\,dy = -\int_{\partial\Omega} g\ ds. \tag{7.46}$$

(b) Consider the problem (7.45) for two given functions f and g satisfying (7.46). Show that there is at most one solution of this problem such that

$$\iint_{\Omega} u \, dx \, dy = 0.$$

Why do we need this extra condition ?

EXERCISE 7.10 Consider a Poisson problem with Robin boundary conditions of the form

$$\begin{aligned} -\Delta u &= f \quad \text{in} \quad \Omega, \\ \frac{\partial u}{\partial \mathbf{n}} + \alpha u &= g \quad \text{on} \quad \partial\Omega, \end{aligned}$$

where $\alpha > 0$ is a constant.

Show that this problem has at most one solution.

EXERCISE 7.11 The strong maximum principle states that if a harmonic function u reaches its maximum in the interior of the domain Ω, then u is necessarily constant on Ω.

Explain how this property follows from the mean value property for harmonic functions. Does this argument put restrictions on the domain Ω?

EXERCISE 7.12 Use the identity (2.31) on page 60 to establish the identity (7.34).

EXERCISE 7.13 Let Ω be the unit square. Use the result of Lemma 7.4 to show that a discrete Poisson problem

$$\begin{aligned} (L_h v)(x, y) &= f(x, y), \qquad (x, y) \in \Omega_h, \\ v(x, y) &= g(x, y), \qquad (x, y) \in \partial\Omega_h, \end{aligned}$$

has a unique solution.

EXERCISE 7.14 Use the bound (2.12) on page 46 to prove Lemma 7.5.

EXERCISE 7.15 Assume that Ω is the unit square. State and prove a strong maximum principle for discrete harmonic functions (see Problem 7.11).

EXERCISE 7.16 Consider the Poisson problem

$$-\Delta u = 2x(1 - x) + 2y(1 - y) \tag{7.47}$$

with $u = 0$ on $\partial\Omega$.

(a) Show that the exact solution is given by

$$u = x(1-x)y(1-y).$$

(b) Use the numerical approximation (7.32) to establish a linear system

$$Aw = b \tag{7.48}$$

of the form considered in Example 7.3.

(c) Implement Algorithm 7.4 for the linear system (7.48).

(d) Use the program developed in (c) to analyze the sharpness of the estimate given in Theorem 7.2 for the problem (7.47).

EXERCISE 7.17 Assume that Ω is the unit square and consider and eigenvalue problem of the form

$$\begin{aligned} -\Delta u &= \lambda u \quad \text{in} \quad \Omega, \\ u &= 0 \quad \text{on} \quad \partial\Omega, \end{aligned} \tag{7.49}$$

where the eigenfunction $u \not\equiv 0$.

(a) Use Green's first identity to show that $\lambda > 0$.

(b) Assume that for each $x \in (0,1)$, $u(x,\cdot)$ can be written in a sine series with respect to y of the form

$$u(x,y) = \sum_{k=1}^{\infty} a_k(x) \sin(k\pi y).$$

Explain that this leads to eigenfunctions of the form

$$u(x,y) = \sin(j\pi x)\sin(k\pi y) \quad \text{for} \quad j,k = 1,2,\ldots$$

and eigenvalues

$$\lambda_j = (j\pi)^2 + (k\pi)^2.$$

EXERCISE 7.18 Let Ω be the unit square and consider an inhomogeneous Poisson problem of the form

$$\begin{aligned} -\Delta u &= f \quad \text{in} \quad \Omega, \\ u &= 0 \quad \text{on} \quad \partial\Omega. \end{aligned}$$

Assume that the right-hand side f has a representation of the form

$$f(x,y) = \sum_{j,k=1}^{\infty} a_{j,k} \sin(j\pi x)\sin(k\pi y),$$

where $\{a_{j,k}\}$ are suitable constants. Find a formal solution $u(x,y)$ of the form

$$u(x,y) = \sum_{j,k=1}^{\infty} b_{j,k} \sin(j\pi x)\sin(k\pi y).$$

EXERCISE 7.19 Consider the heat equation in two space variables x and y. Hence, for a given domain $\Omega \subset \mathbb{R}^2$ we consider

$$\begin{aligned} u_t &= \Delta u && \text{in} \quad \Omega, t > 0, \\ u(x,y,t) &= 0 && \text{for} \quad (x,y) \in \partial\Omega, \\ u(x,y,0) &= f(x,y) && \text{for} \quad (x,y) \in \Omega. \end{aligned} \tag{7.50}$$

(a) Use energy arguments to show that any solution of this problem satisfies

$$\iint_\Omega u^2(x,y)\,dx\,dy \le \iint_\Omega f^2(x,y)\,dx\,dy.$$

(Hint: In the same way as in Section 3.7, consider $\frac{d}{dt}\int\int u^2(x,y,t)dxdy$. In addition, Green's first identity (7.23) will probably be useful.)

(b) Explain why the initial and boundary value problem (7.50) has at most one solution.

(c) Assume that Ω is the unit square. Try to use the results of Exercise 7.17 above to construct a formal solution of (7.50).

EXERCISE 7.20 Assume that Ω is the unit square. Consider the following eigenvalue problem:

Find $v \in D_{h,0}$ such that

$$L_h v = \lambda v.$$

Find all eigenvalues and eigenvectors. (Hint: The problem has at most n^2 eigenvalues. Try to consider eigenfunctions v of the form $\sin(j\pi x)\sin(k\pi y)$.)

EXERCISE 7.21 Let $u \in C^2(B_a(\mathbf{0})) \cap C(\overline{B_a(\mathbf{0})})$ be a solution of the problem

$$\begin{aligned} -\Delta u &= 0 \quad \text{in} \quad B_a(\mathbf{0}), \\ u &= g \quad \text{on} \quad \partial B_a(\mathbf{0}), \end{aligned}$$

for $a > 0$. Here g is a continuous function on $\partial B_a(\mathbf{0})$. The purpose of this exercise is to derive Poisson's formula for this solution; the formula states that

$$u(\mathbf{x}) = \frac{a^2 - |\mathbf{x}|^2}{2\pi a} \int_{\partial B_a(\mathbf{0})} \frac{g(\mathbf{y})}{|\mathbf{x} - \mathbf{y}|^2} ds. \tag{7.51}$$

Observe that this formula reduces to the mean value property (7.27) when $\mathbf{x} = \mathbf{0}$.

Define

$$v(\mathbf{x}) = \frac{a^2 - |\mathbf{x}|^2}{2\pi a} \int_{\partial B_a(\mathbf{0})} \frac{g(\mathbf{y})}{|\mathbf{x} - \mathbf{y}|^2} ds.$$

(a) Assume we can show that v is harmonic in $B_a(\mathbf{0})$ and that $v\Big|_{\partial B_a(\mathbf{0})} = g$. Explain why this implies that (7.51) holds.

(b) Show by direct computations that the function

$$\frac{a^2 - |\mathbf{x}|^2}{|\mathbf{x} - \mathbf{y}|^2}$$

is harmonic in $B_a(\mathbf{0})$ as a function of $\mathbf{x}$ if $|\mathbf{y}| = a$.

(c) Use Proposition 3.1 on page 107 to show that v is harmonic in $B_a(\mathbf{0})$.

Let

$$w(\mathbf{x}) = \frac{a^2 - |\mathbf{x}|^2}{2\pi a} \int_{\partial B_a(\mathbf{0})} \frac{ds}{|\mathbf{x} - \mathbf{y}|^2}.$$

Hence w corresponds to the function v with $g \equiv 1$.

(d) Show that w is rotation invariant, i.e. $w = w(|\mathbf{x}|) = w(r)$.

(e) Use the fact that w is harmonic and rotation invariant to show that $w \equiv 1$.

(f) Let $\mathbf{z} \in \partial B_a(\mathbf{0})$. Use the fact that $g(\mathbf{z}) = w(\mathbf{x})g(\mathbf{z})$ to show that

$$\lim_{\mathbf{x} \to \mathbf{z}} v(\mathbf{x}) = g(\mathbf{z}).$$

Exercise 7.22 Show that an upper triangular matrix of the form (7.39) is nonsingular if and only if the diagonal elements $a_{i,i} \neq 0$ for $i = 1, 2, \ldots, n$.

Exercise 7.23 Give an example of a nonsingular linear system which has the property that Gaussian elimination will fail, i.e Algortithm 7.2 will break down.

Exercise 7.24

(a) Consider the Algorithm 7.2. Show that the required number of operations is approximately $2n^3/3 + O(n^2)$.

Hint: You can use the identities

$$\sum_{k=1}^{n-1} k = \frac{1}{2}n(n-1) \quad \text{and} \quad \sum_{k=1}^{n-1} k^2 = \frac{1}{6}n(n-1)(2n-1).$$

(b) Consider the Algorithm 7.4 for a banded matrix. Show that the number of operations is approximately $2nd^2$.

Exercise 7.25 Consider the recurrence relations (7.42) and assume that $a_{i,i}^{(i)} \neq 0$ for $i = 1, 2, \ldots, k-1$.

(a) Show the identity (7.43).

(b) Show a similar relation for the qth columns of A and $A^{(k)}$, i.e.

$$a_{i,q} = a_{i,q}^{(k)} + \sum_{j=1}^{\min(i-1,k-1)} m_{i,j} a_{j,k}.$$

(c) Assume that $a_{k,k}^{(k)} \neq 0$ for $k = 1, 2, \ldots, n-1$. Show that $A = LU$, where L is a nonsingular lower triangular matrix and U is upper triangular. (This factorization is frequently referred to as an *LU*-factorization of A).

8
Orthogonality and General Fourier Series

In the previous chapters Fourier series have been the main tool for obtaining formal solutions of partial differential equations. The purpose of the present chapter and the two following chapters is to give a more thorough analysis of Fourier series and formal solutions. The Fourier series we have encountered in earlier chapters can be thought of as examples of a more general class of orthogonal series, and many properties of Fourier series can be derived in this general context. In the present chapter we will study Fourier series from this point of view. The next chapter is devoted to convergence properties of Fourier series, while we return to partial differential equations in Chapter 10. There the goal is to show that the formal solutions are in fact rigorous solutions in a strict mathematical sense.

Let us first recall some of our earlier experiences with Fourier series. For example, in Sections 3.3–3.4 we expanded the initial function f of the initial and boundary value problem (3.1)–(3.3) for the heat equation in a Fourier sine-series of the form (cf. (3.25))

$$f(x) = \sum_{k=1}^{\infty} c_k \sin(k\pi x) \tag{8.1}$$

to obtain the formal solution

$$u(x,t) = \sum_{k=1}^{\infty} c_k e^{-(k\pi)^2 t} \sin(k\pi x).$$

A similar procedure was carried out in Section 3.6, where the corresponding Neumann problem was solved by expanding the initial function f in a

Fourier cosine series of the form

$$f(x) = \frac{c_0}{2} + \sum_{k=1}^{\infty} c_k \cos(k\pi x), \tag{8.2}$$

while in Exercise 3.15 we used a full Fourier series of the form

$$f(x) = \frac{a_0}{2} + \sum_{k=1}^{\infty} \left(a_k \cos(k\pi x) + b_k \sin(k\pi x)\right) \tag{8.3}$$

to solve a corresponding problem with periodic boundary conditions.

In all the three examples mentioned above, a key property is that we are expanding the function f as a linear combination of orthogonal basis functions. For example, in (8.1) the set of basis functions, $\{\sin(k\pi x)\}_{k=1}^{\infty}$, is an orthogonal set in the sense that

$$\int_0^1 \sin(k\pi x)\sin(m\pi x)dx = \begin{cases} 0 & k \neq m, \\ 1/2 & k = m, \end{cases}$$

(cf. Lemma 2.8). This immediately leads to the formula

$$c_k = 2\int_0^1 f(x)\sin(k\pi x)dx$$

for the coefficients. Similar formulas hold for the coefficients in (8.2) and (8.3).

Below we shall see that the Fourier sine series (8.1) and the Fourier cosine series (8.2) are in fact special cases of the full Fourier series (8.3).

8.1 The Full Fourier Series

Let f be a function defined on $[-1,1]$. The full Fourier series of f is given by

$$f(x) = \frac{a_0}{2} + \sum_{k=1}^{\infty} \left(a_k \cos(k\pi x) + b_k \sin(k\pi x)\right), \tag{8.4}$$

where

$$\begin{aligned} a_k &= \int_{-1}^{1} f(x)\cos(k\pi x)dx, \qquad k = 0, 1, \dots, \\ b_k &= \int_{-1}^{1} f(x)\sin(k\pi x)dx, \qquad k = 1, 2, \dots. \end{aligned} \tag{8.5}$$

We will assume that f is a piecewise continuous function on $[-1,1]$.

Definition 8.1 *A function f is called piecewise continuous on an interval $[a,b]$ if it is continuous in all but a finite number of interior points $\{x_j\}$, where $\lim_{x\searrow x_j} f(x)$ and $\lim_{x\nearrow x_j} f(x)$ both exist*[1] *(see Fig. 8.1)* ■

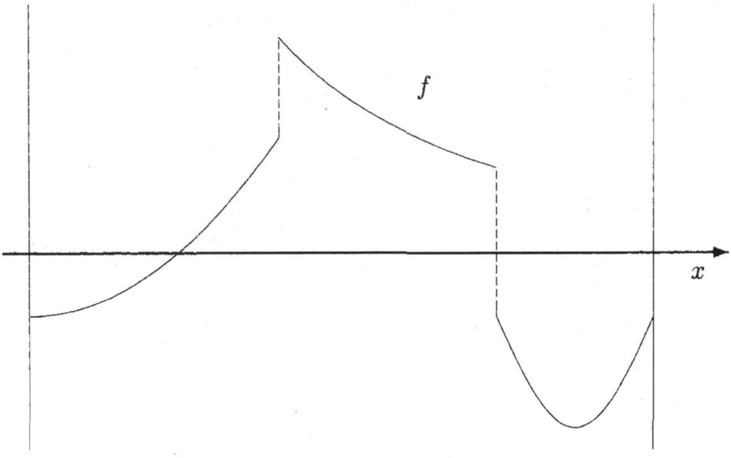

FIGURE 8.1. *A piecewise continuous function.*

Since f is assumed to be piecewise continuous on $[-1, 1]$, the coefficients a_k and b_k given by (8.5) are well defined. However, at this point we would like to be more careful with the equality sign used in (8.4). Since we have an infinite sum on the right-hand side, this equality sign indicates that the partial sums

$$S_N(f)(x) = \frac{a_0}{2} + \sum_{k=1}^{N} (a_k \cos(k\pi x) + b_k \sin(k\pi x))$$

converge to $f(x)$ as $N \to \infty$. However, so far we have not discussed this convergence. All we have argued is that if f can be represented in the form (8.4), then the coefficients must be given by (8.5). The question of convergence of Fourier series will be discussed in detail in the next chapter. With the purpose of being more precise, we shall therefore write

$$f(x) \sim \frac{a_0}{2} + \sum_{k=1}^{\infty} (a_k \cos(k\pi x) + b_k \sin(k\pi x)) \qquad (8.6)$$

instead of (8.4). Here the symbol $\sim$ should be read as "has the Fourier series."

---

[1]Here $\lim_{x \searrow x_j} = \lim_{x \to x_j, x > x_j}$ and $\lim_{x \nearrow x_j} = \lim_{x \to x_j, x < x_j}$.

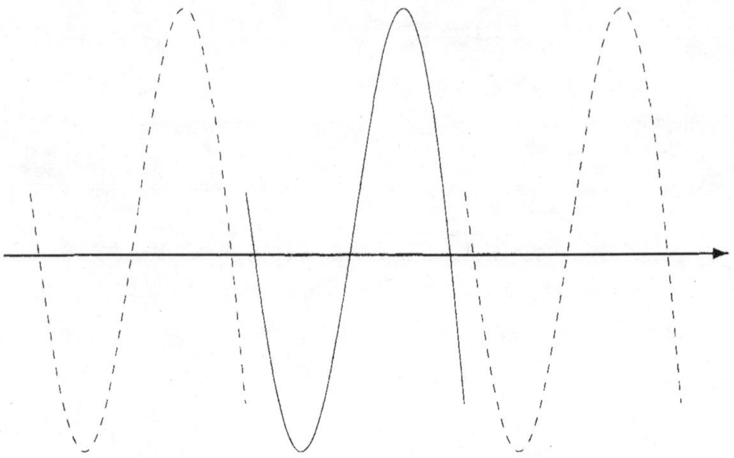

FIGURE 8.2. *Periodic extension of a function.*

Definition 8.2 *Let f be piecewise continuous*[2] *on $[-1, 1]$ and let the coefficients a_k and b_k be given by (8.5). The infinite series*

$$\frac{a_0}{2} + \sum_{k=1}^{\infty} (a_k \cos(k\pi x) + b_k \sin(k\pi x))$$

is referred to as the full Fourier series of f. The notation in (8.6) will be used to indicate the relation between f and its Fourier series. ■

Let us observe that since the trigonometric functions sine and cosine are defined on all of $\mathbb{R}$, the partial sum $S_N(f)$ can naturally be considered to be functions on all of $\mathbb{R}$. Furthermore, since the trigonometric functions are 2π-periodic, the functions $S_N(f)$ are 2-periodic. Here, we recall that a function defined on $\mathbb{R}$ is called p-periodic, with period p, if

$$g(x+p) = g(x). \tag{8.7}$$

On the other hand, if g is defined on an interval of length p then it can be uniquely extended to a p-periodic function defined on all of $\mathbb{R}$ (see Fig. 8.2) by enforcing (8.7). This function is called *the p-periodic extension* of g.

Let us return to the situation for Fourier series. If $S_N(f)$ converges to f on $[-1, 1]$, it will in fact converge to the 2-periodic extension of f on all of $\mathbb{R}$. Therefore the full Fourier series can either be considered as an expansion

[2]Throughout this book we will discuss Fourier series for piecewise continuous functions. Fourier series for less regular functions will not be considered.

of a function defined on $[-1,1]$ or as an expansion of a 2-periodic function defined on all of $\mathbb{R}$.

8.1.1 Even and Odd Functions

Before we consider an example of a full Fourier series, let us recall that a function f is called an even function if $f(-x) = f(x)$. Similarly, f is called an odd function if $f(-x) = -f(x)$. Typical examples of even functions are x^2 and $\cos(x)$, while x^3 and $\sin(x)$ are odd. Furthermore, the product of two even or odd functions is even, while the product of an even and an odd function is odd. Finally, if f is odd, then

$$\int_{-1}^{1} f(x)dx = 0,$$

and if f is even, then

$$\int_{-1}^{1} f(x)dx = 2\int_{0}^{1} f(x)dx.$$

You are asked to discuss these properties of even and odd functions in Exercise 8.2.

EXAMPLE 8.1 Let $f(x) = x$ for $x \in [-1,1]$. In order to find the full Fourier series of f, we have to compute the coefficients a_k, b_k. Since f is an odd function and $\cos(k\pi x)$ is an even function, we conclude that $a_k = 0$ for $k = 0, 1, 2, \ldots$.
Furthermore, integration by parts implies that for $k \geq 1$,

$$\begin{aligned}
b_k &= \int_{-1}^{1} x \sin(k\pi x)dx \\
&= -\frac{1}{k\pi}\left[x\cos(k\pi x)\right]_{-1}^{1} + \frac{1}{k\pi}\int_{-1}^{1} \cos(k\pi x)dx \\
&= -\frac{1}{k\pi}\left(\cos(k\pi) + \cos(-k\pi)\right) \\
&= \frac{2}{k\pi}(-1)^{k+1}.
\end{aligned}$$

Hence, we have

$$x \sim \sum_{k=1}^{\infty} \frac{2}{k\pi}(-1)^{k+1}\sin(k\pi x).$$

The 2-periodic extension of x is plotted on Fig. 8.3. Hence, if the Fourier series converges to x on $[-1,1]$, it converges to this extension on all of $\mathbb{R}$. ■

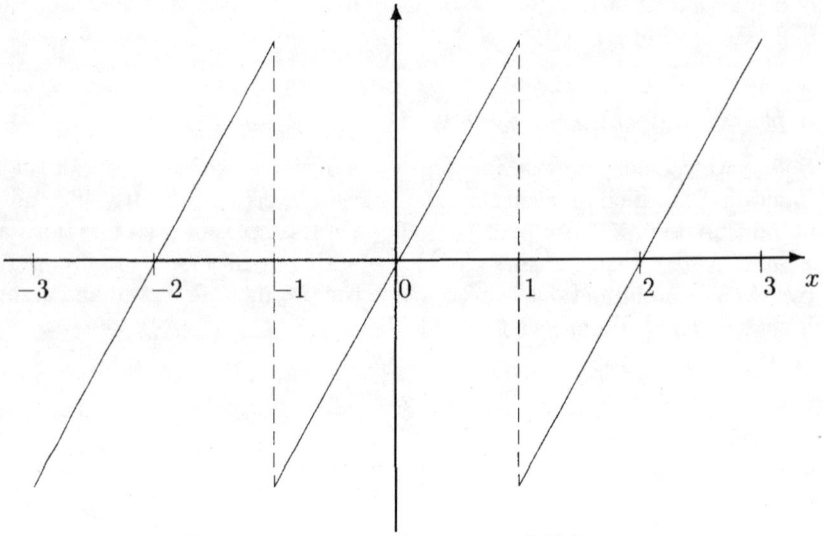

FIGURE 8.3. *Periodic extension of* $f(x) = x$.

In the example above we found that the full Fourier series of the function $f(x) = x$ is the same as the Fourier sine series of this function (cf. Example 3.4 on page 97). In fact, this will be the case for any odd function.

Lemma 8.1 *If* f *is an odd function defined on* $[-1, 1]$, *then*

$$f(x) \sim \sum_{k=1}^{\infty} b_k \sin(k\pi x),$$

where

$$b_k = 2\int_0^1 f(x)\sin(k\pi x)dx.$$

Proof: Since $f(x)$ is odd and $\cos(k\pi x)$ is even, $f(x)\cdot\cos(k\pi x)$ is odd and

$$a_k = \int_{-1}^{1} f(x)\cos(k\pi x)dx = 0.$$

Furthermore, since $f(x)\cdot\sin(k\pi x)$ is even, we have

$$b_k = \int_{-1}^{1} f(x)\sin(k\pi x)dx = 2\int_0^1 f(x)\sin(k\pi x)dx. \qquad \blacksquare$$

We observe that the coefficients b_k given above coincide with the coefficients in the Fourier sine series. (see (3.29)). Hence, we have established that full Fourier series of an odd function is the same as its Fourier sine

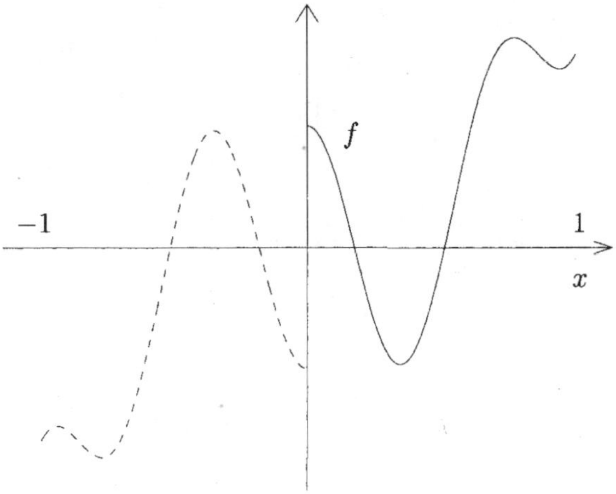

FIGURE 8.4. *Odd extension of a function.*

series. On the other hand, if f is a function defined on $[0,1]$, then f can be extended to an odd function on $[-1,1]$ by letting

$$f(-x) = -f(x);$$

see Fig. 8.4.

Furthermore, the full Fourier series of the odd extension is exactly the sine series of f.

Similar observations can be made for even functions. The full Fourier series of an even function is exactly the cosine series of f. Furthermore, the cosine series of any function f defined on $[0,1]$ is the full Fourier of its even extension defined by

$$f(-x) = f(x);$$

see Fig. 8.5.

EXAMPLE 8.2 Consider the function $f(x) = 1$. Note that this function already has the form of a full Fourier series. By directly computing the coefficients a_k and b_k in (8.5), we easily conclude that

$$1 \sim 1;$$

i.e. the full Fourier series has only one term. Since f is an even function, the full Fourier series is also the cosine series of f. However, in Example 3.3 on page 96 we also computed the sine series of this function. Hence, since $\operatorname{sign}(x)$ is the odd extension of f, it follows that

$$\operatorname{sign}(x) \sim \frac{4}{\pi} \sum_{k=1}^{\infty} \frac{1}{2k-1} \sin\left((2k-1)\pi x\right).$$

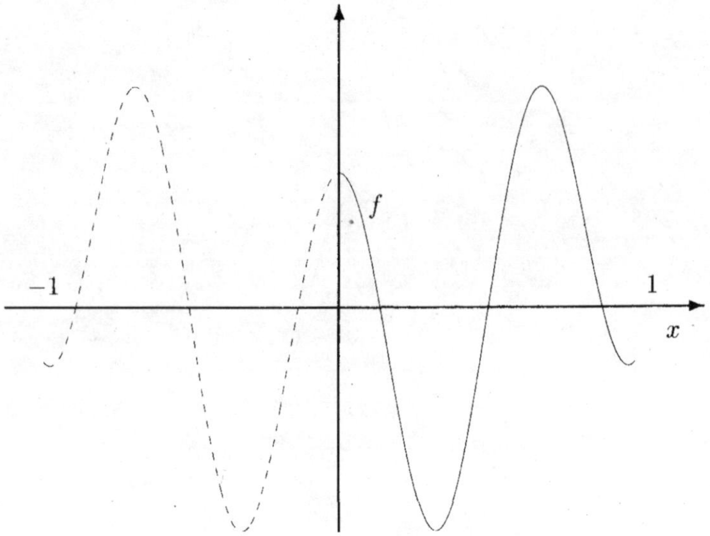

FIGURE 8.5. *Even extension of a function.*

Here

$$\operatorname{sign}(x) = \begin{cases} -1 & \text{for} \quad x < 0, \\ 0 & \text{for} \quad x = 0, \\ 1 & \text{for} \quad x > 0. \end{cases}$$

This Fourier series will potentially converge to the 2-periodic extension of sign(x) plotted in Fig. 8.6. ■

EXAMPLE 8.3 In Example 3.6 on page 102 we found that the function $f(x) = x$ had the cosine series

$$\frac{1}{2} - \frac{4}{\pi^2}\sum_{k=1}^{\infty}\left(\frac{1}{2k-1}\right)^2 \cos\left((2k-1)\pi x\right).$$

Since the even extension of f is $|x|$, we have

$$|x| \sim \frac{1}{2} - \frac{4}{\pi^2}\sum_{k=1}^{\infty}\left(\frac{1}{2k-1}\right)^2 \cos\left((2k-1)\pi x\right).$$

■

8.1.2 Differentiation of Fourier Series

One of the important applications of Fourier series is solving differential equations. In such applications we typically like to express the coefficients

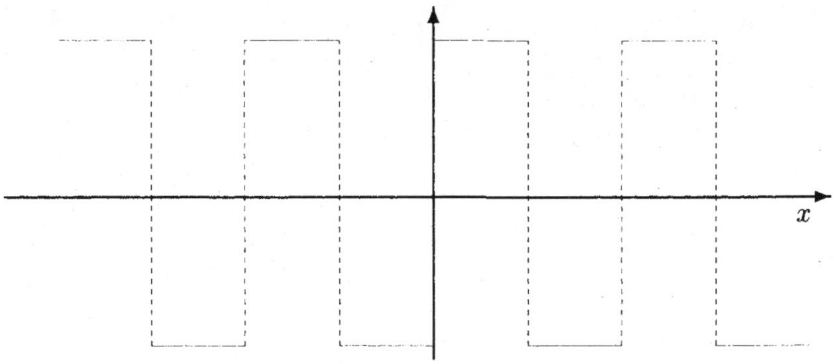

FIGURE 8.6. *Periodic extension of sign(x).*

of the Fourier series of $f'(x)$ by the coefficients of the Fourier series of $f(x)$. Let us assume that $f'(x)$ is piecewise continuous and that

$$f'(x) \sim \frac{\alpha_0}{2} + \sum_{k=1}^{\infty} \left(\alpha_k \cos(k\pi x) + \beta_k \sin(k\pi x)\right). \tag{8.8}$$

Similarly

$$f(x) \sim \frac{a_0}{2} + \sum_{k=1}^{\infty} \left(a_k \cos(k\pi x) + b_k \sin(k\pi x)\right). \tag{8.9}$$

Hence, from the definition of the full Fourier series we have

$$\alpha_k = \int_{-1}^{1} f'(x)\cos(k\pi x)dx, \qquad \beta_k = \int_{-1}^{1} f'(x)\sin(k\pi x)dx,$$

and a_k, b_k are given by (8.5).

Assume that we have ordinary equality (instead of $\sim$) in (8.9) and that we can differentiate the series term by term. Then we obtain from (8.9) that

$$f'(x) = \sum_{k=1}^{\infty} \left(-k\pi a_k \sin(k\pi x) + k\pi b_k \cos(k\pi x)\right),$$

or

$$\alpha_k = k\pi b_k \quad \text{and} \quad \beta_k = -k\pi a_k. \tag{8.10}$$

However, in general these identities are not true. Assume for example that $f(x) = x$. From Example 8.1 we have

$$x \sim \sum_{k=1}^{\infty} \frac{2}{k\pi}(-1)^{k+1}\sin(k\pi x).$$

Therefore, if (8.10) were true, it would follow that

$$1 \sim 2\sum_{k=1}^{\infty}(-1)^{k+1}\cos(k\pi x).$$

However, this is not true since $1 \sim 1$.

On the other hand, recall that in Example 8.3 we derived

$$|x| \sim \frac{1}{2} - \frac{4}{\pi^2}\sum_{k=1}^{\infty}\left(\frac{1}{2k-1}\right)^2 \cos((2k-1)\pi x).$$

Furthermore, $\frac{d}{dx}(|x|) = \text{sign}(x)$ for $x \neq 0$, and from Example 8.2 we have

$$\text{sign}(x) \sim \frac{4}{\pi}\sum_{k=1}^{\infty}\frac{1}{2k-1}\sin((2k-1)\pi x),$$

which is exactly in agreement with formula (8.10).

The following result gives a criteria for when the Fourier coefficients of f' can be determined from term-by-term differentiation of the Fourier series of f.

Theorem 8.1 *Assume that f is continuous on $[-1,1]$, f' is piecewise continuous,*[3] *and $f(-1) = f(1)$. If the Fourier series of f' and f are given by (8.8) and (8.9), then $\alpha_0 = 0$ and*

$$\alpha_k = k\pi b_k \quad \text{and} \quad \beta_k = -k\pi a_k$$

for $k = 1, 2, \ldots$.

We observe that the condition $f(1) = f(-1)$ is satisfied for the function $f(x) = |x|$, but not for $f(x) = x$. Therefore, this theorem is consistent with what we observed above for these two examples.

Proof of Theorem 8.1: Since $f(-1) = f(1)$, we have

$$\alpha_0 = \int_{-1}^{1} f'(x)dx = f(1) - f(-1) = 0.$$

[3]Here the phrase "f' piecewise continuous" means that f is differentiable everywhere, except for a finite number of points $\{x_j\}$ where the one-sided derivatives both exist. Furthermore, the function f' is required to be piecewise continuous.

Furthermore, by using integration by parts we obtain for $k \geq 1$

$$\begin{aligned}\alpha_k &= \int_{-1}^{1} f'(x)\cos(k\pi x)dx \\ &= [f(x)\cos(k\pi x)]_{-1}^{1} + k\pi\int_{-1}^{1} f(x)\sin(k\pi x)dx \\ &= k\pi\int_{-1}^{1} f(x)\sin(k\pi x)dx \\ &= k\pi b_k.\end{aligned}$$

The formula for β_k follows by a completely similar argument. ■

8.1.3 The Complex Form

The full Fourier series can be written in a more elegant, and slightly more compact, form by introducing the complex exponential function. Recall that if $y \in \mathbb{R}$ then

$$e^{iy} = \cos(y) + i\sin(y),$$

where $i = \sqrt{-1}$. Since $\cos(y)$ is an even function and $\sin(y)$ is odd, this also gives

$$e^{-iy} = e^{i(-y)} = \cos(y) - i\sin(y),$$

and hence we obtain

$$\cos(y) = \frac{1}{2}\left(e^{iy} + e^{-iy}\right) \quad \text{and} \quad \sin(y) = \frac{1}{2i}\left(e^{iy} - e^{-iy}\right). \tag{8.11}$$

Consider now the full Fourier series (8.4). By using the complex representation (8.11), we obtain

$$f(x) \sim \frac{a_0}{2} + \sum_{k=1}^{\infty}\left(\frac{a_k}{2}\left(e^{ik\pi x} + e^{-ik\pi x}\right) + \frac{b_k}{2i}\left(e^{ik\pi x} - e^{-ik\pi x}\right)\right) = \sum_{k=-\infty}^{\infty} c_k e^{ik\pi x},$$

where

$$\begin{aligned} c_k &= (a_k - ib_k)/2 \quad \text{for} \quad k > 0, \\ c_0 &= a_0/2, \\ c_{-k} &= (a_k + ib_k)/2 \quad \text{for} \quad k > 0. \end{aligned} \tag{8.12}$$

It is straightforward to show (see Exercise 8.7) that the coefficients c_k can alternatively be expressed as

$$c_k = \frac{1}{2}\int_{-1}^{1} f(x)e^{-ik\pi x}dx.$$

Furthermore, from the expressions for c_k above it follows that if f is a real-valued even function, then

$$c_k = c_{-k} = \frac{a_k}{2}.$$

In particular, the coefficients c_k are all real. On the other hand, if f is a real-valued odd function then $c_0 = 0$ and

$$c_k = -i\frac{b_k}{2} = -c_{-k}.$$

Hence, in this case all the coefficients c_k are purely imaginary, i.e. of the form ir, where r is real.

8.1.4 Changing the Scale

An obvious question to ask is how do we define the Fourier series of functions defined on intervals other than $[-1, 1]$. For example, let $l > 0$ be arbitrary and assume that f is a given piecewise continuous function on $[-l, l]$. In fact, the Fourier series of f can be defined by a simple rescaling of the x-axis. Define a new function $\tilde{f}$ on $[-1, 1]$ by

$$\tilde{f}(y) = f(yl).$$

Hence, we can use Definition 8.2 to define the Fourier series of $\tilde{f}$ in the form

$$\tilde{f}(y) \sim \frac{a_0}{2} + \sum_{k=1}^{\infty} \left(a_k \cos(k\pi y) + b_k \sin(k\pi y)\right).$$

Introducing x by $y = x/l$ and $f(x) = \tilde{f}(x/l)$, we obtain

$$f(x) \sim \frac{a_0}{2} + \sum_{k=1}^{\infty} \left(a_k \cos(k\pi x/l) + b_k \sin(k\pi x/l)\right). \tag{8.13}$$

Expressed in terms of the function f, the coefficients a_k and b_k will be given by (see Exercise 8.8)

$$\begin{aligned} a_k &= \frac{1}{l}\int_{-l}^{l} f(x)\cos(k\pi x/l)dx, \\ b_k &= \frac{1}{l}\int_{-l}^{l} f(x)\sin(k\pi x/l)dx. \end{aligned} \tag{8.14}$$

We should note that the functions $\{\cos(k\pi x/l)\}_{k=0}^{\infty}$ and $\{\sin(k\pi x/l)\}_{k=1}^{\infty}$ are orthogonal with respect to the natural inner product for functions defined on $[-l, l]$. This follows by the corresponding property for $l = 1$ and a simple change of variable. We should also note that these functions correspond to eigenfunctions of the eigenvalue problem

$$\begin{aligned} -X''(x) &= \lambda X(x), & x \in (-l,l) \\ X(-l) &= X(l), & X'(-l) = X'(l) \end{aligned} \tag{8.15}$$

i.e. the periodic problem for the operator $-\frac{d^2}{dx^2}$ on the interval $[-l,l]$.

The complex form of the Fourier series on $[-l,l]$ is

$$f(x) \sim \sum_{k=-\infty}^{\infty} c_k e^{ik\pi x/l},$$

where

$$c_k = \frac{1}{2l}\int_{-l}^{l} f(x)e^{-ik\pi x/l}dx.$$

8.2 Boundary Value Problems and Orthogonal Functions

In the previous section we studied the full Fourier series of piecewise continuous functions. A key tool for obtaining the formulas (8.5) for the coefficients is the orthogonality property for the trigonometric functions $\{\cos(k\pi x)\}_{k=0}^{\infty}$ and $\{\sin(k\pi x)\}_{k=1}^{\infty}$ on $[-1,1]$ (cf. Exercise 3.15). At first sight this property may seem to be a mere coincidence. However, already in Section 2.4.1 it was indicated that this is not an accident, but is closely related to the fact that the trigonometric functions are eigenfunctions of a proper boundary value problem. In this section we will study this connection more systematically.

8.2.1 Other Boundary Conditions

So far in this chapter we have essentially studied three sets of orthogonal functions $\{X_k\}$. If $X_k(x) = \sin(k\pi x)$, then the set $\{X_k\}_{k=1}^{\infty}$ is orthogonal with respect to the inner product on the interval $[0,1]$. Furthermore, these functions are the eigenfunctions of the eigenvalue problem

$$-X'' = \lambda X \tag{8.16}$$

with homogeneous Dirichlet boundary conditions. Similary, the orthogonal set $\{X_k\}_{k=0}^{\infty}$, where $X_k(x) = \cos(k\pi x)$ for $x \in [0,1]$, corresponds to the eigenfunctions of (8.16) with Neumann boundary conditions. Finally, the set

$$\{X_k\}_{k=0}^{\infty} = \{1, \cos(\pi x), \sin(\pi x), \cos(2\pi x), \sin(2\pi x), \dots\}$$

consists of all the eigenfunctions of (8.16) with periodic boundary conditions with respect to the interval $[-1, 1]$. Hence, all these three orthogonal sets are generated by the eigenvalue problem (8.16), but with different boundary conditions. However, there are more possible boundary conditions.

EXAMPLE 8.4 Consider the eigenvalue problem (8.16) with "mixed boundary conditions" of the form

$$X(0) = X'(1) = 0. \tag{8.17}$$

We will show that the problem (8.16)–(8.17) generates a set of orthogonal eigenfunctions $\{X_k\}$. It is straightforward to check by integration by parts that if $X(x)$ and $Y(x)$ both satisfy the boundary conditions (8.17), then the symmetry relation

$$\langle LX, Y\rangle = \int_0^1 X'(x)Y'(x)dx = \langle X, LY\rangle \tag{8.18}$$

holds, where $L = -\frac{d^2}{dx^2}$ and the inner product is given by

$$\langle X, Y\rangle = \int_0^1 X(x)Y(x)dx.$$

If $\lambda \neq \mu$ are two distinct eigenvalues of the problem (8.16)–(8.17), with corresponding eigenfunctions $X(x)$ and $Y(x)$, then (8.18) implies that

$$\lambda\langle X, Y\rangle = \langle LX, Y\rangle = \langle X, LY\rangle = \mu\langle X, Y\rangle,$$

or, since $\lambda \neq \mu$,

$$\langle X, Y\rangle = 0.$$

Hence, two eigenfunctions corresponding to different eigenvalues must be orthogonal. Furthermore, (8.18) implies that

$$\lambda\langle X, X\rangle = \int_0^1 (X'(x))^2\, dx > 0$$

for any eigenvalue λ with associated eigenfunction X. Here the strict inequality follows since the eigenfunctions $X(x)$ must satisfy $X(0) = 0$ and $X \not\equiv 0$, which implies $X' \not\equiv 0$. Therefore all eigenvalues are positive.

All eigenvalues and eigenfunctions can now be determined by calculations similar to those we have performed earlier. If $\lambda = \beta^2$, where $\beta > 0$, the equation (8.16) implies that $X(x)$ takes the form

$$X(x) = c_1 \cos(\beta x) + c_2 \sin(\beta x),$$

while the condition $X(0) = 0$ forces c_1 to be zero. Hence, up to multiplication by a constant, X has to be given by

$$X(x) = \sin(\beta x).$$

The second boundary condition $X'(1) = 0$ will be satisfied if $\beta = \left(k + \frac{1}{2}\right)\pi$, where $k \geq 0$ is an integer. Hence, we conclude that the set of functions $\{X_k\}_{k=0}^{\infty}$, where $X_k(x) = \sin\left(\left(k + \frac{1}{2}\right)\pi x\right)$, are eigenfunctions of the problem (8.16)–(8.17) with eigenvalues $\lambda_k = \left(k + \frac{1}{2}\right)^2 \pi^2$. Furthermore, since eigenfunctions corresponding to different eigenvalues are orthogonal, we conclude that the set $\{X_k\}_{k=0}^{\infty}$ is orthogonal, i.e.

$$\langle X_k, X_m \rangle = 0 \quad \text{for} \quad k \neq m.$$

■

EXAMPLE 8.5 In this example we consider the eigenvalue problem

$$LX = -X'' = \lambda X \tag{8.19}$$

with the boundary conditions

$$X'(0) = X(0), \qquad X'(1) = -X(1). \tag{8.20}$$

These conditions are an example of what is usually referred to as Robin boundary conditions. Using integration by parts, we have

$$\langle LX, Y \rangle = -X'(x)Y(x)\Big|_0^1 + \int_0^1 X'(x)Y'(x)dx. \tag{8.21}$$

However, if the functions X and Y both satisfy (8.20), then

$$X'(x)Y(x)\Big|_0^1 = X(x)Y'(x)\Big|_0^1.$$

From this relation and (8.21) we again obtain

$$\langle LX, Y \rangle = \langle X, LY \rangle. \tag{8.22}$$

By arguing exactly as in the previous example, the symmetry relation (8.22) implies that eigenfunctions corresponding to different eigenvalues must be orthogonal. Furthermore, from (8.20) and (8.21) we find that

$$\lambda\langle X, X \rangle = (X(1))^2 + (X(0))^2 + \int_0^1 (X'(x))^2\, dx > 0$$

for any eigenfunction X with eigenvalue λ, and hence all eigenvalues must be positive.

If $\lambda = \beta^2$, where $\beta > 0$, then (8.16) implies that the eigenfunction X has the form

$$X(x) = c_1 \cos(\beta x) + c_2 \sin(\beta x).$$

Hence,

$$X'(x) = -c_1 \beta \sin(\beta x) + c_2 \beta \cos(\beta x).$$

The condition $X(0) = X'(0)$ therefore implies that $c_1 = c_2\beta$. Hence, up to multiplication by a constant, we must have

$$X(x) = \beta \cos(\beta x) + \sin(\beta x). \tag{8.23}$$

With this expression for X, the second boundary condition $X'(1) = -X(1)$ can be written as

$$\tan(\beta) = \frac{2\beta}{\beta^2 - 1}. \tag{8.24}$$

Hence, solving the eigenvalue problem (8.19)–(8.20) is equivalent to finding the positive roots β of the nonlinear equation (8.24).

If numerical values for these roots are needed, we must solve the equation (8.24) by a numerical method, e.g. the bisection method or Newton's method.[4] However, qualitative information can be derived from a graphical analysis. In Fig. 8.7 we have plotted the functions $\tan(\beta)$ and $2\beta/(\beta^2 - 1)$ for $\beta \geq 0$. It follows from these plots that there is a sequence of roots $\{\beta_k\}_{k=0}^{\infty}$ of the equation (8.24) such that

$$\beta_k \in \left(k\pi, \left(k + \frac{1}{2}\right)\pi\right).$$

Hence, there is an increasing sequence of eigenvalues $\lambda_k = \beta_k^2$ with corresponding eigenfunctions

$$X_k(x) = \beta_k \cos(\beta_k x) + \sin(\beta_k x).$$

Furthermore, the orthogonality property derived above for the eigenfunctions implies that the set $\{X_k\}_{k=0}^{\infty}$ is an orthogonal set of functions on $[0, 1]$.

We finally remark that the results of this example can be used to derive formal solutions of the heat equation with Robin boundary conditions. This discussion is left as an exercise; see Exercise 8.13. ■

---

[4] Newton's method for solving nonlinear algebraic equations is discussed in Exercise 8.9 For the bisection method, consult any introductory text in numerical analysis.

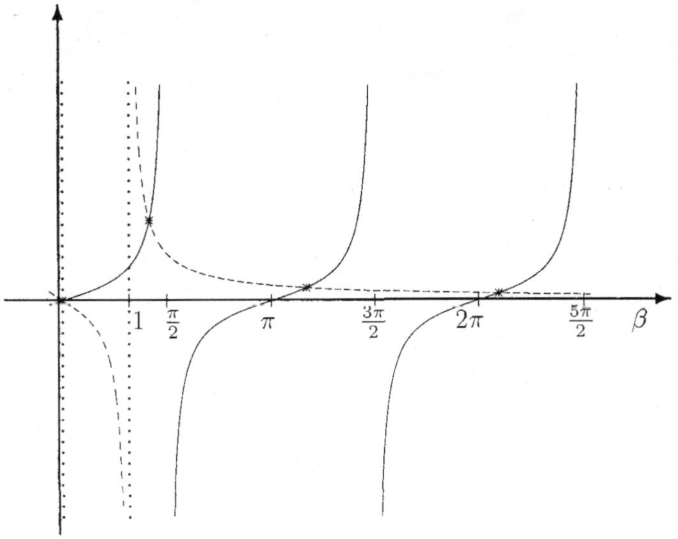

FIGURE 8.7. *The stars "∗" denote the eigenvalues determined as solutions of the equation* $\tan(\beta) = \frac{2\beta}{\beta^2-1}$. *In the plot we have used the notation:* — : $\tan(\beta)$ *and* - - -: $\frac{2\beta}{\beta^2-1}$

8.2.2 Sturm-Liouville Problems

In the previous section we derived different sets of orthogonal functions by studying the eigenvalue problem (8.16) with different boundary conditions. However, in many applications we are forced to study more complicated problems than the simple equation (8.16).

Assume that we want to solve an initial and boundary value problem of the form

$$\begin{aligned} u_t &= (pu_x)_x - qu, && x \in (0,1),\ t > 0, \\ u(0,t) &= u(1,t) = 0, && t > 0, \\ u(x,0) &= f(x), && x \in (0,1), \end{aligned} \tag{8.25}$$

where $p = p(x)$ and $q = q(x)$ are functions of x. If $p \equiv 1$ and $q \equiv 0$, the differential equation in (8.25) reduces to the heat equation. However, in many physical applications we are forced to study more general problems, where p and/or q are not constants.

In order to derive formal solutions of (8.25), we have to consider a more general eigenvalue problem than (8.16).

Let L denote the differential operator of the form

$$(Lu)(x) = -(p(x)u'(x))' + q(x)u(x). \tag{8.26}$$

Here $p \in C^1([0,1])$ and $q \in C([0,1])$ are given functions of x. Furthermore, the function p is strictly positive, while q is nonnegative, i.e. there exists a positive number α such that

$$p(x) \geq \alpha > 0 \quad \text{for all} \quad x \in [0, 1] \tag{8.27}$$

and

$$q(x) \geq 0 \quad \text{for all} \quad x \in [0, 1]. \tag{8.28}$$

The operator L is often referred to as a *Sturm-Liouville operator.*

We observe that if $p \equiv 1$ and $q \equiv 0$, the operator L reduces to the operator $-\frac{d^2}{dx^2}$.

An eigenvalue problem of the form

$$(LX)(x) = \lambda X(x), \qquad X(0) = X(1) = 0 \tag{8.29}$$

is referred to as a Sturm-Liouville problem with Dirichlet boundary conditions.

The eigenvalues and eigenfunctions of the problem (8.29) can be used to find formal solutions of the initial and boundary value problem (8.25). The details of this discussion are outlined in Exercise 8.14. Here, we shall restrict ourselves to deriving some of the fundamental properties of the Sturm-Liouville problem (8.29).

A fundamental property of the problem (8.29) is that eigenfunctions corresponding to different eigenvalues are orthogonal. This result will be established in Corollary 8.1 below, and is in fact a simple consequence of the symmetry property for the Sturm-Liouville operator L. As above, let $\langle \cdot, \cdot \rangle$ denote the inner product

$$\langle u, v \rangle = \int_0^1 u(x)v(x)dx.$$

We now have the following generalizations of the results given in Lemmas 2.3 and 2.4:

Lemma 8.2 *The operator L is symmetric and positive definite in the sense that for any $u, v \in C_0^2\big((0,1)\big)$,*[5]

$$\langle Lu, v \rangle = \langle u, Lv \rangle$$

and

$$\langle Lu, u \rangle \geq 0,$$

with equality only if $u \equiv 0$.

---

[5]The space $C_0^2\big((0,1)\big)$ is introduced in Section 2.1.2 on page 44.

Proof: As above, these properties follow from integration by parts. For $u, v \in C_0^2\big((0,1)\big)$ we derive

$$\begin{aligned}\langle Lu, v\rangle &= \int_0^1 \{-(p(x)u'(x))' + q(x)u(x)\}v(x)dx \\ &= \int_0^1 \{p(x)u'(x)v'(x) + q(x)u(x)v(x)\}dx \\ &= \langle u, Lv\rangle,\end{aligned}$$

where all the boundary terms have disappeared due to the boundary conditions on u and v. In particular, this implies that

$$\langle Lu, u\rangle = \int_0^1 \{p(x)\left(u'(x)\right)^2 + q(x)\left(u(x)\right)^2\}dx \geq \alpha \int_0^1 (u'(x))^2 dx,$$

where we have used (8.27) and (8.28). Hence, if $\langle Lu, u\rangle = 0$, we must have $u' \equiv 0$. Therefore u is constant, and since $u(0) = 0$, we get $u \equiv 0$. ■

Corollary 8.1 *If X and Y are eigenfunctions of (8.29), corresponding to distinct eigenvalues λ and μ, then $\langle X, Y\rangle = 0$.*

Proof: Since L is symmetric, we have

$$\lambda\langle X, Y\rangle = \langle LX, Y\rangle = \langle X, LY\rangle = \mu\langle X, Y\rangle$$

or

$$(\lambda - \mu)\langle X, Y\rangle = 0.$$

Since $\lambda - \mu \neq 0$, we have $\langle X, Y\rangle = 0$. ■

From the positive definite property of the operator L, we also obtain that all eigenvalues of (8.29) are positive. For if λ is an eigenvalue with corresponding eigenfunction X, then

$$\lambda\langle X, X\rangle = \langle LX, X\rangle > 0,$$

and since $\langle X, X\rangle > 0$ for an eigenfunction, we conclude that $\lambda > 0$.

Corollary 8.2 *All eigenvalues of the problem (8.29) are positive.*

For most problems of the form (8.29), where p or q are not constants, it is impossible to derive analytical expressions for the eigenfunctions. However, for a few problems this can be done.

EXAMPLE 8.6 Consider the eigenvalue problem (8.29) with $p(x) = (1+x)^2$ and $q(x) = 0$. Hence, we consider the eigenvalue problem

$$\begin{aligned}-\big((1+x)^2 X'(x)\big)' &= \lambda X(x), \\ X(0) = X(1) &= 0.\end{aligned} \tag{8.30}$$

Let

$$X_k(x) = \frac{1}{\sqrt{1+x}} \sin\left(k\pi \frac{\log(1+x)}{\log(2)}\right) \quad \text{for} \quad k = 1, 2, \ldots. \tag{8.31}$$

It is straightforward to check that these functions are eigenfunctions of problem (8.30) with eigenvalues $\lambda_k = \left(\frac{k\pi}{\log(2)}\right)^2 + \frac{1}{4}$. In fact, in Exercise 8.15 you are asked to show that all the eigenfunctions are given by (8.31). In particular, it follows that the set $\{X_k\}_{k=1}^{\infty}$ is an orthogonal set of functions with respect to the inner product $\langle \cdot, \cdot \rangle$. ■

8.3 The Mean Square Distance

The purpose of the next section is to start the discussion of convergence of Fourier series. However, first we will introduce the mean square distance function.[6] The convergence of Fourier series is intimately related to properties of this distance function.

Let f and g be two piecewise continuous functions defined on an interval $[a, b]$. Already in Chapter 2 (see page 58) we introduced an inner product of the form

$$\langle f, g \rangle = \int_a^b f(x)g(x)dx.$$

The corresponding distance function, or norm, is defined by

$$\|f\| = \langle f, f \rangle^{1/2} = \left(\int_a^b f^2(x)dx\right)^{1/2}.$$

In general, the quantity

$$\|f - g\| = \left(\int_a^b (f(x) - g(x))^2 \, dx\right)^{1/2}$$

is a measure of the distance between two functions f and g. We will refer to $\|f - g\|$ as the mean square distance between f and g. In particular, $\|f\|$ is a measure of the size of f, or the distance from f to the zero function. The quantity $\|f\|$ has properties which resemble corresponding properties of the absolute value of real numbers, or more generally the Euclidian norm of a vector (cf. Project 1.2). For example, $\|f\| \geq 0$, with equality if and only if f is identically zero. Furthermore, if $\alpha \in \mathbb{R}$ then

$$\|\alpha f\| = |\alpha| \|f\|.$$

[6] In more advanced courses in analysis or partial differential equations, this distance function will usually be referred to as the L^2-norm.

We shall also see below that the mean square distance satisfies a triangle inequality of the form

$$\|f+g\| \leq \|f\| + \|g\|. \tag{8.32}$$

This will in fact be a consequence of the following version of Cauchy-Schwarz inequality.

Lemma 8.3 *If f and g are piecewise continuous functions on $[a,b]$, then*

$$|\langle f,g\rangle| \leq \|f\|\|g\|.$$

Proof: If $f \equiv 0$ then we have zero on both sides of the desired inequality, which therefore holds. Since this case is covered, we will assume in the rest of the proof that $\|f\| > 0$. For any $t \in \mathbb{R}$ consider

$$\begin{aligned} p(t) &= \int_a^b (tf-g)^2\,dx \\ &= t^2\int_a^b f^2(x)dx - 2t\int_a^b f(x)g(x)dx + \int_a^b g^2(x)dx \\ &= t^2\|f\|^2 - 2t\langle f,g\rangle + \|g\|^2. \end{aligned}$$

Hence, $p(t)$ is a second-order polynomial with respect to t which has the property that

$$p(t) \geq 0 \quad \text{for all} \quad t \in \mathbb{R}.$$

In particular, $p(t_0) \geq 0$, where t_0 is chosen such that $p'(t_0) = 0$, i.e.

$$t_0 = \frac{\langle f,g\rangle}{\|f\|^2}.$$

Hence,

$$0 \leq p(t_0) = \frac{\langle f,g\rangle^2}{\|f\|^2} - 2\frac{\langle f,g\rangle^2}{\|f\|^2} + \|g\|^2 = -\frac{\langle f,g\rangle^2}{\|f\|^2} + \|g\|^2$$

or

$$\langle f,g\rangle^2 \leq \|f\|^2\|g\|^2.$$

The desired inequality now follows by taking the square roots. ■

Observe that

$$\begin{aligned} \|f+g\|^2 &= \int_a^b \left(f^2(x) + 2f(x)g(x) + g^2(x)\right)dx \\ &= \|f\|^2 + 2\langle f,g\rangle + \|g\|^2 \end{aligned}$$

Therefore, it follows from the Cauchy-Schwarz inequality above that

$$\begin{aligned}\|f+g\|^2 &\le \|f\|^2 + 2\|f\|\|g\| + \|g\|^2 \\ &= (\|f\| + \|g\|)^2,\end{aligned}$$

and hence the desired triangle inequality (8.32) follows by taking square roots. The inequality of Lemma 8.3 also implies the inequality

$$\int_a^b |f(x)||g(x)|dx \le \|f\|\|g\|, \tag{8.33}$$

which appears to be a stronger inequality since, in general, $|\int fg| \le \int |f||g|$. However, (8.33) follows if we apply Lemma 8.3 to the functions $|f|$ and $|g|$.

A useful generalization of Cauchy-Schwarz inequality is Hölder's inequality, which states that

$$\int_a^b |f(x)||g(x)|dx \le \left(\int_a^b |f(x)|^p dx\right)^{1/p} \left(\int_a^b |g(x)|^q dx\right)^{1/q}, \tag{8.34}$$

where p, q are real numbers such that $p, q > 1$ and $\frac{1}{p} + \frac{1}{q} = 1$. Note that the choice $p = q = 2$ gives (8.33). A proof of Hölder's inequality is outlined in Exercise 8.16.

Let us recall that $\|f - g\|$ can be interpreted as the distance between the two functions f and g. This distance can therefore be used to define the concept of mean square convergence.

Definition 8.3 *A sequence $\{f_N\}_{N=1}^{\infty}$ of piecewise continuous functions on $[a,b]$ is said to converge in the mean square sense to a piecewise continuous function f if*

$$\lim_{N\to\infty} \|f_N - f\| = 0.$$

In the next chapter we will discuss the convergence of sequences of functions in more detail. In particular, we will compare mean square convergence to other concepts of convergence. However, in the present chapter we will restrict ourselves to mean square convergence, which in some sense is the natural notion of convergence for Fourier series.

EXAMPLE 8.7 Let $f_N(x) = x^N$ for $x \in [0,1]$. Then

$$\|f_N\|^2 = \int_0^1 x^{2N} dx = \frac{1}{2N+1} \to 0$$

as $N \to \infty$. Hence $\{f_N\}$ converges to the function $f \equiv 0$ in the mean square sense. Observe that $f_N(1) = 1$ for all N. Hence, $\{f_N(x)\}$ does not converge to $f(x)$ for all $x \in [0,1]$. The mean square convergence simply means that the area bounded by $x = 0$, $x = 1$, $y = 0$, and the curve $y = x^{2N}$ tends to zero as N tends to infinity (see Fig. 8.8).

■

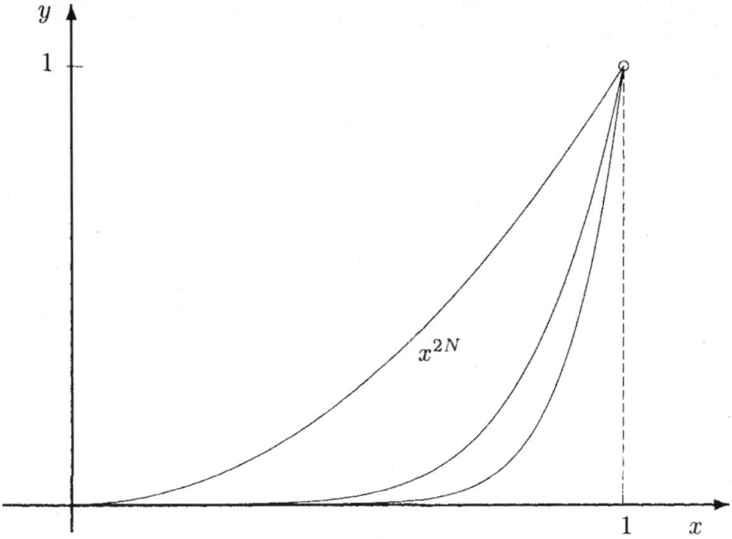

FIGURE 8.8. *Plot of* x^{2N} *for* $N = 1, 3$ *and* 5.

EXAMPLE 8.8 Let $f_N(x) = N/(1 + N^2x^2)$ for $x \in [0, 1]$. Hence

$$\lim_{N\to\infty} f_N(x) = 0 \quad \text{for} \quad x > 0,$$

while $f_N(0) = N \to \infty$.

Does $\{f_N\}$ converge to zero in the mean square sense? We have

$$\begin{aligned}\int_0^1 (f_N(x))^2\,dx &= \int_0^1 \left(\frac{N}{1+N^2x^2}\right)^2 dx \\ &= N\int_0^1 \left(\frac{1}{1+N^2x^2}\right)^2 N\,dx \\ &= N\int_0^N \left(\frac{1}{1+y^2}\right)^2 dy \quad \to \quad \infty\end{aligned}$$

as $N \to \infty$. Hence, we do not have mean square convergence to the zero function. ■

8.4 General Fourier Series

In this section we will start the discussion of convergence of Fourier series. If $S_N(f)$ is the partial sum

$$S_N(f) = \frac{a_0}{2} + \sum_{k=1}^{N} \left(a_k \cos(k\pi x) + b_k \sin(k\pi x)\right)$$

corresponding to the full Fourier series (8.6) of f, we like to know if the sequence of functions $\{S_N(f)\}$ converges to f in the mean square sense. Below we will derive some partial answers to this question. However, the discussion here will only depend on basic orthogonality properties of Fourier series. Therefore, the results will be true for a more general class of orthogonal series which is introduced below. In the next chapter we will return to the more specific Fourier series of the form (8.6).

As above, let $\langle \cdot, \cdot \rangle$ denote the inner product

$$\langle f, g \rangle = \int_a^b f(x)g(x)dx$$

and $\|\cdot\|$ the corresponding norm

$$\|f\| = \langle f, f \rangle^{1/2}.$$

Throughout this section $\{X_k\}_{k=1}^{\infty}$ will be an orthogonal set of piecewise continuous functions with respect to the inner product $\langle \cdot, \cdot \rangle$. Furthemore, none of the functions X_k are allowed to be identical to the zero function, i.e. $\|X_k\| > 0$ for all k.

We recall that in Sections 8.1 and 8.2 above we discussed several examples of sets of orthogonal functions. The theory below will only depend on the assumption that the set $\{X_k\}$ is orthogonal, and hence the theory applies to all the examples discussed above.

If f is a piecewise continuous function on $[a, b]$, then the general Fourier series of f with respect to the orthogonal set $\{X_k\}$ are series of the form $\sum_{k=1}^{\infty} c_k X_k$. We note that if the identity $f = \sum_{k=1}^{\infty} c_k X_k$ holds, then it follows from the orthogonality property of the set $\{X_k\}$ that

$$\langle f, X_k \rangle = c_k \|X_k\|^2.$$

Hence, we have motivated the following definition:

Definition 8.4 *The infinite series $\sum_{k=1}^{\infty} c_k X_k(x)$, where $c_k = \frac{\langle f, X_k \rangle}{\|X_k\|^2}$, is called the general Fourier series of f with respect to the orthogonal set $\{X_k\}$.*

The most fundamental problem for Fourier series is the question of convergence. Will the partial sums $S_N(f)$, where

$$S_N(f) = \sum_{k=1}^{N} c_k X_k\,, \qquad c_k = \frac{\langle f, X_k \rangle}{\|X_k\|^2}, \tag{8.35}$$

converge to the function f?

In general the answer to this question will depend on the choice of orthogonal set $\{X_k\}$. The purpose of the discussion here is to give an important partial answer, (see Theorem 8.2 below), which will be used in the next chapter to derive complete convergence results.

We first establish the following formula for the norm of a finite linear combination of the basis functions $\{X_k\}$.

Lemma 8.4 *Let P_N be any function of the form $P_N = \sum_{k=1}^{N} \alpha_k X_k$, $\alpha_k \in \mathbb{R}$. Then*

$$||P_N||^2 = \sum_{k=1}^{N} \alpha_k^2 \, ||X_k||^2. \tag{8.36}$$

Proof: From the orthogonality property of the set $\{X_k\}$ we get

$$||P_N||^2 = \int_a^b P_N^2(x)\, dx = \sum_{k=1}^{N} \sum_{m=1}^{N} \alpha_k \alpha_m \langle X_k, X_m \rangle = \sum_{k=1}^{N} \alpha_k^2 \, ||X_k||^2. \quad \blacksquare$$

Next we derive an orthogonality property for the difference between f and its finite general Fourier series $S_N(f)$.

Lemma 8.5 *If f is piecewise continuous on $[a, b]$ and $S_N(f)$ is given by (8.35), then*

$$\langle f - S_N(f), P_N \rangle = 0$$

for any function P_N of the form $P_N = \sum_{k=1}^{N} \alpha_k \, X_k$, $\alpha_k \in \mathbb{R}$.

Proof: If $1 \leq k \leq N$, then

$$\langle S_N(f), X_k \rangle = \sum_{m=1}^{N} c_m \langle X_m, X_k \rangle = c_k ||X_k||^2 = \langle f, X_k \rangle.$$

Therefore,

$$\langle f - S_N(f), X_k \rangle = 0 \qquad \text{for } k = 1, 2, \ldots, N.$$

By linearity of the inner product we therefore obtain

$$\begin{aligned} \langle f - S_N(f), P_N \rangle &= \int_a^b (f - S_N(f)) \Big(\sum_{k=1}^{N} \alpha_k X_k \Big) \, dx \\ &= \sum_{k=1}^{N} \alpha_k \langle f - S_N(f), X_k \rangle = 0. \quad \blacksquare \end{aligned}$$

The two simple results above have immediate consequences. First, by letting $P_N = S_N(f)$ in Lemma 8.5, we get

$$\langle f, S_N(f) \rangle = ||S_N(f)||^2.$$

Hence, by using (8.36) we derive the identity

$$||S_N(f)||^2 = \langle f, S_N(f)\rangle = \sum_{k=1}^{N} c_k^2 \, ||X_k||^2, \tag{8.37}$$

where the coefficients c_k are given by (8.35).

Note that any piecewise continuous function f can be written as the sum of the finite Fourier series $S_N(f)$ and the error $f - S_N(f)$, i.e.

$$f = S_N(f) + \big(f - S_N(f)\big).$$

Furthermore, it is a direct consequence of Lemma 8.5 that this decomposition is orthogonal, i.e.

$$\langle f - S_N(f), S_N(f)\rangle = 0.$$

This orthogonality property is illustrated in Fig. 8.9. As a consequence of this property, the decomposition will satisfy a "Pythagoras theorem" of the form

$$||f||^2 = ||S_N(f)||^2 + ||f - S_N(f)||^2. \tag{8.38}$$

This identity follows since

$$\begin{aligned} ||f - S_N(f)||^2 &= \langle f - S_N(f), f - S_N(f)\rangle = \langle f - S_N(f), f\rangle - \langle f - S_N(f), S_N(f)\rangle \\ &= \langle f - S_N(f), f\rangle = ||f||^2 - ||S_N(f)||^2, \end{aligned}$$

where the final equality follows from (8.37). Hence, (8.38) is established. By using (8.37) to express $||S_N(f)||^2$ with respect to the Fourier coefficients, this final identity can be rewritten in the form

$$||f - S_N(f)||^2 = ||f||^2 - \sum_{k=1}^{N} c_k^2 \, ||X_k||^2. \tag{8.39}$$

This formulation of the Pythagoras identity is very useful.

Since $||f - S_N(f)||^2 \geq 0$ it follows directly from (8.39) that

$$\sum_{k=1}^{N} c_k^2 ||X_k||^2 \leq ||f||^2$$

for all N. Hence, since the partial sums only contain positive terms, the series $\sum_{k=1}^{\infty} c_k^2 ||X_k||^2$ converges and satisfies

$$\sum_{k=1}^{\infty} c_k^2 ||X_k||^2 \leq ||f||^2. \tag{8.40}$$

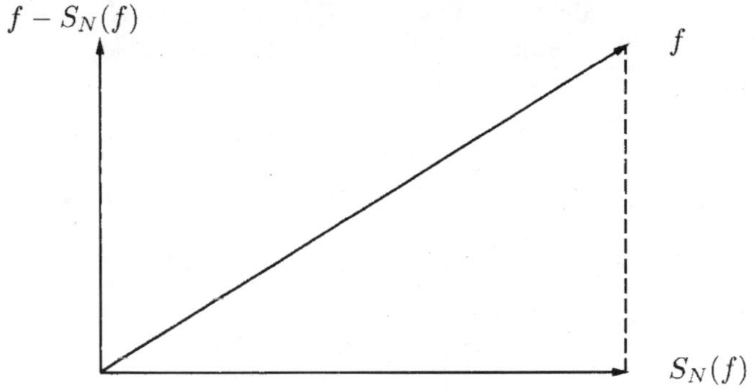

FIGURE 8.9. *An orthogonal decomposition of f.*

This inequality holds for any orthonormal set $\{X_k\}$ and any piecewise continuous[7] function f. The inequality (8.40) is usually referred to as *Bessel's inequality.* If this inequality becomes an identity, i.e. if

$$\sum_{k=1}^{\infty} c_k^2 \|X_k\|^2 = \|f\|^2, \tag{8.41}$$

then this identity is called *Parseval's identity.* From the identity (cf. (8.39))

$$\|f - S_N(f)\|^2 = \|f\|^2 - \sum_{k=1}^{N} c_k^2 \|X_k\|^2,$$

we see that Parseval's identity will hold, if $S_N(f)$ converges to f in the mean square sense. Alternatively, if Parseval's identity holds, then this implies the mean square convergence of $S_N(f)$ to f. We summarize this important result:

Theorem 8.2 *Let f be a piecewise continuous function on $[a, b]$. A generalized Fourier series of f with respect to an orthogonal set $\{X_k\}$ converges to f in the mean square sense if and only if the corresponding Parseval's identity (8.41) holds.*

A set of orthogonal functions $\{X_k\}_{k=1}^{\infty}$, defined on $[a, b]$, is called *complete* if Parseval's identity holds for all piecewise continuous functions on $[a, b]$. We will show below (see Corollary 9.1) that the orthogonal set

$$\{1, \cos(\pi x), \sin(\pi x), \cos(2\pi x), \sin(2\pi x), \dots\},$$

---

[7]In fact, it holds as long as $\int_a^b f^2(x)dx$ is finite.

corresponding to the full Fourier series, is complete with respect to the interval $[-1, 1]$. Similarly, the sets $\{\sin(k\pi x)\}_{k=1}^{\infty}$ and $\{\cos(k\pi x)\}_{k=0}^{\infty}$ are complete on $[0, 1]$. However, these sets of functions are not complete on $[-1, 1]$ (see Exercise 8.19).

EXAMPLE 8.9 Assume that f is a piecewise continuous function on $[-1, 1]$ and let

$$\frac{a_0}{2} + \sum_{k=1}^{\infty} (a_k \cos(k\pi x) + b_k \sin(k\pi x))$$

be the full Fourier series of f. What do we obtain from Bessel's inequality in this case?

In order to see this, we define

$$X_0(x) \equiv 1$$

and

$$X_{2k}(x) = \cos(k\pi x), \quad X_{2k-1}(x) = \sin(k\pi x) \quad \text{for} \quad k \geq 1.$$

Hence $\|X_0\|^2 = 2$, while

$$\|X_k\|^2 = \int_{-1}^{1} X_k^2(x)dx = 1 \quad \text{for} \quad k \geq 1.$$

Furthermore, by letting $c_0 = \frac{a_0}{2}$ and

$$c_{2k} = a_k, \quad c_{2k-1} = b_k \quad \text{for} \quad k \geq 1,$$

the full Fourier series is rewritten as $\sum_{k=0}^{\infty} c_k X_k$. Hence, Bessel's inequality gives

$$2c_0^2 + \sum_{k=1}^{\infty} c_k^2 \leq \|f\|^2$$

or

$$\frac{a_0^2}{2} + \sum_{k=1}^{\infty} (a_k^2 + b_k^2) \leq \int_{-1}^{1} f^2(x)dx. \tag{8.42}$$

■

EXAMPLE 8.10 Let f be piecewise continuous on $[0, 1]$ with Fourier sine series

$$\sum_{k=1}^{\infty} c_k \sin(k\pi x).$$

Since the sine series is the full Fourier series of the odd extension of f, we obtain from the previous example that

$$\sum_{k=1}^{\infty} c_k^2 \leq 2\int_0^1 f^2(x)dx. \tag{8.43}$$

This is the proper form of Bessel's inequality in this case. ■

EXAMPLE 8.11 Recall from Example 8.1 that

$$x \sim \frac{2}{\pi}\sum_{k=1}^{\infty} \frac{(-1)^{k+1}}{k} \sin{(k\pi x)}.$$

Since $\int_{-1}^1 x^2 dx = 2\int_0^1 x^2 dx = 2/3$, we obtain from (8.42) (or (8.43)) that

$$\frac{4}{\pi^2}\sum_{k=1}^{\infty}\frac{1}{k^2} \leq \frac{2}{3}$$

or

$$\sum_{k=1}^{\infty}\frac{1}{k^2} \leq \frac{\pi^2}{6}.$$

The results of the next chapter will in fact imply that this inequality is an identity. ■

8.5 A Poincaré Inequality

We will end this chapter by deriving a simple version of what is usually referred to as a *Poincaré inequality.* The main tool for obtaining this inequality is the Cauchy-Schwarz inequality (8.33). As an application of the Poincaré inequality, we will improve the energy estimates for the heat equation derived in Chapter 3.7.

Let f be a function defined on $[a,b]$ such that f is differentiable and f' is piecewise continuous. Furthermore, assume that $f(a) = 0$. Note that this last condition implies that if $f' \equiv 0$, then $f \equiv 0$. More generally, since $|f(x)|$ can be large only if $|f'|$ is large, it is reasonable to believe that an inequality of the form

$$\int_a^b f^2(x)dx \leq \lambda \int_a^b \left(f'(x)\right)^2 dx$$

holds, for a suitable positive constant λ. In fact, we have the following result:

Lemma 8.6 *Let $f : [a,b] \to \mathbb{R}$ be such that $f(a) = 0$, f' exists and is piecewise continuous. Then*

$$\|f\| \leq \frac{(b-a)}{\sqrt{2}} \|f'\|.$$

Proof: Since $f(a) = 0$ we have from the fundamental theorem of calculus that

$$f(x) = \int_a^x f'(y)dy.$$

Hence, we obtain from the Cauchy-Schwarz inequality (8.33) that

$$|f(x)|^2 \leq \int_a^x 1dy \int_a^x |f'(y)|^2 dy \leq (x-a)\|f'\|^2$$

or

$$\|f\|^2 \leq \int_a^b (x-a)dx \|f'\|^2 \leq \frac{(b-a)^2}{2} \|f'\|^2.$$

The desired result is obtained by taking sqare roots. ■

EXAMPLE 8.12 As an illustration of the use of Poincaré's inequality we will reconsider the study of energy arguments for the heat equation (cf. Chapter 3.7). Recall that we studied the solution $u(x,t)$ of the heat equation on the interval $[0,1]$, with Dirichlet boundary conditions. The idea was to study the dynamics of the scalar variable

$$E(t) = \int_0^1 u^2(x,t)dx.$$

Above we found that $E(t)$ is nonincreasing with time. Here we shall show that

$$E(t) \leq e^{-4t} E(0) \quad \text{for} \quad t \geq 0. \tag{8.44}$$

In particular, this will imply that $\lim_{t \to \infty} E(t) = 0$.

In Chapter 3.7 (see page 103) we showed that

$$E'(t) = -2 \int_0^1 u_x^2(x,t)dx \leq 0.$$

However, from the Poincaré inequality given in Lemma 8.6, we have

$$\int_0^1 u^2(x,t)dx \leq \frac{1}{2} \int_0^1 u_x^2(x,t)dx,$$

and hence we obtain

$$E'(t) = -2\int_0^1 u_x^2(x,t)dx \leq -4E(t). \tag{8.45}$$

If instead of this inequality we have the equality $E'(t) = -4E(t)$, then we immediately obtain

$$E(t) = e^{-4t}E(0).$$

Hence, it seems reasonable that the inequality (8.45) implies (8.44). In order to see this, let

$$z(t) = E(t)e^{4t}.$$

Then

$$z'(t) = e^{4t}\left(E'(t) + 4E(t)\right) \leq 0,$$

where the last inequality follows from (8.45). Hence, $z(t)$ is nonincreasing, i.e.

$$z(t) \leq z(0) \quad \text{for} \quad t \geq 0.$$

From the definition of $z(t)$ we therefore obtain (8.44). Since the definition of $E(t)$ implies that $E(t)$ is nonnegative, we therefore have

$$0 \leq E(t) \leq e^{-4t}E(0) \quad \text{for} \quad t \geq 0,$$

which implies that $\lim_{t\to\infty} E(t) = 0$. ■

An even sharper upper bound for $E(t)$ will be derived in Chapter 10 from an improved Poincaré inequality (see Example 10.2).

A main part of the discussion above was how the estimate (8.44) was obtained from the differential inequality (8.45). Such a result is usually referred to as *Gronwall's inequality.* For later references we state the following result.

Lemma 8.7 *Let $y : [0,b] \to \mathbb{R}$ be continuous, differentiable, and satisfy*

$$y'(t) \leq \alpha y(t), \quad t \in (0,b),$$

for a suitable $\alpha \in \mathbb{R}$. Then

$$y(t) \leq e^{\alpha t}y(0) \quad \text{for} \quad t \in [0,b].$$

Proof: We repeat the argument used in Example 8.12. Let $z(t) = e^{-\alpha t}y(t)$. Then

$$z'(t) = e^{-\alpha t}\left(y'(t) - \alpha y(t)\right) \leq 0.$$

Hence, $z(t) \leq z(0)$ or

$$y(t) \leq e^{\alpha t}y(0) \quad \text{for} \quad t \in [0,b].$$

■

8.6 Exercises

EXERCISE 8.1 Sketch the periodic extension of the function

(a) $f(x) = |x|$ defined on $[-1, 1]$

(b) $g(x) = \sin(x)$ defined on $[0, \pi]$

(c) $h(x) = x^2$ defined on $[0, 1]$.

EXERCISE 8.2 (a) Show that the product of two functions which are either both even or both odd is even.

(b) Show that the product of an even and an odd function is odd.

(c) Show that if f is an odd function defined on $[-1, 1]$, then

$$\int_{-1}^{1} f(x)dx = 0.$$

EXERCISE 8.3 Find the full Fourier series of the functions

(a) $f(x) = x^2$

(b) $f(x) = e^x$

defined on $[-1, 1]$.

EXERCISE 8.4 Find the full Fourier series of the functions

(a) $f(x) = \sin^2(x)$

(b) $f(x) = \cos^2(x)$

defined on $[-\pi, \pi]$.

EXERCISE 8.5 Let $f(x)$ be piecewise continuous on $[-1, 1]$ and assume that

$$f(x) \sim \sum_{k=1}^{\infty} \left(\alpha_k \cos(k\pi x) + \beta_k \sin(k\pi x)\right).$$

Let $g(x)$ satisfy $g'(x) = f(x)$; i.e. g is an integral of f. Show that the Fourier series of g can be obtained from term-by-term integration, i.e.

$$g(x) \sim \frac{a_0}{2} + \sum_{k=1}^{\infty} \left(\frac{\alpha_k}{k\pi} \sin(k\pi x) - \frac{\beta_k}{k\pi} \cos(k\pi x) \right),$$

where $a_0 = \int_{-1}^{1} g(x)dx$.

EXERCISE 8.6 Let $f(x) = \frac{1}{2} - x$ for $x \in [0, 1]$.

(a) Find the Fourier sine series of f.

(b) Find the Fourier cosine series of f.

(c) Use the results above and the result of Exercise 8.5 to find the Fourier sine series of

$$g(x) = \frac{1}{2}(1 - x)x.$$

EXERCISE 8.7 Consider the complex form of the Fourier series:

$$f(x) \sim \sum_{k=-\infty}^{\infty} c_k e^{ik\pi x}.$$

Derive the formula

$$c_k = \frac{1}{2} \int_{-1}^{1} f(x) e^{-ik\pi x} \, dx$$

from (8.12).

EXERCISE 8.8 Derive the formulas (8.14) from the corresponding formulas (8.5).

EXERCISE 8.9 The purpose of this exercise is to derive Newton's method for solving nonlinear algebraic equations, and to apply the method to the equation (8.24), i.e.

$$\tan(\beta) = \frac{2\beta}{\beta^2 - 1}.$$

But we begin by considering a general equation $f = f(\beta)$, and we want to find β^* such that

$$f(\beta^*) = 0. \qquad (1)$$

Newton's method is an iterative procedure for computing numerical approximations of solutions of (1). The first step in the procedure is to guess an initial value β_0. Now we can use a Taylor series to obtain that

$$f(\beta) = f(\beta_0) + (\beta - \beta_0) f'(\beta_0) + O((\beta - \beta_0)^2).$$

(a) Explain why it is reasonable to choose

$$\beta_1 = \beta_0 - \frac{f(\beta_0)}{f'(\beta_0)}.$$

Newton's method is defined by the following algorithm:

1. Choose ϵ, β_0 and put $n = 0$.
2. While $|f(\beta_n)| > \epsilon$ do
   $\beta_{n+1} = \beta_n - f(\beta_n)/f'(\beta_n)$
   $n = n + 1$.

Here $\epsilon > 0$ is a given tolerence.

(b) Implement the scheme above and apply it to solve the equation

$$f(\beta) = e^{\beta} - e = 0$$

using $\beta_0 = 3/4$ and $\epsilon = 10^{-6}$.

(c) Apply the same procedure for

$$f(\beta) = x^2 - 4$$

with $\beta_0 = 3$ and $\epsilon = 10^{-6}$.

(d) Explain the geometric interpretation of Newton's method given in Fig. 8.10, and use this interpretation to discuss the speed of convergence observed in the two examples above.

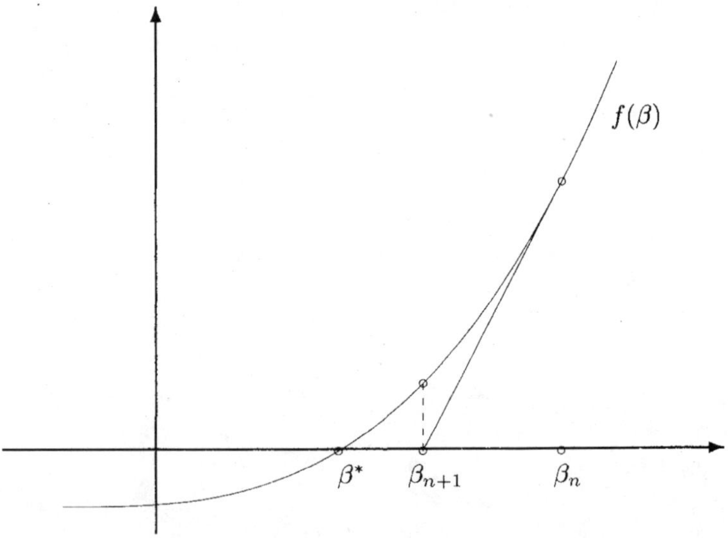

FIGURE 8.10. *A geometric view of Newton's method.*

(e) Use the program developed above to compute the first 100 eigenvalues of the operator considered in Example 8.5. In other words, compute the first 100 roots of the equation

$$f(\beta) = \tan(\beta) - \frac{2\beta}{\beta^2 - 1}.$$

Here the first root β_0^* is in the interval $[1, \pi/2]$, β_1^* is in $[\pi/2, \pi]$, and in general

$$\beta_k^* \in \left(k\pi, \left(k + \frac{1}{2}\pi\right)\right) \quad \text{for} \quad k \geq 1.$$

Discuss the following choices of initial values:

$$(i) \quad \beta_0^0 = 1.3, \quad \beta_k^0 = \left(k + \frac{1}{4}\right)\pi, \qquad k \geq 1$$

$$(ii) \quad \beta_0^0 = 1.3, \quad \beta_1^0 = 5\pi/4, \quad \beta_k^0 = \beta_{k-1}^* + \pi, \quad k \geq 2.$$

Again you can choose $\epsilon = 10^{-6}$.

EXERCISE 8.10 Consider the eigenvalue problem

$$\begin{aligned} -X''(x) &= \lambda X(x), & 0 < x < 1, \\ X'(0) &= 2X(0), & X'(1) = X(1). \end{aligned} \tag{8.46}$$

(a) Show that if λ is an eigenvalue with eigenfunction X, then

$$\lambda \int_0^1 X^2(x)dx = \int_0^1 (X'(x))^2 \, dx + 2(X(0))^2 - (X(1))^2.$$

Can we conclude from this that all eigenvalues are positive?

(b) Show that $\lambda = -\mu^2 < 0$ is a negative eigenvalue of (8.46) if and only if $\mu > 0$ satisfies

$$\tanh(\mu) = \frac{\mu}{2 - \mu^2}.$$

(c) Show that the problem (8.46) has exactly one negative eigenvalue.

(d) Compute this negative eigenvalue by Newton's method.

EXERCISE 8.11 Consider the initial and boundary value problem

$$\begin{aligned} u_t &= u_{xx}, & x \in (0,1),\ t > 0, \\ u_x(0,t) &= 2u(0,t), \quad u_x(1,t) = u(1,t),\ t > 0, & \\ u(x,0) &= f(x) & \end{aligned}$$

Explain why the solution of this problem does not satisfy an energy estimate of the form

$$\int_0^1 u^2(x,t)\,dx \le \int_0^1 f^2(x)\,dx \qquad \text{for } t > 0.$$

(Hint: Use the results of Exercise 8.10.)

EXERCISE 8.12 Consider the eigenvalue problem

$$\begin{aligned} X'(x) &= \lambda X(x), \qquad 0 < x < 1, \\ X(0) &= X(1). \end{aligned}$$

(a) Show that $\lambda = 0$ is the only real eigenvalue of this problem.

(b) Derive complex eigenvalues of this problem by considering eigenfunctions of the form

$$X(x) = e^{i\beta x} = \cos(\beta x) + i \sin(\beta x),$$

where $\beta \in \mathbb{R}$.

EXERCISE 8.13 Consider the heat equation with Robin boundary conditions, i.e.

$$\begin{aligned} u_t &= u_{xx}, & x \in (0,1),\ t > 0, \\ u_x(0,t) &= u(0,t), \quad u_x(1,t) = -u(1,t),\ t > 0, & \\ u(x,0) &= f(x). & \end{aligned} \tag{8.47}$$

(a) For each $t \ge 0$ let

$$E(t) = \int_0^1 u^2(x,t)\,dx.$$

Use energy arguments to show that

$$E(t) \le E(0) \quad \text{for } t \ge 0.$$

Here you are allowed to assume that

$$E'(t) = \int_0^1 \frac{\partial}{\partial t}\left(u^2(x,t)\right) dx; \quad \text{cf. Proposition 3.1.}$$

(b) Discuss how you can use the eigenvalues and eigenfunctions of the problem (8.19)–(8.20) to find a representation of the solution u.

EXERCISE 8.14 Consider the initial and boundary value problem (8.25), where we assume that $p \in C([0,1])$ and $q \in C([0,1])$ satisfy the conditions (8.27) and (8.28).

(a) Use energy arguments to show that

$$\int_0^1 u^2(x,t)\,dx \le \int_0^1 f^2(x)\,dx \quad \text{for } t \ge 0.$$

(b) Discuss how you can use the eigenvalues and eigenfunctions of the problem (8.29) to find a formal representation of the solution u.

(Hint: Use the ansatz $u(x,t) = \sum_k T_k(t)X_k(x)$, where $\{X_k\}$ are the eigenfunctions of (8.29).)

EXERCISE 8.15 Consider the second-order differential equation

$$\big((1+x)X'(x)\big)' = \beta^2 X(x)$$

for $x > 0$, where $\beta > 0$ is a parameter.

(a) Show that the functions $X_1(x) = \frac{1}{\sqrt{1+x}}\cos\big(\beta\log(1+x)\big)$ and $X_2(x) = \frac{1}{\sqrt{1+x}}\sin\big(\beta\log(1+x)\big)$ are both solutions of this equation.

(b) Explain why any solution of the equation is of the form

$$c_1X_1(x) + c_2X_2(x),$$

where $c_1, c_2 \in \mathbb{R}$.

(c) Show that all the eigenfunctions of problem (8.30) are given by (8.31).

EXERCISE 8.16 The purpose of this exercise is to prove Hölder's inequality (8.34). Let $p, q > 1$ be real numbers such that

$$\frac{1}{p} + \frac{1}{q} = 1$$

and consider the function

$$\phi(x) = \frac{\lambda^p}{p} + \frac{x^q}{q} - \lambda x$$

for $x \ge 0$, where $\lambda \ge 0$ is a parameter.

(a) Show that $\phi'(x) = 0$ if and only if $x = \lambda^{\frac{1}{q-1}}$.

(b) Show that the inequality

$$\lambda\mu \le \frac{\lambda^p}{p} + \frac{\mu^q}{q} \tag{8.48}$$

holds for any λ, $\mu \ge 0$.

(c) Apply the inequality (8.48), with

$$\lambda = |f(x)| \left(\int_a^b |f(x)|^p dx \right)^{-1/p} \quad \text{and} \quad \mu = |g(x)| \left(\int_a^b |g(x)|^q dx \right)^{-1/q}$$

to establish Hölder's inequality.

EXERCISE 8.17 Consider the sequence of functions $\{f_N\}_{N=1}^{\infty}$, where $f_N(x) = N^{3/2} x e^{-(Nx)^2}$ for $x \in [-1, 1]$.

(a) Show that $f_N(x) \to 0$ for all $x \in [-1, 1]$.

(b) Show that f_N does not converge to zero in the mean square sense.

EXERCISE 8.18 Use Bessel's inequality and the full Fourier series for the function sign(x) to derive the inequality

$$\sum_{k=1}^{\infty} \left(\frac{1}{2k-1} \right)^2 \le \frac{\pi^2}{8}.$$

EXERCISE 8.19

(a) Show that

$$\int_{-1}^{1} \sin(k\pi x)\, dx = 0 \quad \text{for} \quad k = 1, 2, \ldots$$

and explain why this implies that the set $\{\sin(k\pi x)\}_{k=1}^{\infty}$ is not complete on $[-1, 1]$.

(b) Show that the set $\{\cos(k\pi x)\}_{k=0}^{\infty}$ is not complete on $[-1, 1]$.

EXERCISE 8.20 The purpose of this exercise is to derive a result which is usually referred to as "the best approximation property" for general Fourier series. Let $\{X_k\}_{k=1}^{\infty}$ be an orthogonal set of piecewise continuous functions defined on $[a,b]$. Furthermore, let f be an arbitrary piecewise continuous function with general Fourier series with respect to $\{X_k\}$ given by

$$f \sim \sum_{k=1}^{\infty} c_k X_k.$$

For $N \geq 1$, let $S_N(f)$ be the partial sums

$$S_N(f) = \sum_{k=1}^{N} c_k X_k$$

and P_N an arbitrary function of the form $P_N = \sum_{k=1}^{N} \alpha_k X_k$. Show that

$$||f - S_N(f)|| \leq ||f - P_N||,$$

where $||\cdot||$ is the mean square distance function with respect to the interval $[a,b]$. Give a geometric interpretation of this result.

EXERCISE 8.21 The purpose of this exercise is to derive a generalization of Gronwall's inequality given in Lemma 8.7.

Let $y(t)$ be a continuous and differentiable function defined for $t \geq 0$ which satisfies

$$y'(t) \leq ay(t) + b \quad \text{for } t > 0,$$

where $a, b \in \mathbb{R}$, $a \neq 0$. Show that

$$y(t) \leq e^{at}\left(y(0) + \frac{b}{a}\right) - \frac{b}{a} \quad \text{for } t \geq 0.$$

EXERCISE 8.22 Consider a two-point boundary value problem of the form

$$\begin{aligned} Lu &= f \quad \text{for} \quad 0 < x < 1, \\ u(0) &= u(1) = 0, \end{aligned} \tag{8.49}$$

where $f \in C([0,1])$ is given. Here the differential operator L is the Sturm-Liouville operator (8.26), and we assume that the conditions (8.27) and (8.28) are satisfied.

(a) Explain why Lemma 8.2 implies that the problem (8.49) has at most one solution.

(b) Consider the initial value problem

$$Lw = 0, \quad w(0) = 0, \quad w'(0) = 1.$$

Show that the solution of this problem is strictly positive in $[0, 1]$.

(c) Consider initial value problems of the form

$$Lu = f, \quad u(0) = 0, \quad u'(0) = z,$$

where $z \in \mathbb{R}$. Show that there exists a unique choice of z such that u solves (8.49).

EXERCISE 8.23 Throughout this exercise we assume that the functions $p(x)$ and $q(x)$ satisfy the conditions (8.27) and (8.28). The Sturm-Liouville operator L, given by (8.26), can be approximated by the finite difference operator

$$(L_h v)(x) = \frac{p(x+\frac{h}{2})\left(\frac{v(x+h)-v(x)}{h}\right) - p(x-\frac{h}{2})\left(\frac{v(x)-v(x-h)}{h}\right)}{h} + q(x)v(x),$$

where $h > 0$. Consider the finite difference approximation of the two-point boundary value problem (8.49) given by

$$\begin{aligned} (L_h v)(x_j) &= f(x_j) \quad \text{for} \quad j = 1, 2, \ldots, n, \\ v(0) &= v(1) = 0, \end{aligned} \tag{8.50}$$

where $n \geq 1$ is an integer, $h = \frac{1}{(n+1)}$ and where $\{x_j = jh\}_{j=0}^{n+1}$ are the grid points.

(a) Let $v, b \in \mathbb{R}^n$ be the vectors given by

$$v = \Big(v(x_1), v(x_2), \ldots, v(x_n)\Big)^T \text{ and } b = h^2\Big(f(x_1), f(x_2), \ldots f(x_n)\Big)^T.$$

Identify an $n \times n$ matrix A such that the problem (8.50) can be written in the form $Av = b$. Is the matrix A symmetric and tridiagonal?

(b) Show that the matrix A is positive definite.

(c) Explain why all eigenvalues of A are real and positive.

(d) Explain why the system (8.50) always has a unique solution.

(e) Can we use Algorithm 2.1 on page 53 to compute the solution of the systen (8.50)? Justify your answer.

9
Convergence of Fourier Series

Let f be a piecewise continuous function defined on $[-1,1]$ with a full Fourier series given by

$$\frac{a_0}{2} + \sum_{k=1}^{\infty} \left(a_k \cos\left(k\pi x\right) + b_k \sin\left(k\pi x\right)\right).$$

The main purpose of this chapter is to discuss the convergence question for Fourier series, i.e. "Do the partial sums

$$S_N(f) = \frac{a_0}{2} + \sum_{k=1}^{N} \left(a_k \cos\left(k\pi x\right) + b_k \sin\left(k\pi x\right)\right)$$

converge to the function f ?" If we here refer to convergence in the mean square sense, then a partial answer to this question is already established by Theorem 8.2. At least we have seen that we have convergence if and only if the corresponding Parseval's identity holds. However, we like to establish convergence under assumptions which are easier to check. Also, frequently we are interested in notions of convergence other than convergence in the mean.

9.1 Different Notions of Convergence

In Chapter 8 we defined mean square convergence of a sequence of piecewise continuous functions. Before we continue the study of convergence of Fourier series, a few different concepts of convergence for sequences of functions will be discussed.

Let $\{f_N\}_{N=1}^{\infty}$ be a sequence of piecewise continuous functions defined on $[a,b]$. Recall that the sequence converges to a piecewise continuous function f in the mean square sense if

$$\lim_{N\to\infty} \|f_N - f\| = 0,$$

where $\|\cdot\|$ is the mean square norm given by

$$\|f\| = \left(\int_a^b f^2(x)dx\right)^{1/2}.$$

Another norm, or distance function, which we encountered several times before is *the uniform norm*, $\|\cdot\|_\infty$, given by

$$\|f\|_\infty = \sup_{x\in[a,b]} |f(x)|.$$

This distance function leads to the notion of uniform convergence.

Definition 9.1 *The sequence $\{f_N\}$ converges to f uniformly in $[a,b]$ if*

$$\lim_{N\to\infty} \|f_N - f\|_\infty = 0.$$

In addition to the two notions of convergence described above, we also mention *pointwise convergence.* A sequence $\{f_N\}$ defined on an interval I (closed or open) is said to converge pointwise to f on I if

$$\lim_{N\to\infty} f_N(x) = f(x)$$

for all $x \in I$.

EXAMPLE 9.1 In Example 8.8 we studied the sequence $\{f_N\}$ given by

$$f_N(x) = \frac{N}{1+N^2x^2} \quad \text{for} \quad x \in [0,1].$$

We found that $\{f_N\}$ converges to the zero function pointwise in $(0,1]$, but not in the mean square sense on $[0,1]$. Furthermore, $f_N(0) = N$. Therefore,

$$\|f_N\|_\infty \geq N,$$

and we do not have uniform convergence to the zero function on $[0,1]$. ■

EXAMPLE 9.2 Let

$$f_N(x) = \begin{cases} -1 + (1+x)^N & \text{for} \quad x \in [-1,0], \\ 1 - (1-x)^N & \text{for} \quad x \in [0,1]. \end{cases}$$

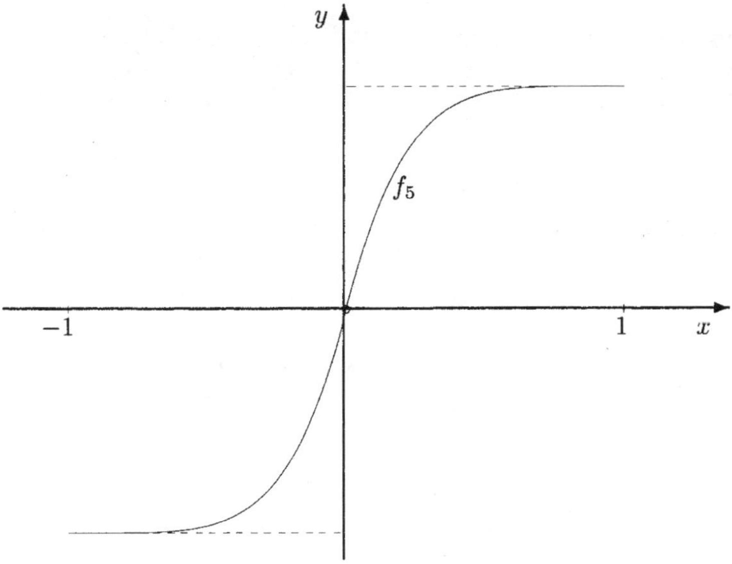

FIGURE 9.1. $f_5(x)$ *and* $f(x)$.

Hence, $\{f_N\}$ is a sequence of continuous functions defined on $[-1,1]$, with values in $[-1,1]$. From the fact that

$$\lim_{N\to\infty} y^N = 0$$

for $|y| < 1$, it follows that $\{f_N\}$ will converge pointwise to the function

$$f(x) = \begin{cases} -1 & \text{for} \quad x \in [-1,0), \\ 0 & \text{for} \quad x = 0, \\ 1 & \text{for} \quad x \in (0,1]. \end{cases}$$

(see Fig. 9.1).

Furthermore, a straightforward calculation shows that

$$\|f_N - f\|^2 = 2\int_0^1 (1-x)^{2N} dx = 2\int_0^1 z^{2N} dz = \frac{2}{2N+1},$$

and hence mean square convergence follows. However, for each N we have

$$\|f_N - f\|_\infty = \sup_{x\in[-1,1]} |f_N(x) - f(x)| = 1.$$

Therefore, we do not have uniform convergence.

We conclude that $\{f_N\}$ converges to f pointwise and in the mean square sense, but not uniformly on $[-1,1]$. ■

The following result shows that uniform convergence will always imply pointwise and mean square convergence.

Proposition 9.1 *Assume that* $\{f_N\}_{N=1}^{\infty}$ *is a sequence of piecewise continuous functions which converges uniformly to a piecewise continuous function* f *on* $[a,b]$*. Then the sequence also converges pointwise and in the mean square sense on* $[a,b]$*.*

Proof: Recall that uniform convergence means

$$\lim_{N\to\infty} \|f_N - f\|_\infty = 0.$$

However, for each $x \in [a,b]$ the inequality

$$|f_N(x) - f(x)| \le \|f - f_N\|_\infty$$

holds. Therefore,

$$\lim_{N\to\infty} f_N(x) = f(x)$$

for all $x \in [a,b]$, i.e. $\{f_N\}$ converges to f pointwise. Similarly, mean square convergence follows since

$$\|f_N - f\|^2 = \int_a^b |f_N(x) - f(x)|^2\, dx \le (b-a)\,\|f_N - f\|_\infty^2.$$

■

The next result shows that if a sequence $\{f_N\}$ converges to f uniformly, and all the functions f_N are continuous, then the limit function f also has to be continuous. We note that such a result will not contradict the results of Example 9.2. There the sequence $\{f_N\}$ converges pointwise (and in mean) to a discontinuous function f. However, we observed that the convergence is not uniform.

Proposition 9.2 *Assume that* $\{f_N\}_{N=1}^{\infty}$ *is a sequence of continuous functions on* $[a,b]$ *which converges uniformly to* f*. Then* f *is continuous on* $[a,b]$*.*

Proof: Let $x \in [a,b]$ be arbitrary. In order to show that f is continuous at x, we have to show that for each $\epsilon > 0$ there is a $\delta > 0$ such that

$$|x-y| < \delta \quad \text{implies} \quad |f(x) - f(y)| < \epsilon.$$

Let $\epsilon > 0$ be given. Since $\{f_N\}$ converges uniformly to f, there exists a function f_{N_0} such that

$$\|f_{N_0} - f\|_\infty < \epsilon/3.$$

Furthermore, since f_{N_0} is continuous there is a $\delta > 0$ such that

$$|x-y| < \delta \quad \text{implies} \quad |f_{N_0}(x) - f_{N_0}(y)| < \epsilon/3.$$

Now, for $|x - y| < \delta$ we have

$$\begin{aligned} |f(x) - f(y)| &\le |f(x) - f_{N_0}(x)| + |f_{N_0}(x) - f_{N_0}(y)| + |f_{N_0}(y) - f(y)| \\ &\le 2\|f - f_{N_0}\|_\infty + |f_{N_0}(x) - f_{N_0}(y)| \\ &< \epsilon. \end{aligned}$$

■

The result above states that a uniform limit of continuous functions is continuous. The next result, which will be useful in the application of Fourier series to differential equations, states that a proper uniform limit of continuous differentiable functions is continuously differentiable.

Proposition 9.3 *Assume that $\{f_N\}_{N=1}^\infty$ is a sequence of continuously differentiable functions on $[a, b]$ which converges uniformly to a continuous function f. Suppose furthermore that $\{f_N'\}$ converges uniformly to a continuous function g. Then f is continuously differentiable and $f' = g$.*

Proof: From the fundamental theorem of integral calculus we have

$$f_N(x) = f_N(a) + \int_a^x f_N'(y)dy \tag{9.1}$$

for each $x \in [a, b]$. Our purpose is to take the limit of this identity. From the uniform convergence of $\{f_N\}$ it follows that

$$\lim_{N\to\infty} f_N(x) = f(x) \quad \text{and} \quad \lim_{N\to\infty} f_N(a) = f(a).$$

Furthermore, from the uniform convergence of $\{f_N'\}$ it follows that

$$\begin{aligned} \left|\int_a^x f_N'(y)dy - \int_a^x g(y)dy\right| &\le \int_a^x |f_N'(y) - g(y)|dy \\ &\le (b-a)\|f_N' - g\|_\infty \\ &\longrightarrow 0 \end{aligned}$$

as N tends to infinity. Therefore

$$\lim_{N\to\infty} \int_a^x f_N'(y)dy = \int_a^x g(y)dy.$$

By letting N tend to infinity in (9.1), we now obtain

$$f(x) = f(a) + \int_a^x g(y)dy,$$

and hence $f' = g$. ■

9.2 Pointwise Convergence

Let f be a piecewise continuous function on $[-1,1]$ with full Fourier series

$$\frac{a_0}{2} + \sum_{k=1}^{\infty} \left(a_k \cos(k\pi x) + b_k \sin(k\pi x)\right).$$

Hence, the coefficients a_k and b_k are given by

$$\begin{aligned} a_k &= \int_{-1}^{1} f(y)\cos(k\pi y)dy, \\ b_k &= \int_{-1}^{1} f(y)\sin(k\pi y)dy. \end{aligned} \tag{9.2}$$

We will start by investigating pointwise convergence of the Fourier series. More precisely, if

$$S_N(f) = \frac{a_0}{2} + \sum_{k=1}^{N} \left(a_k \cos(k\pi x) + b_k \sin(k\pi x)\right) \tag{9.3}$$

are the partial sums, then we will give conditions which guarantee that $\{S_N(f)\}$ converges pointwise to f on $[-1,1]$.

As a preliminary step in this discussion we will derive an alternative representation of the functions $S_N(f)$. By combining (9.2) and (9.3), we obtain

$$S_N(f)(x) = \int_{-1}^{1} \left[\frac{1}{2} + \sum_{k=1}^{N} \left(\cos(k\pi x)\cos(k\pi y) + \sin(k\pi x)\sin(k\pi y)\right)\right] f(y)dy.$$

Furthermore, by using the trigonometric identity

$$\cos(u)\cos(v) + \sin(u)\sin(v) = \cos(u-v),$$

this expression simplifies to

$$S_N(f)(x) = \frac{1}{2}\int_{-1}^{1} \left[1 + 2\sum_{k=1}^{N} \cos(k\pi(x-y))\right] f(y)dy.$$

Hence, if we let

$$K_N(z) = 1 + 2\sum_{k=1}^{N} \cos(k\pi z), \tag{9.4}$$

we obtain the representation

$$S_N(f)(x) = \frac{1}{2}\int_{-1}^{1} K_N(x-y)f(y)dy \tag{9.5}$$

for the partial sums $S_N(f)$. The function $K_N(z)$ is called the *Dirichlet kernel.* The next step in our convergence analysis is to study the properties of this function.

Observe that the periodicity of the trigonometric functions implies that

$$\int_{-1}^{1} \cos(k\pi z)dz = 0 \quad \text{for} \quad k \geq 1.$$

Hence, we obtain from (9.4) that

$$\frac{1}{2}\int_{-1}^{1} K_N(z)dz = 1. \tag{9.6}$$

In addition to this the series for $K_N(z)$ can be summed.

Lemma 9.1 *The function $K_N(z)$ has the alternative representation*

$$K_N(z) = \frac{\sin\left(\left(N + \frac{1}{2}\right)\pi z\right)}{\sin\left(\frac{\pi z}{2}\right)}. \tag{9.7}$$

Proof: We use the complex representation (8.11) of sine and cosine. Furthermore, let $\theta = \pi z$. If $i = \sqrt{-1}$, we obtain from (9.4) that

$$\begin{aligned} K_N(z) &= 1 + 2\sum_{k=1}^{N} \cos(k\theta) = 1 + \sum_{k=1}^{N} \left(e^{ik\theta} + e^{-ik\theta}\right) \\ &= \sum_{k=-N}^{N} e^{ik\theta} = \sum_{k=-N}^{N} \left(e^{i\theta}\right)^k \end{aligned}$$

or

$$K_N(z) = e^{-i\theta N} \sum_{k=0}^{2N} \left(e^{i\theta}\right)^k.$$

However, the sum of this finite geometric series is given by

$$\begin{aligned} K_N(z) &= e^{-i\theta N}\frac{e^{i(2N+1)\theta} - 1}{e^{i\theta} - 1} = \frac{e^{i\left(N+\frac{1}{2}\right)\theta} - e^{-i\left(N+\frac{1}{2}\right)\theta}}{e^{\frac{i\theta}{2}} - e^{-\frac{i\theta}{2}}} \\ &= \frac{\sin\left(\left(N + \frac{1}{2}\right)\pi z\right)}{\sin\left(\frac{\pi z}{2}\right)}. \end{aligned}$$

■

In addition to the properties (9.6) and (9.7), it is also easy to see that K_N is a 2-periodic function. It will also be convenient to assume that the function f, which we have assumed to be given on $[-1, 1]$, is extended to a 2-periodic function on $\mathbb{R}$. Hence, substituting $z = y - x$, we have

$$\begin{aligned} S_N(f)(x) &= \frac{1}{2}\int_{-1}^{1} K_N(x-y)f(y)dy \\ &= \frac{1}{2}\int_{-1-x}^{1-x} K_N(-z)f(x+z)dz. \end{aligned}$$

However, since K_N is an even function and since $K_N{\cdot}f$ is 2-periodic, this can be written as

$$S_N(f)(x) = \frac{1}{2}\int_{-1}^{1} K_N(z)f(x+z)dz. \tag{9.8}$$

Furthermore, by (9.6) we can rewrite $f(x)$ as

$$f(x) = f(x){\cdot}1 = \frac{1}{2}\int_{-1}^{1} K_N(z)f(x)dz.$$

Therefore, the error $S_N(f) - f$ admits the representation

$$S_N(f)(x) - f(x) = \frac{1}{2}\int_{-1}^{1} K_N(z)\left(f(x+z) - f(x)\right)dz. \tag{9.9}$$

From this error representation it is straightforward to establish pointwise convergence of the Fourier series under proper assumptions on the function f. In order to avoid unnecessary technical difficulties, we will first assume a rather strong condition on f, i.e. we will assume that the periodic extension of f is continuous and differentiable. Later the conditions on f will be relaxed.

Theorem 9.1 *Let f be a function defined on $[-1, 1]$ such that its 2-periodic extension is continuous and differentiable for all $x \in \mathbb{R}$. Then $\{S_N(f)\}$ converges pointwise to f on $[-1, 1]$, and hence to the periodic extension of f on $\mathbb{R}$.*

Proof: Although it may seem unlikely, this theorem will be derived from Bessel's inequality (8.40).

Let $x \in [-1, 1]$ be fixed. We have to show that

$$\lim_{N\to\infty} S_N(f)(x) = f(x).$$

From (9.7) and (9.9) we obtain that the error can be written in the form

$$S_N(f)(x) - f(x) = \frac{1}{2}\int_{-1}^{1} g(z)\sin\left(\left(N + \frac{1}{2}\right)\pi z\right)dz, \tag{9.10}$$

where

$$g(z) = \frac{f(x+z) - f(x)}{\sin\left(\frac{\pi z}{2}\right)}.$$

Of course, in addition to z, g also depends on x. However, since x is fixed throughout the proof, this dependence is suppressed. Note that the function g is obviously continuous at all points in $[-1, 1]$, with the exception of the origin, where it is not defined. However, since

$$\lim_{z \to 0} g(z) = \lim_{z \to 0} \frac{2}{\pi} \frac{f(x+z) - f(x)}{z} \frac{\pi z/2}{\sin(\pi z/2)}$$
$$= \frac{2}{\pi} f'(x),$$

it follows that g can be defined to be continuous in all of $[-1, 1]$. Hence, g is bounded and, in particular,

$$\|g\|^2 = \langle g, g \rangle = \int_{-1}^{1} g^2(z)dz < \infty. \tag{9.11}$$

Next let us consider the functions

$$Z_k(z) = \sin\left(\left(k + \frac{1}{2}\right)\pi z\right) \quad \text{for } k = 1, 2, \ldots$$

Recall from Example 8.4 that these functions are orthogonal on $[0, 1]$. Hence, since these functions are odd they will also be orthogonal on $[-1, 0]$ and, as a consequence of this, they are orthogonal on $[-1, 1]$. Furthermore, a straightforward calculation using the formula

$$\sin^2(\alpha) = \frac{1}{2}(1 - \cos(2\alpha))$$

shows that $\|Z_k\|^2 = 1$. You are asked to verify this in Exercise 9.9. Hence, from Bessel's inequality (8.40) with respect to the function g and the orthogonal set $\{Z_k\}$, we derive from (9.11)

$$\sum_{k=1}^{\infty} \langle g, Z_k \rangle^2 \leq \|g\|^2 < \infty.$$

In particular, this means that

$$\langle g, Z_N \rangle = \int_{-1}^{1} g(z) \sin\left(\left(N + \frac{1}{2}\right)\pi z\right) dz \longrightarrow 0$$

as N tends to infinity, and by (9.10) this implies that

$$\lim_{N \to \infty} S_N(f)(x) = f(x).$$

■

We would like to extend the argument above such that it also applies when the 2-periodic extension of f is only piecewise continuous. For such functions let[1]

$$f(x-) = \lim_{h \searrow 0} f(x-h) \quad \text{and} \quad f(x+) = \lim_{h \searrow 0} f(x+h).$$

Hence, f is continuous at x if $f(x-) = f(x+)$.

Definition 9.2 *A piecewise continuous function f is said to be "one-sided differentiable" at x if the two limits*

$$\lim_{h \searrow 0} \frac{f(x-) - f(x-h)}{h} \quad \text{and} \quad \lim_{h \searrow 0} \frac{f(x+h) - f(x+)}{h}$$

both exist.

EXAMPLE 9.3 The function $f(x) = |x|$ is one-sided differentiable at $x = 0$ since

$$\lim_{h \searrow 0} \frac{|0| - |-h|}{h} = -1 \quad \text{and} \quad \lim_{h \searrow 0} \frac{|h| - |0|}{h} = 1.$$

■

EXAMPLE 9.4 The function $f(x) = \text{sign}(x)$ is one-sided differentiable at $x = 0$ since

$$\lim_{h \searrow 0} \frac{\text{sign}(0-) - \text{sign}(-h)}{h} = 0 \quad \text{and} \quad \lim_{h \searrow 0} \frac{\text{sign}(h) - \text{sign}(0+)}{h} = 0.$$

■

We now have the following stronger pointwise convergence theorem:

Theorem 9.2 *Let f be a piecewise continuous function on $[-1, 1]$ such that its 2-periodic extension is one-sided differentiable for all $x \in \mathbb{R}$. Then the sequence $\{S_N(f)(x)\}$ converges pointwise to $\frac{1}{2}[f(x-) + f(x+)]$ for all $x \in \mathbb{R}$.*

Proof: We will do a proper modification of the proof of Theorem 9.1 above. First we write (9.8) as

$$S_N(f) = \frac{1}{2}\left[\int_{-1}^{0} K_N(z) f(x+z)dz + \int_{0}^{1} K_N(z) f(x+z)dz\right]. \tag{9.12}$$

Furthermore, since $K_N(z)$ is an even function, it follows from (9.6) that

$$\frac{1}{2}\int_{-1}^{0} K_N(z)dz = \frac{1}{2}\int_{0}^{1} K_N(z)dz = \frac{1}{2}.$$

[1] Here the symbol $\lim_{h \searrow 0}$ means $\lim_{h \to 0,\, h>0}$.

Let $\overline{f(x)}$ be the average value

$$\overline{f(x)} = \frac{1}{2}[f(x-) + f(x+)].$$

We obtain

$$\overline{f(x)} = \frac{1}{2}\left[\int_{-1}^{0} K_N(z)f(x-)dz + \int_{0}^{1} K_N(z)f(x+)dz\right].$$

Together with (9.12) this means that we can rewrite (9.10) in the form

$$S_N(f)(x) - \overline{f(x)} = \frac{1}{2}\left[\int_{-1}^{0} g^-(z)Z_N(z)dz + \int_{0}^{1} g^+(z)Z_N(z)dz\right], \quad (9.13)$$

where

$$g^{\pm}(z) = \frac{f(x+z) - f(x\pm)}{\sin\left(\frac{\pi z}{2}\right)}.$$

However, since f is one-sided differentiable at x, the two functions g^- and g^+ are continuous on the two intervals $[-1,0]$ and $[0,1]$, respectively. Therefore, we have

$$\int_{-1}^{0} |g^-(z)|^2 dz, \quad \int_{0}^{1} |g^+(z)|^2 dz \quad < \quad \infty.$$

Furthermore, we recall that the functions $\{Z_k(x)\}$ are orthogonal on each of the two intervals $[-1,0]$ and $[0,1]$. Hence, by applying Bessel's inequality with respect to each interval, we conclude, as above, that each of the two integrals on the right-hand side of (9.13) tends to zero as N tends to infinity. ■

EXAMPLE 9.5 Recall from Example 8.3 on page 252 that

$$|x| \sim \frac{1}{2} - \frac{4}{\pi^2}\sum_{k=1}^{\infty}\left(\frac{1}{2k-1}\right)^2 \cos\left((2k-1)\pi x\right).$$

Since the 2-periodic extension of $|x|$ is continuous and one-sided differentiable, the Fourier series will converge to $|x|$ for each $x \in [-1,1]$. Hence, by letting $x = 0$ we have

$$0 = \frac{1}{2} - \frac{4}{\pi^2}\sum_{k=1}^{\infty}\left(\frac{1}{2k-1}\right)^2$$

or

$$\sum_{k=1}^{\infty}\left(\frac{1}{2k-1}\right)^2 = 1 + \frac{1}{9} + \frac{1}{25} + \cdots = \frac{\pi^2}{8}.$$

■

Example 9.6 The Fourier series of sign(x) is derived in Example 8.2 on page 251. We have

$$\operatorname{sign}(x) \sim \frac{4}{\pi} \sum_{k=1}^{\infty} \frac{1}{2k-1} \sin\left((2k-1)\pi x\right).$$

Since sign(x) is piecewise continuous and one-sided differentiable, we obtain for $x = \frac{1}{2}$ that

$$1 = \frac{4}{\pi} \sum_{k=1}^{\infty} \frac{1}{2k-1} (-1)^{k-1}$$

or

$$\sum_{k=1}^{\infty} \frac{(-1)^{k-1}}{2k-1} = 1 - \frac{1}{3} + \frac{1}{5} - \frac{1}{7} + \cdots = \frac{\pi}{4}.$$

■

9.3 Uniform Convergence

In this section we shall establish necessary conditions which will guarantee that the Fourier series converges uniformly to f. Recall from Proposition 9.1 above that uniform convergence always implies pointwise convergence. Therefore, we would expect that the conditions required on f to guarantee uniform convergence must be at least as strong as those required for pointwise convergence. The conditions below are slightly stronger than those assumed in Theorem 9.2.

Theorem 9.3 *Let f be a function defined on $[-1,1]$ such that its periodic extension is continuous and let f' be piecewise continuous. Then $S_N(f)$ converges uniformly to f on $[-1,1]$.*

Proof: Let us first observe that since the conditions on f above are stronger than the ones given in Theorem 9.2, we have, for any $x \in [-1,1]$, that

$$f(x) = \lim_{N\to\infty} S_N(f)(x) = \frac{a_0}{2} + \sum_{k=1}^{\infty} \left(a_k \cos(k\pi x) + b_k \sin(k\pi x)\right),$$

where the Fourier coefficients are given by

$$a_k = \int_{-1}^{1} f(y)\cos(k\pi y)dy, \qquad b_k = \int_{-1}^{1} f(y)\sin(k\pi y)dy.$$

Therefore,

$$\|f - S_N(f)\|_\infty = \sup_{x\in[-1,1]} \left| \sum_{k=N+1}^{\infty} a_k \cos(k\pi x) + b_k \sin(k\pi x) \right|$$
$$\leq \sum_{k=N+1}^{\infty} (|a_k| + |b_k|)\,.$$

The proof will be completed by showing that the right-hand side of this inequality tends to zero as N tends to infinity. In fact, we will show that

$$\sum_{k=1}^{\infty} (|a_k| + |b_k|) < \infty, \tag{9.14}$$

i.e. this series converges to a finite number. This will immediatly imply the desired convergence.

In order to establish (9.14) let

$$\alpha_k = \int_{-1}^{1} f'(y)\cos(k\pi y)dy \quad \text{and} \quad \beta_k = \int_{-1}^{1} f'(y)\sin(k\pi y)dy$$

be the Fourier coefficients of f'. Since f' is assumed to be piecewise continuous, it follows from Bessel's inequality (8.42) that

$$\frac{\alpha_0^2}{2} + \sum_{k=1}^{\infty} (\alpha_k^2 + \beta_k^2) \leq \|f'\|^2 = \int_{-1}^{1} (f'(x))^2\, dx < \infty. \tag{9.15}$$

Furthermore, since $f(1) = f(-1)$, it follows from Theorem 8.1 that $\alpha_0 = 0$ and

$$\alpha_k = k\pi b_k\,, \qquad \beta_k = -k\pi a_k.$$

For any integer $N > 0$ we therefore have

$$\sum_{k=1}^{N} (|a_k| + |b_k|) = \frac{1}{\pi}\sum_{k=1}^{N} \frac{1}{k}(|\alpha_k| + |\beta_k|)\,.$$

By applying the Cauchy-Schwarz inequality, (cf. Project 1.2 in Chapter 1), we obtain

$$\sum_{k=1}^{N} \frac{1}{k}|\alpha_k| \leq \left(\sum_{k=1}^{N} \frac{1}{k^2}\right)^{1/2} \left(\sum_{k=1}^{N} \alpha_k^2\right)^{1/2}$$
$$\leq \left(\sum_{k=1}^{N} \frac{1}{k^2}\right)^{1/2} \left(\sum_{k=1}^{N} (\alpha_k^2 + \beta_k^2)\right)^{1/2}.$$

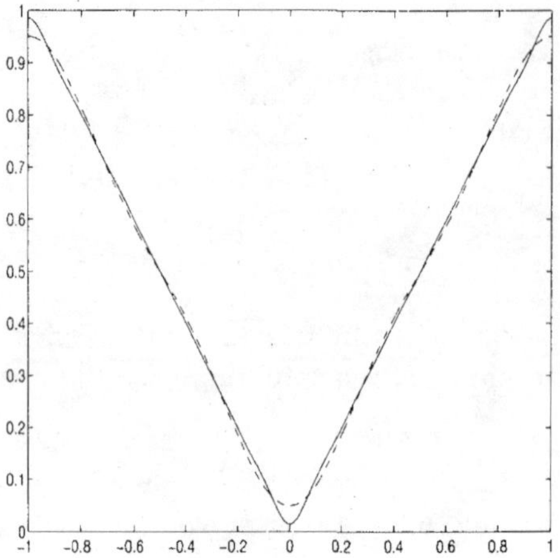

FIGURE 9.2. $S_2(f)$ *(dashed) and* $S_7(f)$ *when* $f(x) = |x|$.

Together with a similar inequality for $\sum_{k=1}^N \frac{1}{k}|\beta_k|$ we now have

$$\begin{aligned}\sum_{k=1}^N (|a_k| + |b_k|) &\leq \frac{2}{\pi}\left(\sum_{k=1}^N \frac{1}{k^2}\right)^{1/2}\left(\sum_{k=1}^N (\alpha_k^2 + \beta_k^2)\right)^{1/2} \\ &\leq \frac{2}{\pi}\frac{\pi}{\sqrt{6}}\|f'\| \\ &\leq \|f'\|,\end{aligned} \tag{9.16}$$

where we have used (9.15) and the inequality

$$\sum_{k=1}^N \frac{1}{k^2} \leq \sum_{k=1}^\infty \frac{1}{k^2} \leq \frac{\pi^2}{6}$$

derived in Example 8.11. Hence, by letting N tend to infinity in (9.16), we have established (9.14). The proof is therefore completed. ■

EXAMPLE 9.7 Let $f(x) = |x|$ for $x \in [-1, 1]$. Then the periodic extension of f is continuous and $f' = \text{sign}(x)$ is piecewise continuous. Theorem 9.3 therefore implies that $S_N(f)$ converges uniformly to f on $[-1, 1]$. This convergence is illustrated in Fig. 9.2, where we have plotted $S_N(f)$ for $N = 2, 7$. ■

EXAMPLE 9.8 Let $f(x) = \text{sign}(x)$. Since this function is not continuous, we cannot conclude that $S_N(f)$ converges to f uniformly on $[-1, 1]$. In fact, from Proposition 9.2 we know that it is impossible that $S_N(f)$ converges uniformly to f, since this would imply that f itself is continuous. Still, it may seem reasonable to belive that $S_N(f)(x)$ always takes values in the

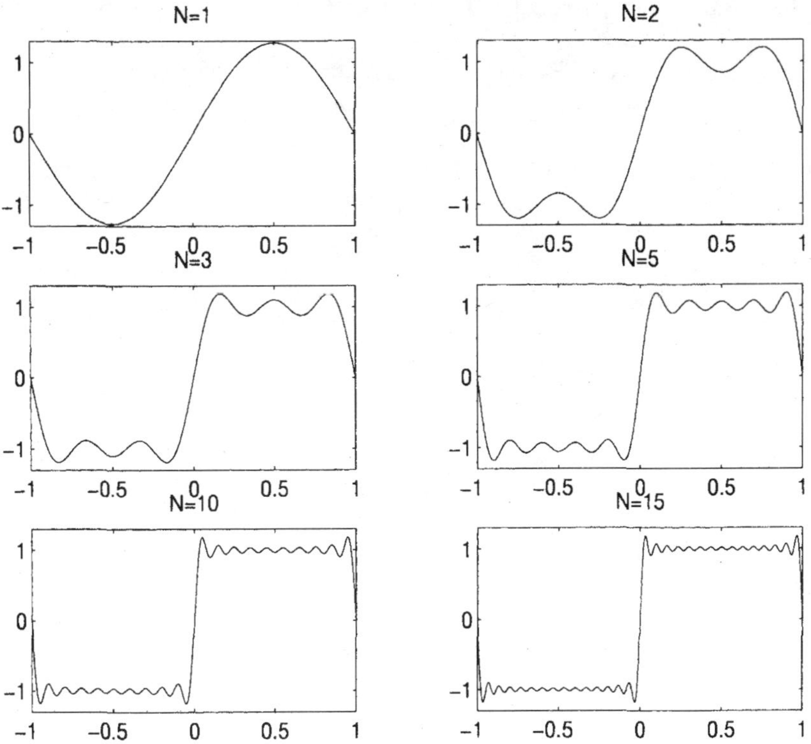

FIGURE 9.3. S_{2N-1} *for different values of* N *when* $f(x) = \text{sign}(x)$.

interval $[-1,1]$. However, this is not the case. In Fig. 9.3 we plotted the functions

$$S_{2N-1}(f)(x) = \frac{4}{\pi}\sum_{k=1}^{N}\frac{1}{2k-1}\sin\left((2k-1)\pi x\right)$$

for $N = 1, 2, 3, 5, 10, 15$.

We see that $S_N(f)(x)$ takes values larger than 1 and smaller than -1 for x close to zero. This overshoot is usually referred to as *the Gibbs phenomenon.* It can in fact be shown that there is a $\delta > 0$ such that

$$\lim_{N\to\infty}\sup \|S_N(f)\|_\infty \geq 1 + \delta,$$

i.e. $\|S_N(f)\|_\infty$ does not converge to 1. We refer for example to Strauss [25] for a further discussion of this phenomenon. ■

9.4 Mean Square Convergence

Finally, we shall consider mean square convergence of Fourier series. Recall from Proposition 9.1 that uniform convergence implies mean square convergence. Hence, if the periodic extension of f is continuous, with f' piecewise continuous, it follows from Theorem 9.3 that $S_N(f)$ converges in the mean square sense to f. However, these conditions are unnecessarily strong. In fact, we have mean square convergence for any piecewise continuous function f.

Theorem 9.4 *Let f be a piecewise continuous function on $[-1,1]$. Then $S_N(f)$ converges to f in the mean square sense.*

Proof: The idea is to approximate f by a smoother function f_δ which satisfies the hypothesis of Theorem 9.3.

First we extend f to a 2-periodic function on $\mathbb{R}$. For each $\delta > 0$ let $f_\delta(x)$ be defined by averaging the original function f around the point x. More precisely,

$$f_\delta(x) = \frac{1}{2\delta}\int_{x-\delta}^{x+\delta} f(y)dy.$$

Since f is 2-periodic, f_δ will be 2-periodic. The fundamental theorem of integral calculus implies that

$$f'_\delta(x) = \frac{1}{2\delta}\left[f((x+\delta)-) - f((x-\delta)+)\right].$$

Hence, f'_δ is piecewise continuous, and since any differentiable function is continuous, f_δ is continuous. We have therefore verified that f_δ satisfies the hypothesis of Theorem 9.3, i.e. $S_N(f_\delta)$ converges uniformly to f_δ as N tends to infinity. Since uniform convergence implies mean square convergence (see Proposition 9.1), we therefore obtain

$$\lim_{N\to\infty} ||S_N(f_\delta) - f_\delta|| = 0. \tag{9.17}$$

Furthermore, it can be shown that

$$\lim_{\delta\to 0} ||f_\delta - f|| = 0. \tag{9.18}$$

In fact, you are asked to establish this convergence in Exercise 9.17.

Observe now that the Fourier series $S_N(f)$ depends linearly on f. This follows since the Fourier coefficients depend linearly on f. Therefore,

$$S_N(f) - S_N(f_\delta) = S_N(f - f_\delta).$$

From the Pythagoras identity (8.38) it follows that $||S_n(f)|| \le ||f||$ for any piecewise continuous function f. In particular,

$$||S_N(f - f_\delta)|| \le ||f - f_\delta||. \tag{9.19}$$

Now write

$$S_N(f) - f = S_N(f - f_\delta) + (S_N(f_\delta) - f_\delta) + (f_\delta - f).$$

By using the triangle inequality for the mean square norm and (9.19), we therefore obtain

$$||S_N(f) - f|| \leq 2||f - f_\delta|| + ||S_N(f_\delta) - f_\delta||. \tag{9.20}$$

In order to show that $S_N(f)$ converges to f in the mean square sense, we have to show that $||S_N(f) - f||$ can be made arbitrarily small by choosing N sufficiently large.

Let $\epsilon > 0$ be given. Since f_δ converges to f (see (9.18)), it follows that we can choose a δ such that

$$||f - f_\delta|| < \frac{\epsilon}{3}.$$

Furthermore, with δ fixed, it follows from (9.17) that we can choose N_0 such that

$$||S_N(f_\delta) - f_\delta|| < \frac{\epsilon}{3} \quad \text{for} \quad N \geq N_0.$$

Hence, by (9.20)

$$||S_N(f) - f|| < \epsilon \quad \text{for} \quad N \geq N_0.$$

Since $\epsilon > 0$ is arbitrary, this shows that

$$\lim_{N\to\infty} ||S_N(f) - f|| = 0.$$

■

We recall from Theorem 8.2 that mean square convergence of the Fourier series implies Parseval's identity

$$\frac{a_0^2}{2} + \sum_{k=1}^{\infty} (a_k^2 + b_k^2) = ||f||^2. \tag{9.21}$$

Hence, the following result is a simple consequence of Theorem 9.4 above.

Corollary 9.1 *If f is piecewise continuous on $[-1, 1]$, then Parseval's identity (9.21) holds.*

Example 9.9 In Example 8.11 on page 273 we studied the full Fourier series of $f(x) = x$ and we concluded from Bessel's inequality that

$$\sum_{k=1}^{\infty} \frac{1}{k^2} \leq \frac{\pi^2}{6}.$$

By Corollary 9.1 this inequality is an identity. ■

EXAMPLE 9.10 Recall from Example 8.2 on page 251 that the full Fourier series of $f(x) = \text{sign}(x)$ is given by

$$\text{sign}(x) \sim \frac{4}{\pi}\sum_{k=1}^{\infty} \frac{1}{2k-1}\sin((2k-1)\pi x).$$

Hence, since f is piecewise continuous and $\|f\|^2 = 2$, it follows from Parseval's identity that

$$\frac{16}{\pi^2}\sum_{k=1}^{\infty}\left(\frac{1}{2k-1}\right)^2 = 2$$

or

$$\sum_{k=1}^{\infty}\left(\frac{1}{2k-1}\right)^2 = \frac{\pi^2}{8}.$$

In fact, this formula was also derived in Example 9.5 above as a consequence of the pointwise convergence of the Fourier series of $|x|$ at $x = 0$. ■

9.5 Smoothness and Decay of Fourier Coefficients

We will end this chapter with a short discussion of the relation between the smoothness of a function f and how fast its Fourier coefficients a_k and b_k tend to zero as k tends to infinity. Recall from Corollary 9.1 that if f is piecewise continuous, then

$$\frac{a_0^2}{2} + \sum_{k=1}^{\infty}\left(a_k^2 + b_k^2\right) = \|f\|^2 < \infty.$$

Since the infinite series converges this implies, in particular, that

$$a_k\,,\; b_k \longrightarrow 0 \quad \text{as} \quad k \to \infty.$$

We shall see below that the smoother f is the faster the Fourier coefficients will converge to zero. Here, the rate at which the Fourier coefficients tend to zero will be measured by checking if

$$\sum_{k=1}^{\infty} k^{2m}\left(a_k^2 + b_k^2\right) < \infty$$

for positive integers m. Larger values of m indicate faster convergence to zero for the Fourier coefficients.

Throughout this section we let C_p^m denote the set of functions on $\mathbb{R}$ such that $f, f', \ldots, f^{(m)}$ are all continuous and 2-periodic. Hence, if $f \in C_p^m$, then

$$f^{(j)}(-1) = f^{(j)}(1) \quad \text{for} \quad j = 0, 1, \ldots, m. \tag{9.22}$$

In fact, since any 2-periodic function is uniquely determined by its values in $[-1, 1]$, the space C_p^m can be alternatively defined as all functions $f \in C^m([-1, 1])$ which satisfy (9.22).

Assume first that $f \in C_p^0$ and that f' is piecewise continuous. Let α_k and β_k denote the Fourier coefficients of f'. From Theorem 8.1 we have

$$\alpha_k = k\pi b_k \quad \text{and} \quad \beta_k = -k\pi a_k. \tag{9.23}$$

Furthermore, Parseval's identity (9.21) implies that

$$\sum_{k=1}^{\infty} \left(\alpha_k^2 + \beta_k^2\right) = \|f'\|^2,$$

or by using (9.23),

$$\sum_{k=1}^{\infty} k^2 \left(a_k^2 + b_k^2\right) = \frac{\|f'\|^2}{\pi^2}.$$

The following theorem is a generalization of this identity.

Theorem 9.5 *Let $m \geq 1$ be an integer. Assume that $f \in C_p^{m-1}$ and $f^{(m)}$ is piecewise continuous. Then*

$$\sum_{k=1}^{\infty} k^{2m} \left(a_k^2 + b_k^2\right) = \pi^{-2m} \|f^{(m)}\|^2,$$

where a_k and b_k are the Fourier coefficients of f.

Proof: We prove this by induction on m. For $m = 1$ the result was established above. Assume the result holds for m. If $f \in C_p^m$ with $f^{(m+1)}$ piecewise continuous, then $f' \in C_p^{m-1}$ with $\frac{d^m}{dx^m} f' = f^{(m+1)}$ piecewise continuous. Therefore, the induction hypothesis applied to f' gives

$$\sum_{k=1}^{\infty} k^{2m} \left(\alpha_k^2 + \beta_k^2\right) = \pi^{-2m} \|f^{(m+1)}\|^2,$$

where α_k and β_k are the Fourier coefficients of f'. From the identities (9.23) we obtain

$$\pi^2 \sum_{k=1}^{\infty} k^{2(m+1)} \left(a_k^2 + b_k^2\right) = \pi^{-2m} \|f^{(m+1)}\|^2,$$

which is the desired result for $m + 1$. ■

EXAMPLE 9.11 Let $f(x) = \text{sign}(x)$. Recall from Example 9.10 above that the Fourier series is given by

$$\frac{4}{\pi}\sum_{k=1}^{\infty}\frac{1}{2k-1}\sin((2k-1)\pi x).$$

We have already checked Parseval's identity (9.21) in Example 9.10. However, we observe that the series

$$\sum_{k=1}^{\infty} k^2 b_k^2 = \sum_{k=1}^{\infty}\left(\frac{4}{\pi}\right)^2$$

diverges. But this does not contradict Theorem 9.5 above since $f \notin C_p^0$. ■

EXAMPLE 9.12 Let $f(x) = |x|$. Then $f \in C_p^0$ and $f'(x) = \text{sign}(x)$ is piecewise continuous. Hence, Theorem 9.5 predicts that

$$\sum_{k=1}^{\infty} k^2\left(a_k^2 + b_k^2\right) = \pi^{-2}\|f'\|^2 = \frac{2}{\pi^2}.$$

On the other hand we recall from Example 9.5 on page 295 that

$$|x| \sim \frac{1}{2} - \frac{4}{\pi^2}\sum_{k=1}^{\infty}\left(\frac{1}{2k-1}\right)^2\cos\left((2k-1)\pi x\right),$$

and that

$$\sum_{k=1}^{\infty}\left(\frac{1}{2k-1}\right)^2 = \frac{\pi^2}{8}.$$

This gives

$$\sum_{k=1}^{\infty} k^2\left(a_k^2 + b_k^2\right) = \frac{16}{\pi^4}\sum_{k=1}^{\infty}\left(\frac{1}{2k-1}\right)^2 = \frac{2}{\pi^2}.$$

We have therefore confirmed the theorem for this example. ■

The interpretation of Theorem 9.5 is that the smoother f is, the faster the Fourier coefficients will decay to zero. However, we can also show the converse, i.e. that fast decay of the Fourier coefficients implies that f is smooth. In fact, the argument needed to prove this has already been introduced in the proof of Theorem 9.3. Assume first that $\{a_k\}_{k=0}^{\infty}$ and $\{b_k\}_{k=1}^{\infty}$ are real coefficients such that

$$\sum_{k=1}^{\infty} k^2\left(a_k^2 + b_k^2\right) < \infty. \tag{9.24}$$

For $N \geq 1$ let

$$S_N(x) = \frac{a_0}{2} + \sum_{k=1}^{N} \left(a_k \cos(k\pi x) + b_k \sin(k\pi x)\right).$$

We like to show that S_N converges to a function $f \in C_p^0$. Since all the functions $S_N \in C_p^0$, this will follow if we can show that S_N converges uniformly to a function f; see Proposition 9.2.

Let $x \in [-1, 1]$ be arbitrary. If $M > N$, then

$$\begin{aligned}
|S_M(x) - S_N(x)| &\leq \sum_{k=N+1}^{M} (|a_k| + |b_k|) \\
&= \sum_{k=N+1}^{M} \frac{1}{k} \left(k(|a_k| + |b_k|)\right) \\
&\leq 2 \left(\sum_{k=N+1}^{M} \frac{1}{k^2} \right)^{1/2} \left(\sum_{k=N+1}^{M} k^2 \left(a_k^2 + b_k^2\right) \right)^{1/2}.
\end{aligned}$$

Here, the final inequality follows from the Cauchy-Schwarz inequality. Hence,

$$|S_M(x) - S_N(x)| \leq 2 \left(\sum_{k=N+1}^{\infty} \frac{1}{k^2} \right)^{1/2} \left(\sum_{k=1}^{\infty} k^2 \left(a_k^2 + b_k^2\right) \right)^{1/2}.$$

Since $\sum_k 1/k^2 < \infty$ (see Example 9.10 above), it follows that $\lim_{N\to\infty} \sum_{k=N+1}^{\infty} 1/k^2 = 0$. Together with (9.24) this implies that

$$\lim_{M,N\to\infty} |S_M(x) - S_N(x)| = 0.$$

Therefore, since $\{S_N(x)\}$ is a Cauchy sequence, $S_N(x)$ converges. We call the limit $f(x)$. Furthermore, by replacing $S_M(x)$ by $f(x)$ in the calculation above, and by taking supremum over $x \in [-1, 1]$, we derive

$$\begin{aligned}
\|f - S_N\|_\infty &\leq \sum_{k=N+1}^{\infty} (|a_k| + |b_k|) \\
&\leq 2 \left(\sum_{k=N+1}^{\infty} \frac{1}{k^2} \right)^{1/2} \left(\sum_{k=1}^{\infty} k^2 \left(a_k^2 + b_k^2\right) \right)^{1/2} \longrightarrow 0 \text{ as } N \to \infty.
\end{aligned}$$

Therefore S_N converges uniformly to f, and by Proposition 9.2 we can conclude that $f \in C_p^0$. Furthermore, it is straightforward to show that $S_N = S_N(f)$, or equivalently, that

$$f \sim \frac{a_0}{2} + \sum_{k=1}^{\infty} \left(a_k \cos(k\pi x) + b_k \sin(k\pi x)\right).$$

You are asked to establish this in Exercise 9.13.

The following result is a generalization of this observation.

Theorem 9.6 *Let $m \geq 1$ be an integer and assume that $\{a_k\}_{k=0}^{\infty}$ and $\{b_k\}_{k=1}^{\infty}$ are real coefficients such that*

$$\sum_{k=1}^{\infty} k^{2m} \left(a_k^2 + b_k^2\right) < \infty.$$

Let

$$S_N(x) = \frac{a_0}{2} + \sum_{k=1}^{N} \left(a_k \cos(k\pi x) + b_k \sin(k\pi x)\right). \tag{9.25}$$

Then there exists a function $f \in C_p^{m-1}$ such that

$$S_N^{(j)} \longrightarrow f^{(j)} \quad \textit{uniformly as} \quad N \to \infty$$

for $0 \leq j \leq m-1$. Furthermore, the Fourier series of f is given by

$$\frac{a_0}{2} + \sum_{k=1}^{\infty} \big(a_k \cos(k\pi x) + b_k \sin(k\pi x)\big). \tag{9.26}$$

Proof: We will use induction on m to establish the uniform convergence of the functions $S_N^{(j)}$. For $m = 1$ the result is derived above. Assume that the result holds for m, where $m \geq 1$, and that

$$\sum_{k=1}^{\infty} k^{2(m+1)} \left(a_k^2 + b_k^2\right) < \infty.$$

Since this assumption is stronger than (9.24), we can conclude from the discussion above that there is a function $f \in C_p^0$ such that S_N converges uniformly to f. We have to show that $f \in C_p^m$ and that

$$S_N^{(j)} \longrightarrow f^{(j)} \quad \text{uniformly for} \quad 0 \leq j \leq m \quad \text{as} \quad N \to \infty.$$

Let

$$T_N(x) = S_N'(x) = \sum_{k=1}^{N} \left(\alpha_k \cos(k\pi x) + \beta_k \sin(k\pi x)\right),$$

where $\alpha_k = k\pi b_k$ and $\beta_k = -k\pi a_k$. The coefficients α_k and β_k satisfy

$$\sum_{k=1}^{\infty} k^{2m} \left(\alpha_k^2 + \beta_k^2\right) = \pi^2 \sum_{k=1}^{\infty} k^{2(m+1)} \left(a_k^2 + b_k^2\right) < \infty.$$

Hence, the induction hypothesis implies that there is a function $g \in C_p^{m-1}$ such that

$$T_N^{(j)} = S_N^{(j+1)} \longrightarrow g^{(j)} \quad \text{for} \quad 0 \leq j \leq m-1.$$

From Proposition 9.3 on page 289 we can also conclude that $g = f'$ and hence the desired uniform convergence is established. Finally, we have to show that the Fourier series of f is given by (9.26). However, this is a consequence of the fact that the functions S_N, given by (9.25), converge uniformly to f. You are asked to verify this in Exercise 9.13. ■

Note that the two theorems established in this section are close to providing an equivalence between the property that the Fourier coefficients of f satisfies

$$\sum_{k=1}^{\infty} k^{2m} \left(a_k^2 + b_k^2\right) < \infty \tag{9.27}$$

and the property that $f \in C_p^{m-1}$. In fact, Theorem 9.6 states that the convergence of the series (9.27) implies that $f \in C_p^{m-1}$. On the other hand, Theorem 9.5 implies that if $f \in C_p^{m-1}$, and if in addition $f^{(m)}$ is piecewise continuous, then the series (9.27) converges. However, strict equivalence between convergence of the series (9.27) and smoothness properties of f would require the introduction of Lebesgue integration and Sobolev spaces. This is beyond the scope of this book.

9.6 Exercises

EXERCISE 9.1 Let $f_N(x) = 1/(N+x)$ for $x \in [0,1]$. Show that $\{f_N\}$ converges uniformly to zero as N tends to infinity.

EXERCISE 9.2 Let $f_N(x) = e^{-x/N}$ for $x \in [-1,1]$. Show that $f_N \to 1$ uniformly as $N \to \infty$.

EXERCISE 9.3 Let $f_N(x) = e^{-|x|N}$.

(a) Show that $f_N(x) \to 0$ as $N \to \infty$ for all $x \neq 0$.

(b) Consider the functions f_N on $[-1,1]$. Does $\{f_N\}$ converge to zero in the mean square sense?

(c) Does $\{f_N\}$ converge uniformly to zero on $[-1,1]$?

Exercise 9.4

(a) Let $f_N(x) = 1/(Nx+1)$ for $x \in [0,1]$. Show that $f_N(x) \to 0$ for all $x \in (0,1]$, but that the convergence is not uniform on $[0,1]$.

(b) Show that $\{f_N\}$ converges to zero in the mean square sense.

(c) Let $g_N(x) = x/(Nx+1)$. Show that $\{g_N\}$ converges uniformly to zero on $[0,1]$.

Exercise 9.5 Let $f_N(x) = x/(1+Nx^2)$ for $x \in \mathbb{R}$.

(a) Find the two pointwise limits

$$f(x) = \lim_{N\to\infty} f_N(x) \quad \text{and} \quad g(x) = \lim_{N\to\infty} f_N'(x).$$

(b) Show that $f'(x)$ exists for all x, but that $f'(0) \neq g(0)$.

(c) Explain why this does not contradict Proposition 9.3.

Exercise 9.6 Let $f_N(x) = N^{-1}e^{-N^2x^2}$ for $x \in \mathbb{R}$.

(a) Show that $f_N(x) \to 0$ as N tends to infinity and that the convergence is uniform on any closed interval.

(b) Show that $f_N'(x) \to 0$ for all $x \in \mathbb{R}$, but that the convergence is not uniform on any closed interval containing the origin.

Exercise 9.7 Let $||f||_\infty$ be the uniform norm given by

$$||f||_\infty = \sup_{x\in[a,b]} |f(x)|$$

for $f \in C([a,b])$.

(a) Establish the triangle inequality

$$||f+g||_\infty \le ||f||_\infty + ||g||_\infty$$

for $f, g \in C([a,b])$.

(b) Use the triangle inequality to show that

$$\Big| ||f||_\infty - ||g||_\infty \Big| \le ||f-g||_\infty$$

for $f, g \in C([a,b])$.

(c) Show that $||\cdot||_\infty$ is continuous with respect to uniform convergence, i.e if $\{f_n\} \subset C([a,b])$ converges uniformly to f, then

$$\lim_{N\to\infty} ||f_N||_\infty = ||f||_\infty.$$

(d) Let $\{f_N\}$ be a sequence of piecewise continuous functions on an interval $[a,b]$ which converges to a piecewise continuous function f in the mean square sense. Show that

$$\lim_{N\to\infty} ||f_N|| = ||f||,$$

where $||f|| = \left(\int_a^b f^2(x)\,dx\right)^{1/2}$.

EXERCISE 9.8 Use a computer program to plot the Dirichlet kernel $K_N(z)$ for increasing values of N.

EXERCISE 9.9 Consider the functions

$$Z_k(z) = \sin\big((k+\frac{1}{2})\pi z\big) \quad \text{for } k = 1, 2, \ldots$$

appearing in the proofs of Theorems 9.1 and 9.2. Use the trigonometric identity

$$\sin^2(\alpha) = \frac{1}{2}(1-\cos(2\alpha))$$

to show that

$$||Z_k||^2 = \int_{-1}^{1} Z_k^2(z)\,dz = 1.$$

EXERCISE 9.10

(a) Use the full Fourier series of $f(x) = x^2$ and Parseval's identity to show that

$$\sum_{k=1}^{\infty} \frac{1}{k^4} = \frac{\pi^4}{90}.$$

You can use the result from Exercise 3.2 that

$$x^2 \sim \frac{1}{3} + \frac{4}{\pi^2}\sum_{k=1}^{\infty} \frac{(-1)^k}{k^2}\cos(k\pi x).$$

(b) Use the full Fourier series of $f(x) = |x|$ and Parseval's identity to compute

$$\sum_{k=1}^{\infty} \left(\frac{1}{2k-1}\right)^4.$$

EXERCISE 9.11

(a) Recall from Example 8.1 on page 249 that the Fourier series of $f(x) = x$ is

$$x \sim \sum_{k=1}^{\infty} b_k \sin(k\pi x),$$

where $b_k = \frac{2}{k\pi}(-1)^{k+1}$.

Find the largest integer $m \geq 0$ such that

$$\sum_{k=1}^{\infty} k^{2m} b_k^2 < \infty.$$

Explain how your conclusion relates to the results of Theorems 9.5 and 9.6.

(b) Repeat the problem above with $f(x) = x^2$.

EXERCISE 9.12 Let $f(x)$ be the 2-periodic function defined by

$$f(x) = x^3 - x \quad \text{for } x \in [-1, 1].$$

(a) Show that $f \in C_p^1$, but $f \notin C_p^2$.

(b) Let a_k and b_k denote the Fourier coefficients of f. Explain why

$$\sum_{k=1}^{\infty} k^4 (a_k^2 + b_k^2) < \infty,$$

without calculating a_k and b_k. Is the sum $\sum_{k=1}^{\infty} k^6 (a_k^2 + b_k^2)$ finite?

(c) Use the fact that $f''(x) = 6x$ to compute the Fourier series of f. Compare the results with your conclusions above.

EXERCISE 9.13 Let $\{a_k\}_{k=0}^{\infty}$ and $\{b_k\}_{k=1}^{\infty}$ be real coefficients such that (9.24) holds, and let $f \in C_p^0$ be the function derived from Theorem 9.6, i.e. f is the uniform limit of the sequence $\{S_N\}$ defined by (9.25). Show that

$$f \sim \frac{a_0}{2} + \sum_{k=1}^{\infty} \big(a_k \cos(k\pi x) + b_k \sin(k\pi x)\big).$$

Exercise 9.14

(a) Find the full Fourier series of the function

$$f(x) = \cos(ax)$$

with respect to the interval $[-\pi, \pi]$. Here $a \in \mathbb{R}$, but a is not an integer. You will probably find the formula

$$2\cos u \cos v = \cos(u+v) + \cos(u-v)$$

useful.

(b) Use the Fourier series above to establish the formula

$$\frac{\cos(\pi a)}{\sin(\pi a)} = \frac{1}{\pi a} - 2\frac{a}{\pi}\sum_{k=1}^{\infty}\frac{1}{k^2-a^2}.$$

(c) Use Theorem 8.1 to find the full Fourier series of $g(x) = \sin(ax)$.

(d) Find a formal series for $\cos(ax)$ by differentiating the full Fourier series of g term by term. Compare the result with the full Fourier series for f. Explain what you observe.

Exercise 9.15 Let $f : [-1, 1] \to \mathbb{R}$ be defined by

$$f(x) = \begin{cases} a - |x| & \text{for} \quad |x| < a, \\ 0 & \text{for} \quad |x| \geq a, \end{cases}$$

where $a \in \mathbb{R}$ satisfies $0 < a < 1$.

(a) Find the full Fourier series of f.

(b) Perform a term-by-term differentiation of the Fourier series above. Explain why this series converges for all $x \in \mathbb{R}$. Sketch the graph of the sum $g(x)$ for $x \in [-2, 2]$. What is the value of $g(a)$ and $g(0)$?

Exercise 9.16

(a) Let f be a piecewise continuous function defined on $[0, 1]$ with Fourier sine series

$$\sum_{k=1}^{\infty} c_k \sin(k\pi x).$$

Use Corollary 9.1 to explain why Parseval's identity,

$$\sum_{k=1}^{\infty} c_k^2 = 2\int_0^1 f^2(x)\,dx,$$

holds.

(b) Formulate the corresponding result for Fourier cosine series.

EXERCISE 9.17 The purpose of this exercise is to establish the convergence (9.18). Hence, for a given 2-periodic piecewise continuous function f, we like to show that the functions $f_\delta(x)$ given by

$$f_\delta(x) = \frac{1}{2\delta}\int_{x-\delta}^{x+\delta} f(y)dy \quad \text{for} \quad \delta > 0$$

converge in the mean square sense to f as δ tends to zero.

(a) Let $H(x)$ be the 2-periodic function defined by

$$H(x) = \begin{cases} 1 & \text{for} \quad x \in (0,1] \\ 0 & \text{for} \quad x \in (-1,0]. \end{cases}$$

Note that the function H is piecewise continuous.
Show that $\lim_{\delta\to 0} \|H_\delta - H\| = 0$, where $\|\cdot\|$ denotes the mean square norm on $[-1,1]$.

(b) Let g be a 2-periodic continuous function. Show that $\lim_{\delta\to 0} \|g_\delta - g\| = 0$.

(c) Explain why any 2-periodic piecewise continuous function can be written in the form

$$f(x) = g(x) + \sum_{j=1}^{M} c_j H(x - x_j),$$

where g is continuous, M is a finite positive integer and c_j are suitable coefficients.

(d) Use the results above and the triangle inequality for the mean square norm to show that $\lim_{\delta\to 0} \|f_\delta - f\| = 0$ for any 2-periodic piecewise continuous function f.

10
The Heat Equation Revisited

The two previous chapters have been devoted to Fourier series. Of course, the main motivation for the study of Fourier series was their appearance in formal analytic solutions of various partial differential equations like the heat equation, the wave equation, and Poisson's equation.

In this chapter we will return to partial differential equations. We will reinvestigate formal solutions derived above and discuss the consequence of the results on Fourier series for these formal solutions. The convergence results for Fourier series, derived in the previous chapter, will be used to show that, under proper assumptions, the formal solutions are in fact rigorous solutions in a strict mathematical sense.

In order to avoid this discussion becoming too long, we shall concentrate on the solution of the heat equation in one space dimension with Dirichlet boundary conditions.

Other formal solutions can be treated in a similar manner, and some examples are discussed in the exercises. In the final section of this chapter we shall also derive a rigorous error estimate for a finite difference method for the heat equation.

10.1 Compatibility Conditions

Consider the initial and boundary value problem

$$\begin{aligned} u_t &= u_{xx} \quad \text{for} \quad x \in (0,1), \quad t > 0 \\ u(0,t) &= u(1,t) = 0, \quad t > 0 \\ u(x,0) &= f(x), \quad x \in (0,1). \end{aligned} \tag{10.1}$$

Recall that if

$$f(x) = \sum_{k=1}^{\infty} c_k \sin(k\pi x)$$

is the Fourier sine series of the initial function f, then the formal solution of this problem is given by

$$u(x,t) = \sum_{k=1}^{\infty} c_k e^{-(k\pi)^2 t} \sin(k\pi x). \tag{10.2}$$

The purpose of the next section will be to show that the solution (10.2) is not only a formal solution, but a rigorous mathematical solution. However, before we start this discussion, we will focus attention on a possible difficulty.

Assume that u is a solution of (10.1) which is continuous in the domain $[0,1] \times [0,\infty)$. In particular, this means that u is continuous at the origin $(x,t) = (0,0)$. Hence,

$$0 = \lim_{t \to 0} u(0,t) = u(0,0) = \lim_{x \to 0} u(x,0) = f(0).$$

A similar argument shows that $f(1) = 0$. Therefore, if u is a solution of (10.1) which is continuous in the entire domain $[0,1] \times [0,\infty)$, then the initial function f must satisfy the *compatibility condition*

$$f(0) = f(1) = 0. \tag{10.3}$$

EXAMPLE 10.1 Consider the problem (10.1) with $f \equiv 1$. Note that the function f does not satisfy (10.3). Hence, it follows from the discussion above that there is no solution u which is continuous in the entire domain $[0,1] \times [0,\infty)$.

On the other hand, a formal solution of this problem was derived already in Example 3.2 on page 93. There we computed the formal solution

$$u(x,t) = \frac{4}{\pi} \sum_{k=1}^{\infty} \frac{1}{2k-1} e^{-\left((2k-1)\pi\right)^2 t} \sin((2k-1)\pi x). \tag{10.4}$$

In fact, it will follow from the discussion in the next section (see Exercise 10.11) that the series (10.4) defines a solution of (10.1). However, this solution will not be continuous at the corners $(x,t) = (0,0)$ and $(x,t) = (1,0)$. ■

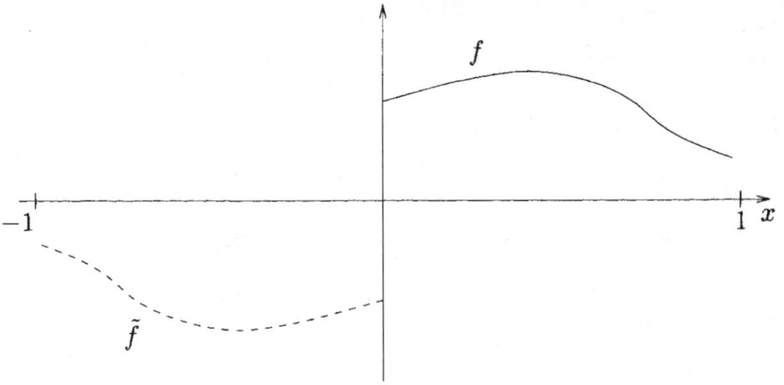

FIGURE 10.1. f *and the odd extension* $\tilde{f}$.

The compatibility condition (10.3) is closely related to the space C_p^0 introduced in Section 9.5 above. We recall that the space C_p^m consists of functions defined on $[-1, 1]$ such that the 2-periodic extension is m-times continuously differentiable, i.e. $f \in C_p^m$ if and only if $f \in C^m\big([-1,1]\big)$ and

$$f^{(j)}(-1) = f^{(j)}(1) \quad \text{for } j = 0, 1, \ldots, m.$$

In order to see the relation between the space C_p^0 and the condition (10.3), let us first recall from Chapter 8 that the Fourier sine series of a function f defined on $[0, 1]$ is simply the full Fourier series of the odd extension of f. Let $\tilde{f}$ denote the odd extension of f, i.e.

$$\tilde{f}(-x) = -f(x) \quad \text{for} \quad x \in (0, 1].$$

Hence, it follows that $\tilde{f} \in C_p^0$ if and only if f is a continuous function on $[0, 1]$ which satisfies (10.3); see Fig. 10.1.

In general, from the chain rule we derive

$$\tilde{f}^{(j)}(-x) = (-1)^{j+1} f^{(j)}(x) \quad \text{for} \quad x \in [0, 1].$$

As a consequence of this identity we conclude that we always have

$$\tilde{f}^{(j)}(-1) = \tilde{f}^{(j)}(1) \quad \text{and} \quad \tilde{f}^{(j)}(0-) = \tilde{f}^{(j)}(0+)$$

if j is odd. On the other hand, if j is even then

$$\tilde{f}^{(j)}(-1) = -\tilde{f}^{(j)}(1) \quad \text{and} \quad \tilde{f}^{(j)}(0-) = -\tilde{f}^{(j)}(0+)$$

Therefore, $\tilde{f} \in C_p^m$ if and only if $f \in C^m\big([0,1]\big)$ with

$$f^{(2j)}(0) = f^{(2j)}(1) = 0 \quad \text{for} \quad 0 \le 2j \le m. \tag{10.5}$$

This result motivates the definition of the space

$$C_{p,o}^m = \left\{ f \in C^m([0,1]) \mid f^{(2j)}(0) = f^{(2j)}(1) = 0 \quad \text{for} \quad 0 \leq 2j \leq m \right\},$$

where $m \geq 0$ is an integer. For example, the space $C_{p,o}^2$ is given by

$$C_{p,o}^2 = \{ f \in C^2([0,1]) | f(0) = f''(0) = f(1) = f''(1) = 0 \}.$$

The discussion above can be summarized as follows:

Lemma 10.1 *Let $\tilde{f}$ be the odd extension of a function f defined on $[0,1]$. Then $\tilde{f} \in C_p^m$ if and only if $f \in C_{p,o}^m$.*

Hence, the space $C_{p,o}^m$ corresponds exactly to all the odd functions in C_p^m.

Recall that in Section 9.5 above we studied the relation between the smoothness of f and decay to zero of the Fourier coefficients. As a consequence of Lemma 10.1 and the fact that the Fourier sine series of a function is the full Fourier series of its odd extension, these relations can be translated to sine series.

The theorem below follows directly from the Theorems 9.5 and 9.6. In this chapter, $\|f\|$ is defined with respect to $[0,1]$, i.e.

$$\|f\|^2 = \int_0^1 f^2(x)dx.$$

Theorem 10.1 *Let f be a piecewise continuous function defined on $[0,1]$ with Fourier sine series*

$$f(x) \sim \sum_{k=1}^{\infty} c_k \sin(k\pi x)$$

and let $m \geq 1$ be an integer.

(i) If $f \in C_{p,o}^{m-1}$ and $f^{(m)}$ is piecewise continuous, then

$$\sum_{k=1}^{\infty} k^{2m} c_k^2 = 2\pi^{-2m} \|f^{(m)}\|^2.$$

(ii) If

$$\sum_{k=1}^{\infty} k^{2m} c_k^2 < \infty$$

then $f \in C_{p,o}^{m-1}$. Furthermore, $S_N^{(j)}(f)$ converges uniformly to $f^{(j)}$ for $0 \leq j \leq m-1$, where

$$S_N(f) = \sum_{k=1}^{N} c_k \sin(k\pi x).$$

Note that the condition (10.5) reduces to (10.3) when $m = 0$. Furthermore, recall that the condition (10.3) has to be satisfied if the solution u of the problem (10.1) is continuous at the corners $(x,t) = (0,0)$ and $(x,t) = (1,0)$. The more general condition (10.5) also arises naturally in connection with the initial and boundary value problem (10.1).

If u_t and u_{xx} are both continuous at the origin, we must have

$$0 = \lim_{t \to 0} u_t(0,t) = \lim_{x,t \to 0} u_t(x,t) = \lim_{x,t \to 0} u_{xx}(x,t) = \lim_{x \to 0} u_{xx}(x,0) = f''(0).$$

In this manner we can argue that if $\frac{\partial^j}{\partial t^j} u$ and $\frac{\partial^{2j}}{\partial x^{2j}} u$ for $0 \le j \le k$ all are continuous at the origin, then

$$f^{(2j)}(0) = 0 \quad \text{for} \quad 0 \le j \le k.$$

A similar discussion applies to the other endpoint $x = 1$. The conditions (10.5) for the initial function f are therefore referred to as *compatibility conditions of order m* for the initial and boundary data of the problem (10.1).

Before we end this section we will reconsider Poincaré's inequality (see Lemma 8.6 on page 274) and the use of this inequality in obtaining energy estimates for the problem (10.1). In fact, from Theorem 10.1 we obtain the following strong version of Poincaré's inequality:

Corollary 10.1 *Assume that $f \in C^0_{p,o}$ with f' piecewise continuous. Then*

$$||f|| \le \frac{1}{\pi} ||f'||. \tag{10.6}$$

Proof: From Parseval's identity (see Exercise 9.16) we have

$$||f||^2 = \frac{1}{2} \sum_{k=1}^{\infty} c_k^2,$$

and from Theorem 10.1

$$||f'||^2 = \frac{1}{2} \pi^2 \sum_{k=1}^{\infty} k^2 c_k^2.$$

Therefore,

$$2\pi^{-2} ||f'||^2 = \sum_{k=1}^{\infty} k^2 c_k^2 \ge \sum_{k=1}^{\infty} c_k^2 = 2||f||^2,$$

which implies the desired inequality. ■

We note that this result represents an improvement over Lemma 8.6, in the sense that the constant appearing in front of $||f'||$ is smaller. Furthermore, by taking $f(x) = \sin(\pi x)$ we obtain equality in (10.6). Therefore, the constant in front of $||f'||$ in (10.6) is the smallest possible.

EXAMPLE 10.2 We will consider energy estimates for the problem (10.1) once more. Recall that this has been studied in Section 3.7, where we derived that the energy

$$E(t) = \int_0^1 u^2(x,t)\,dx$$

is nonincreasing with time, and in Example 8.12 on page 274, where we established the decay estimate

$$E(t) \le e^{-4t}E(0) \quad \text{for} \quad t \ge 0.$$

However, by assuming that $u(\cdot,t) \in C^0_{p,o}$, with $u_x(\cdot,t)$ piecewise continuous for any $t > 0$, and using (10.6), this decay estimate can be improved further. In fact, we have

$$E(t) \le e^{-2\pi^2 t}E(0). \tag{10.7}$$

To see this, we recall from Example 8.12 (see (8.45)) the identity

$$E'(t) = -2\int_0^1 u_x^2(x,t)\,dx.$$

However, (10.6) implies that

$$-2\int_0^1 u_x^2(x,t)\,dx \le -2\pi^2 E(t)$$

and therefore

$$E'(t) \le -2\pi^2 E(t).$$

Hence, the estimate (10.7) follows from Gronwall's inequality (see Lemma 8.7 on page 275). It is also easy to see that the decay estimate (10.7) cannot in general be improved. In fact, if we consider the problem (10.1) with $f(x) = \sin(\pi x)$, then we have equality in (10.7) (see Exercise 10.4). ■

We remark that in the discussion above, the estimate (10.7) is derived by energy arguments, i.e. no representation of the solution u is used. All we have used is the fact that u is a solution of problem (10.1). For the present problem, the estimate (10.7) can also be derived directly from the representation (10.2); see Exercise 10.6. However, the advantage of energy arguments is that they can be used for more complex problems, where no representation of the solution is available. This will for example be illustrated by the study of nonlinear reaction-diffusion equations in Chapter 11.

10.2 Fourier's Method: A Mathematical Justification

The purpose of this section is to show that the formal solution (10.2) is a rigorous solution in a strict mathematical sense. We shall also discuss the smoothing property of the heat equation.

We will assume that the initial function f is just a piecewise continuous function. Compatibility conditions of the form (10.5) will not be assumed. In particular, we will allow $f(0)$ and $f(1)$ to be different from zero. Hence, we will permit the initial function $f \equiv 1$, studied in Example 10.1, and also a discontinuous function like

$$f(x) = \begin{cases} 1 & \text{for} \quad 0 \le x \le 1/2, \\ -1 & \text{for} \quad 1/2 < x \le 1. \end{cases}$$

For this function, f'' is not defined at $x = 1/2$. Hence, for $t = 0$ the differential equation $u_t = u_{xx}$ cannot be satisfied. However, in (10.1) we only require this equation to hold for $t > 0$. We shall see below that this will in fact be the case, even if the initial function f is discontinuous.

Let us also note that the maximum principle stated in Theorem 6.2 on page 182 requires that u be continuous down to $t = 0$. Hence, in general we cannot apply the result of Theorem 6.2 to our solution.

10.2.1 The Smoothing Property

Let $\sum_k c_k \sin(k\pi x)$ be the Fourier sine series of the piecewise continuous function f. It follows from Parseval's identity that

$$\sum_{k=1}^{\infty} c_k^2 = 2\|f\|^2 < \infty. \tag{10.8}$$

Below we will show that even if f is just piecewise continuous, the series (10.2) for $t > 0$ will define a C^∞-function $u(\cdot, t)$ as a function of x. In fact, we will show that $u(\cdot, t) \in C^\infty_{p,o}$, where $C^\infty_{p,o} = \bigcap_{m=0}^{\infty} C^m_{p,o}$. Alternatively, the space $C^\infty_{p,o}$ can be defined by

$$C^\infty_{p,o} = \{g \in C^\infty([0,1]) \mid g^{(2j)}(0) = g^{(2j)}(1) = 0 \quad \text{for } j = 0, 1, 2, \dots\}.$$

The following technical result will be useful:

Lemma 10.2 *Let a and b be positive real numbers. There is a positive constant M, depending on a and b, such that*

$$0 \le x^a e^{-bx} \le M \quad \text{for} \quad x \ge 0.$$

Proof: Let $g(x)$ be the function

$$g(x) = x^a e^{-bx}.$$

We note that $g(x) \geq 0$ for $x \geq 0$ and that

$$g(0) = 0 \quad \text{and} \quad \lim_{x \to \infty} g(x) = 0.$$

Furthermore,

$$g'(x) = x^{a-1} e^{-bx} (a - bx),$$

which implies $g'(x) = 0$ only if $x = a/b$. Hence, we can take

$$M = g\left(\frac{a}{b}\right) = \left(\frac{a}{b}\right)^a e^{-a}.$$

■

Consider the series for $u(x,t)$, i.e.

$$\sum_{k=1}^{\infty} c_k e^{-(k\pi)^2 t} \sin(k\pi x). \tag{10.9}$$

For each fixed $t \geq 0$ we can view this as a Fourier sine series with respect to x, where the Fourier coefficients are given by $c_k e^{-(k\pi)^2 t}$.

Theorem 10.2 *Assume that f is piecewise continuous. For each $t > 0$ the series (10.9) converges uniformly to a function $u(\cdot, t) \in C^{\infty}_{p,o}$.*

Proof: Consider the Fourier sine series (10.9). For each integer $m \geq 1$ and $t > 0$ the Fourier coefficients $c_k e^{-(k\pi)^2 t}$ satisfy

$$\sum_{k=1}^{\infty} k^{2m} c_k^2 e^{-2(k\pi)^2 t} \leq M \sum_{k=1}^{\infty} c_k^2 = 2M||f||^2 < \infty, \tag{10.10}$$

where we used Lemma 10.2 to obtain the bound

$$k^{2m} e^{-2(k\pi)^2 t} \leq M.$$

Here M depends on m and t, but is independent of k. Hence, we can conclude from Theorem 10.1 that for any $t > 0$

$$u(\cdot, t) \in C^{m-1}_{p,o}.$$

However, since $m \geq 1$ is arbitrary, this must imply that $u(\cdot, t) \in C^{\infty}_{p,o}$. ■

Let $u(x,t)$ be the function defined by (10.9), i.e.

$$u(x,t) = \sum_{k=1}^{\infty} c_k e^{-(k\pi)^2 t} \sin(k\pi x). \tag{10.11}$$

It follows from the theorem above that *for any* $t > 0$*,* u *is a* C^∞*-function as a function of* x. Furthermore, for $t > 0$

$$\frac{\partial^{2j} u(0,t)}{\partial x^{2j}} = \frac{\partial^{2j} u(1,t)}{\partial x^{2j}} = 0 \quad \text{for} \quad j = 0, 1, 2, \ldots. \tag{10.12}$$

In particular, this means that u satisfies the boundary conditions

$$u(0,t) = u(1,t) = 0 \quad \text{for} \quad t > 0. \tag{10.13}$$

The property that $u(\cdot, t)$, for $t > 0$, is a C^∞-function, even when the initial function f is just piecewise continuous, is frequently referred to as *the smoothing property of the heat equation.* A similar property for the Cauchy problem for the heat equation was observed in Section 1.4.4 of Chapter 1.

10.2.2 The Differential Equation

Above we observed that the function $u(x, t)$, defined by (10.11), satisfies the boundary conditions $u(0, t) = u(1, t) = 0$ for $t > 0$. Next, we shall show that u satisfies the differential equation in (10.1), i.e we shall show that

$$u_t = u_{xx} \quad \text{for} \quad x \in (0,1), \quad t > 0.$$

We can conclude from Theorem 10.1 and the bound (10.10) that the derivatives of u with respect to x can be obtained as a uniform limit of corresponding derivatives of the partial sums, i.e. for $t > 0$

$$\frac{\partial^{2j}}{\partial x^{2j}} u(x,t) = \sum_{k=1}^{\infty} c_k \left(-(k\pi)^2\right)^j e^{-(k\pi)^2 t} \sin(k\pi x). \tag{10.14}$$

Here, the equality sign means that the right-hand side, as a function of x, converges uniformly to the C^∞-function $\frac{\partial^{2j}}{\partial x^{2j}} u(\cdot, t)$.

Next, we would like to establish that $u_t(x, t)$ exists and that

$$u_t(x,t) = -\sum_{k=1}^{\infty} c_k \, (k\pi)^2 \, e^{-(k\pi)^2 t} \sin(k\pi x).$$

In order to show this, let $u_N(x, t)$ be defined by the finite sum

$$u_N(x,t) = \sum_{k=1}^{N} c_k e^{-(k\pi)^2 t} \sin(k\pi x).$$

It is straightforward to check that $(u_N)_t = (u_N)_{xx}$. In order to establish the corresponding identity for u_N replaced by u, we first show the following preliminary result:

Lemma 10.3 *Let $x \in [0,1]$ and $\delta, T > 0$, $\delta < T$, be arbitrary. For each integer $j \geq 0$ consider the sequence $\{\frac{\partial^j}{\partial t^j} u_N(x,\cdot)\}_{N=1}^{\infty}$ as a function of t. This sequence converges uniformly to $\frac{\partial^{2j}}{\partial x^{2j}} u(x,\cdot)$ in the interval $[\delta, T]$.*

Proof: Throughout the proof x, δ, T, and j will be fixed, and with properties as described in the lemma. In order to simplify the notation we let

$$g(t) = \frac{\partial^{2j}}{\partial x^{2j}} u(x,\cdot)$$

and

$$v_N(t) = \frac{\partial^j}{\partial t^j} u_N(x,\cdot).$$

Our goal is to show that $\{v_N\}_{N=1}^{\infty}$ converges uniformly to g in $[\delta, T]$.

Since uniform convergence implies pointwise convergence, we obtain from (10.14) that

$$g(t) = \sum_{k=1}^{\infty} c_k \left(-(k\pi)^2\right)^j e^{-(k\pi)^2 t} \sin(k\pi x)$$

for any $t > 0$. Furthermore, by differentiating the finite series for u_N with respect to t j-times, we get

$$v_N(t) = \sum_{k=1}^{N} c_k \left(-(k\pi)^2\right)^j e^{-(k\pi)^2 t} \sin(k\pi x)$$

and hence,

$$v_N(t) - g(t) = -\sum_{k=N+1}^{\infty} c_k \left(-(k\pi)^2\right)^j e^{-(k\pi)^2 t} \sin(k\pi x).$$

For any $t \in [\delta, T]$ we therefore have the bound

$$|v_N(t) - g(t)| \leq \sum_{k=N+1}^{\infty} |c_k| \, (k\pi)^{2j} e^{-(k\pi)^2 \delta}.$$

At this point we use Lemma 10.2 to conclude that there is a constant M, independent of k, such that

$$k^{2j+1} e^{-\pi^2 \delta k^2} \leq M \quad \text{for} \quad k \geq 0.$$

Hence, by letting $M_1 = M\pi^{2j}$, we have

$$|v_N(t) - g(t)| \leq M_1 \sum_{k=N+1}^{\infty} \frac{|c_k|}{k}$$

for any $t \in [\delta, T]$. By applying the Cauchy-Schwarz inequality, this implies

$$\sup_{t\in[\delta,T]} |v_N(t) - g(t)| \leq M_1 \Big(\sum_{k=N+1}^{\infty} c_k^2 \Big)^{1/2} \Big(\sum_{k=N+1}^{\infty} k^{-2} \Big)^{1/2}$$

$$\leq M_1 \frac{\pi}{\sqrt{6}} \Big(\sum_{k=N+1}^{\infty} c_k^2 \Big)^{1/2},$$

where we have used the bound $\sum k^{-2} \leq \pi^2/6$ (see Example 8.11 on page 273). However, the identity (10.8) implies that

$$\lim_{N\to\infty} \Big(\sum_{k=N+1}^{\infty} c_k^2 \Big) = 0,$$

and hence the desired uniform convergence is established. ■

An immediate consequence of the lemma above and Proposition 9.3 is the following desired result.

Theorem 10.3 *Let the function $u(x,t)$ be defined by (10.11). For each $t > 0$ and $x \in [0,1]$, the partial derivative $u_t(x,t)$ exists. Furthermore,*

$$u_t(x,t) = u_{xx}(x,t) \quad \text{for} \quad t > 0, \quad x \in [0,1].$$

Proof: Let $x \in [0,1]$ be fixed and consider the sequence $\{u_N(x,\cdot)\}_{N=1}^{\infty}$ as functions of t. It follows from Lemma 10.3, with $j = 0$, that this sequence converges uniformly to $u(x,\cdot)$ in any interval of the form $[\delta, T]$, where $\delta > 0$. Similarly, by applying Lemma 10.3 with $j = 1$ we obtain that $\{(u_N)_t(x,\cdot)\}$ converges uniformly to $u_{xx}(x,\cdot)$ in $[\delta, T]$. However, by Proposition 9.3 this implies that $u_t(x,t)$ exists, and is equal to $u_{xx}(x,t)$ for any $t \in [\delta, T]$. Since $\delta > 0$ can be chosen arbitrarily small and $T > 0$ arbitrarily large, we must have

$$u_t(x,t) = u_{xx}(x,t)$$

for any $t > 0$. ■

10.2.3 The Initial Condition

Recall that the purpose of this section is to show that the formula (10.11) defines a strict mathematical solution of the initial-boundary value problem (10.1). Up to now we have shown that (10.11) defines a function u which solves the differential equation for $t > 0$ (see Theorem 10.3), and by (10.13) u satisfies the boundary conditions for $t > 0$. However, so far we have not discussed the initial condition. Hence, we have to show that

$$u(\cdot,t) \longrightarrow f \quad \text{as} \quad t \to 0, \quad t > 0. \tag{10.15}$$

Recall that we are only assuming that f is piecewise continuous. Therefore, in general, we cannot expect the convergence (10.15) to be uniform for $x \in [0,1]$, since Proposition 9.2 will then imply that f is continuous. Instead we shall establish the convergence (10.15) in the mean square sense.

From Parseval's identity (10.8) we obtain that for $t > 0$

$$\begin{aligned}\|u(\cdot,t) - f\|^2 &= \frac{1}{2}\sum_{k=1}^{\infty} c_k^2 \left(1 - e^{-(k\pi)^2 t}\right)^2 \\ &\leq \frac{1}{2}\left(\sum_{k=1}^{N} c_k^2 \left(1 - e^{-(k\pi)^2 t}\right)^2 + \sum_{k=N+1}^{\infty} c_k^2\right)\end{aligned}$$

for any integer $N \geq 1$. Let $\epsilon > 0$ be given. Since the sum $\sum c_k^2$ converges, by (10.8) we can choose N so large that

$$\sum_{k=N+1}^{\infty} c_k^2 < \epsilon/2.$$

Furthermore, when N is fixed it follows from (10.8) that the finite sum is bounded by

$$\sum_{k=1}^{N} c_k^2 \left(1 - e^{-(k\pi)^2 t}\right)^2 \leq 2\left(1 - e^{-(N\pi)^2 t}\right)^2 \|f\|^2 \longrightarrow 0$$

as t tends to zero. Therefore, by taking $t > 0$ sufficiently small we obtain

$$\|u(\cdot,t) - f\|^2 < \epsilon.$$

Since $\epsilon > 0$ was arbitrary, this implies that

$$\lim_{t \searrow 0} \|u(\cdot,t) - f\| = 0.$$

We summarize the discussion above in the following theorem:

Theorem 10.4 *Assume that the initial function f is piecewise continuous. Let the function $u(x,t)$ be defined by (10.11), where the coefficients c_k are the Fourier coefficients in the sine series of f. Then*

$$\lim_{t \searrow 0} ||u(\cdot,t) - f|| = 0.$$

The results obtained in this section, up to this point, establish that (10.11) defines a solution u of the initial-boundary value problem (10.1) under the weak assumption that f is just piecewise continuous. The differential equation holds as a consequence of Theorem 10.3, the boundary conditions are verified in (10.13) and the initial condition is satisfied in the sense described in Theorem 10.4.

10.2.4 Smooth and Compatible Initial Functions

Before we end our discussion on the justification of Fourier's method, we would like to show that if we assume slightly stronger conditions on the initial function f, then we can show that the convergence (10.15) is uniform. In fact, this will follow from the maximum principle derived for smooth solutions of the heat equation (see Theorem 6.2). Assume that $f \in C^0_{p,o}$ with f' piecewise continuous. We would like to show that

$$\lim_{t \searrow 0} ||u(\cdot, t) - f||_\infty = 0,$$

where $||f||_\infty = \sup_{x \in [0,1]} |f(x)|$. Under the present assumptions on f it follows from Theorem 9.3 that

$$\lim_{N \to \infty} ||S_N(f) - f||_\infty = 0, \qquad (10.16)$$

where the finite series $S_N(f)$ is given by

$$S_N(f) = \sum_{k=1}^{N} c_k \sin(k\pi x).$$

As above, let

$$u_N(x,t) = \sum_{k=1}^{N} c_k e^{-(k\pi)^2 t} \sin(k\pi x),$$

i.e. the solution of (10.1) with initial function $S_N(f)$. We recall from Theorem 10.2 above that $u_N(\cdot, t)$ converges uniformly to $u(\cdot, t)$ for any $t > 0$. Furthermore, since u_N is defined from a finite series, it is easy to see that

$$\lim_{t \to 0} ||u_N(\cdot, t) - S_N(f)||_\infty = 0, \qquad (10.17)$$

for every fixed N. This follows since

$$||u_N(\cdot, t) - S_N(f)||_\infty \leq \left(1 - e^{-(N\pi)^2 t}\right) \sum_{k=1}^{N} |c_k| \longrightarrow 0 \quad \text{as } t \to 0.$$

Next, we apply the maximum principle for the heat equation to the smooth solutions $u_N(x,t)$. Let $N, M \geq 1$ be integers. From Corollary 6.1 we obtain

$$||u_N(\cdot, t) - u_M(\cdot, t)||_\infty \leq ||S_N(f) - S_M(f)||_\infty.$$

By letting M tend to infinity, we can replace $u_M(\cdot, t)$ by $u(\cdot, t)$ and $S_M(f)$ by f in this inequality. This follows since $u_M(\cdot, t)$ and $S_M(f)$ converge uniformly to $u(\cdot,t)$ and f respectively, and since $||\cdot||_\infty$ is continuous with respect to uniform convergence (see Exercise 9.7). Hence, we have

$$\|u_N(\cdot,t) - u(\cdot,t)\|_\infty \le \|S_N(f) - f\|_\infty \qquad (10.18)$$

for $t \ge 0$. From (10.18) and the triangle inequality we now have for $t \ge 0$

$$\begin{aligned}\|u(\cdot,t) - f\|_\infty &\le \|u(\cdot,t) - u_N(\cdot,t)\|_\infty + \|u_N(\cdot,t) - S_N(f)\|_\infty + \|S_N(f) - f\|_\infty \\ &\le \|u_N(\cdot,t) - S_N(f)\|_\infty + 2\|S_N(f) - f\|_\infty.\end{aligned}$$

In order to see that we can get $\|u(\cdot,t)-f\|_\infty$ less than any $\epsilon > 0$ by choosing a small t, first choose N so large that

$$\|S_N(f) - f\|_\infty < \frac{\epsilon}{3}.$$

This is possible by (10.16). But when N is fixed it follows from (10.17) that

$$\|u_N(\cdot,t) - S_N(f)\| < \frac{\epsilon}{3}$$

for t sufficiently small. Hence,

$$\|u(\cdot,t) - f\|_\infty < \epsilon$$

or

$$\lim_{t\searrow 0} \|u(\cdot,t) - f\|_\infty = 0.$$

Hence, we have established that if $f \in C^0_{p,o}$, with f' piecewise continuous, then $u(\cdot,t)$, given by (10.11), converges uniformly to f as t tends to zero.

We summarize this discussion in the following theorem:

Theorem 10.5 *Assume that the initial function $f \in C^0_{p,o}$ and that f' is piecewise continuous. Let the function $u(x,t)$ be defined by (10.11), where the coefficients c_k are the Fourier coefficients in the sine series of f. Then*

$$\lim_{t\searrow 0} \|u(\cdot,t) - f\|_\infty = 0 \qquad (10.19)$$

Before we end this section we shall also note that the discussion above has implications for the maximum principle for the solution u defined by (10.11). The maximum principle for a solution u of (10.1) will imply that

$$\|u(\cdot,t)\|_\infty \le \|f\|_\infty \quad \text{for} \quad t \ge 0 \qquad (10.20)$$

(see Corollary 6.1).

However, the derivation of this inequality relies on the assumption that u is continuous in the domain $\{(x,t) : x \in [0,1], t \ge 0\}$, and, in general, the solution u will not have this property (see Example 10.1 above.) As we observed in Section 10.1, a necessary condition for continuity at the endpoints $x = 0$ and $x = 1$ for $t = 0$ is that the compatibility conditions

$$f(0) = f(1) = 0$$

are satisfied. On the other hand, we have seen above that if $f \in C^0_{p,o}$, with f' piecewise continuous, then (10.19) holds. Hence, u is continuous in $[0,1] \times [0,\infty)$ and the maximum principle (10.20) holds.

In fact, we have the following result:

Corollary 10.2 *Let $j \geq 0$ be an integer and assume that the initial function f in (10.1) is such that $f \in C^{2j}_{p,o}$ with $f^{(2j+1)}$ piecewise continuous. If u is given by (10.11), then the estimate*

$$\|\frac{\partial^j}{\partial t^j}u(\cdot,t)\|_\infty = \|\frac{\partial^{2j}}{\partial x^{2j}}u(\cdot,t)\|_\infty \leq \|f^{(2j)}\|_\infty \quad \text{for} \quad t \geq 0$$

holds.

Proof: For $j = 0$ the estimate corresponds to (10.20), and the result follows from the discussion above. For $j > 0$ we just note that $v = \frac{\partial^j u}{\partial t^j} = \frac{\partial^{2j} u}{\partial x^{2j}}$ is a solution of (10.1) with $v(\cdot,0) = f^{(2j)}$. ■

10.3 Convergence of Finite Difference Solutions

The purpose of the final section of this chapter is to establish a rigorous error bound for finite difference approximations of the initial-boundary value problem (10.1). Recall that we have derived such error bounds in earlier chapters for other problems. In Chapter 2 (see Theorem 2.2 on page 65) we considered the finite difference approximations of the one-dimensional Poisson's equation, while the corresponding problem in two space dimensions was discussed in Chapter 7 (see Theorem 7.2 on page 229).

If you look at the proofs of these theorems, you will discover that they are quite similar. The main strategy in both cases is to characterize the solution of the differential equation as a solution of the difference equation, but with an extra forcing term referred to as "the truncation error." If the truncation error has the property that it tends to zero as the grid becomes finer, we call the difference scheme "consistent" with the differential equation. The desired error bound is then derived from a proper "stability estimate" for the finite difference scheme. To summarize, we have[1]

$$\text{consistency} \quad + \quad \text{stability} \quad \Longrightarrow \quad \text{convergence.}$$

This approach to deriving error estimates is rather general. Here we shall apply this methodology to the heat equation, or more precisely, the system (10.1).

We shall study the implicit finite difference scheme introduced in Chapter 4.4. Assume we consider the initial-boundary value problem (10.1) with a nonhomogeneous forcing term in the differential equation, i.e. we consider

$$\begin{aligned} u_t &= u_{xx} + g(x,t) \quad \text{for} \quad x \in (0,1),\ t > 0, \\ u(0,t) &= u(1,t) = 0, \quad t > 0, \\ u(x,0) &= f(x), \quad x \in (0,1), \end{aligned} \tag{10.21}$$

where g is assumed to be a continuous function in x and t. The corresponding implicit finite difference approximation is given by

$$\begin{aligned} \frac{v_j^{m+1} - v_j^m}{\Delta t} &= \frac{v_{j-1}^{m+1} - 2v_j^{m+1} + v_{j+1}^{m+1}}{(\Delta x)^2} \\ &\quad + g(x_j, t_{m+1}) \quad j = 1, \ldots, n, \quad m \geq 0 \\ v_0^m &= v_{n+1}^m = 0, \quad m \geq 0, \\ v_j^0 &= f(x_j) \quad \text{for} \quad j = 1, 2, \ldots, n, \end{aligned} \tag{10.22}$$

where v_j^m approximates $u(j\Delta x, m\Delta t) = u(x_j, t_m)$ and where $\Delta x = \frac{1}{n+1}$. Our goal is to show that v_j^m tends to $u(x_j, t_m)$ as the grid parameters Δx and Δt tend to zero.

We introduce the notation

$$\|f\|_{\Delta,\infty} = \max_{1 \leq j \leq n} |f(x_j)|\,.$$

The following stability result for the difference scheme (10.22) is closely related to the maximum principle for this scheme discussed in Section 6.2.4.

Lemma 10.4 *A solution of the finite difference scheme (10.22) satisfies the estimate*

$$\|v^m\|_{\Delta,\infty} \leq \|f\|_{\Delta,\infty} + t_m \max_{1 \leq k \leq m} \|g(\cdot, t_k)\|_{\Delta,\infty}.$$

Proof: The difference scheme can be rewritten in the form

$$(1 + 2r)v_j^{m+1} = v_j^m + r(v_{j-1}^{m+1} + v_{j+1}^{m+1}) + \Delta t g_j^{m+1},$$

[1] In fact, if formulated properly this implication can also be reversed, i.e. convergence implies stability and consistency. This result is then known as the Lax equivalence theorem. We refer to [26] and references given there for a further discussion.

where $r = \frac{\Delta t}{(\Delta x)^2}$ and $g_j^m = g(x_j, t_m)$. Hence, if we let

$$V^m = \|v^m\|_{\Delta,\infty} \quad \text{and} \quad G^m = \|g^m\|_{\Delta,\infty},$$

we obtain

$$(1+2r)\left|v_j^{m+1}\right| \leq V^m + 2rV^{m+1} + \Delta t G^{m+1}.$$

However, by taking the maximum with respect to j on the left-hand side, this implies

$$V^{m+1} \leq V^m + \Delta t G^{m+1}.$$

Hence, by repeated use of this inequality,

$$\begin{aligned} V^m &\leq V^0 + \Delta t \sum_{k=1}^{m} G^k \\ &\leq V^0 + t_m \max_{1 \leq k \leq m} G^k. \end{aligned}$$

■

The lemma above contains the stability estimate we will use to derive convergence. Next, we will study the truncation error, or consistency.

Let u be a solution of (10.1), i.e. (10.21) with $g \equiv 0$. Define a grid function $\{u_j^m\}$ by letting $u_j^m = u(x_j, t_m) = u(j\Delta x, m\Delta t)$. Obviously, this function will satisfy the boundary conditions and initial conditions given in (10.22). Furthermore, we recall (see Chapter 2.2 on page 46) that Taylor's theorem implies that

$$\left|\frac{u(x_{j-1},t) - 2u(x_j,t) + u(x_{j+1},t)}{(\Delta x)^2} - u_{xx}(x_j,t)\right| \leq \frac{(\Delta x)^2}{12}\left\|\frac{\partial^4 u(\cdot,t)}{\partial x^4}\right\|_\infty. \tag{10.23}$$

A corresponding result for the time difference is (see formula (1.13) on page 7) that

$$\left|\frac{u(x,t_{m+1}) - u(x,t_m)}{\Delta t} - u_t(x,t_{m+1})\right| \leq \frac{\Delta t}{2} \sup_{t \in [t_m, t_{m+1}]} |u_{tt}(x,t)|. \tag{10.24}$$

Assume now that the initial function $f = u(\cdot, 0)$ is in $C^4_{p,o}$ with $f^{(5)}$ piecewise continuous. From Corollary 10.2 above it then follows that

$$\|u_{tt}(\cdot,t)\|_\infty, \left\|\frac{\partial^4 u(\cdot,t)}{\partial x^4}\right\|_\infty \leq \|f^{(4)}\|_\infty \tag{10.25}$$

for $t \geq 0$. As a consequence of the estimates (10.23)–(10.25) above, we therefore conclude that $\{u_j^m\}$ satisfies a difference equation of the form

$$\frac{u_j^{m+1} - u_j^m}{\Delta t} - \frac{u_{j-1}^{m+1} - 2u_j^{m+1} + u_{j+1}^{m+1}}{(\Delta x)^2} = \tau_\Delta(x_j, t_{m+1}), \tag{10.26}$$

where

$$|\tau_\Delta(x_j, t_m)| \leq \left(\frac{\Delta t}{2} + \frac{(\Delta x)^2}{12}\right) \|f^{(4)}\|_\infty. \tag{10.27}$$

We now have all the information needed to derive the desired error estimate.

Theorem 10.6 *Let $f \in C^4_{p,o}$, let $f^{(5)}$ be piecewise continuous, and let u be the solution of (10.21) with $g \equiv 0$. Furthermore, let v be the corresponding solution of the difference scheme (10.22). Then for any $m \geq 0$*

$$\|u(\cdot, t_m) - v^m\|_{\Delta,\infty} \leq t_m \left(\frac{\Delta t}{2} + \frac{(\Delta x)^2}{12}\right) \|f^{(4)}\|_\infty. \tag{10.28}$$

Proof: This result follows more or less directly from the stability estimate given in Lemma 10.4 and the estimate (10.27) of the truncation error. Let $e_j^m = u_j^m - v_j^m$. Then, by subtracting the first equation of (10.22), with $g \equiv 0$, from (10.26) we obtain that $\{e_j^m\}$ is a solution of the difference scheme

$$\begin{aligned}
\frac{e_j^{m+1} - e_j^m}{\Delta t} &= \frac{e_{j-1}^{m+1} - 2e_j^{m+1} + e_{j+1}^{m+1}}{(\Delta x)^2} + \tau_\Delta(x_j, t_{m+1}), \\
e_0^m &= e_{n+1}^m = 0, \\
e_j^0 &= 0.
\end{aligned}$$

Hence, it follows from Lemma 10.4 and (10.27) that

$$\begin{aligned}
\|e^m\|_{\Delta,\infty} &\leq t_m \max_{1 \leq k \leq m} \|\tau_\Delta(\cdot, t_k)\|_{\Delta,\infty} \\
&\leq t_m \left(\frac{\Delta t}{2} + \frac{(\Delta x)^2}{12}\right) \|f^{(4)}\|_\infty.
\end{aligned}$$

■

The theorem above implies that if t_m is fixed, for example $t_m = 1$, then the error $\|u(\cdot, t_m) - v^m\|_{\Delta,\infty}$ tends to zero as the grid parameters Δt and Δx tend to zero. This is exactly the desired convergence result.

However, there is a weakness in the estimate given above. If Δt and Δx are fixed, the right-hand side of the estimate (10.28) tends to infinity as t_m tends to infinity. This result cannot be sharp, since obviously both $\|u(\cdot, t)\|_\infty$ and $\|v^m\|_{\Delta,\infty}$ are bounded by the maximum principle. In fact, in proper norms both the continuous and discrete solutions tend to zero as t tends to infinity.

An alternative error estimate, which is an improvement on the estimate given above when t is large, will be derived in Exercise 10.14. The main difference in the argument is that we use a discrete mean square norm to measure the error.

10.4 Exercises

EXERCISE 10.1 Let f be a continuous function on an interval $[0,l]$, with $f(0)=f(l)=0$ and f' piecewise continuous. Show the Poincaré inequality

$$\int_0^l f^2(x)\,dx \le \frac{l^2}{\pi^2}\int_0^l [f'(x)]^2\,dx.$$

EXERCISE 10.2 Let f be a continuous 2-periodic function with f' piecewise continuous. Furthermore, assume that

$$\int_{-1}^1 f(x)\,dx = 0.$$

(a) Show that if $f' \equiv 0$, then $f \equiv 0$.

(b) Establish Poincaré's inequality

$$\int_{-1}^1 f^2(x)\,dx \le \frac{1}{\pi^2}\int_{-1}^1 [f'(x)]^2\,dx.$$

EXERCISE 10.3 Find a function $f \in C^1([0,1])$, with $f(0)=0$, such that

$$||f|| > \frac{1}{\pi}||f'||.$$

Explain why your result does not contradict Corollary 10.1. Compare your result with the inequality established in Lemma 8.6 on page 274.

EXERCISE 10.4 Consider the initial-boundary value problem (10.1) with $f(x)=\sin(\pi x)$. Show that in this case the inequality (10.7) becomes an equality.

EXERCISE 10.5 Consider the problem (10.1), but with the Dirichlet boundary conditions replaced by the corresponding Neumann conditions, i.e.

$$u_x(0,t) = u_x(1,t) = 0.$$

Does the decay estimate (10.7) hold in this case? Justify your answer.

EXERCISE 10.6 Consider the initial and boundary value problem (10.1). Assume that the initial function f is piecewise continuous. Use the representation (10.11) and Parseval's identity to establish the energy estimate (10.7).

EXERCISE 10.7 In this problem we study the heat equation with periodic boundary conditions (see Exercise 3.15 on page 111). Hence, we consider

$$\begin{aligned} u_t &= u_{xx} \quad \text{for} \quad x \in (-1,1), \quad t > 0, \\ u(-1,t) &= u(1,t), \quad u_x(-1,t) = u_x(1,t) = 0, \\ u(x,0) &= f(x), \end{aligned}$$

where the initial function f is assumed to be piecewise continuous. If the Fourier series of f is given by

$$\frac{a_0}{2} + \sum_{k=1}^{\infty} \big(a_k \cos(k\pi x) + b_k \sin(k\pi x)\big),$$

then the formal solution is given by

$$u(x,t) = \frac{a_0}{2} + \sum_{k=1}^{\infty} e^{-(k\pi)^2 t} \big(a_k \cos(k\pi x) + b_k \sin(k\pi x)\big). \tag{10.29}$$

(a) Show that (10.29) defines a function u with the property that $u(\cdot,t) \in C_p^\infty$ for any $t > 0$. Here $C_p^\infty = \bigcap_{m=1}^{\infty} C_p^m$, and the spaces C_p^m are defined in Section 9.5.

(b) Let $E(t)$ denote the corresponding energy given by

$$E(t) = \int_{-1}^{1} u^2(x,t)\,dx.$$

Explain why it is not true, in general, that

$$\lim_{t\to\infty} E(t) = 0.$$

(c) Assume that the initial function f is such that

$$\int_{-1}^{1} f(x)\,dx = 0.$$

Show that

$$E(t) \le e^{-2\pi^2 t} E(0).$$

EXERCISE 10.8 In this problem we shall study a procedure for defining C^∞-approximations of a piecewise continuous function f. Let f be a piecewise continuous function defined on $[0,1]$ with Fourier sine series

$$f(x) = \sum_{k=1}^{\infty} c_k \sin(k\pi x).$$

For each $t > 0$ define

$$f_t(x) = \sum_{k=1}^{\infty} c_k e^{-kt} \sin(k\pi x).$$

(a) Show that $f_t \in C_{p,o}^\infty$ for any $t > 0$.

(b) Show that $\lim_{t \searrow 0} ||f_t - f|| = 0$.

The two properties above show that the functions $f_t, t > 0$, are all C^∞-functions, but at the same time they can be arbitrarily close to the piecewise continuous function f in the mean square sense. Another set of functions which has this property is the functions $u(\cdot, t)$ given by (10.11), i.e. the solution of the heat equation. The two properties are in fact a consequence of the Theorems 10.2 and 10.4.

EXERCISE 10.9 In this exercise we shall study the formal solution of the wave equation with Dirichlet boundary conditions. In order to simplify the discussion, we only consider problems with $u_t(\cdot, 0) = 0$. Hence, we consider the initial and boundary value problem

$$\begin{aligned} u_{tt} &= u_{xx} \quad \text{for} \quad x \in (0,1), \quad t > 0, \\ u(0,t) &= u(1,t) = 0, \quad t > 0, \\ u(x,0) &= f(x), \quad u_t(x,0) = 0, \quad x \in (0,1). \end{aligned}$$

If the initial function f has a Fourier sine series given by

$$\sum_{k=1}^{\infty} a_k \sin(k\pi x),$$

then the formal solution is given by

$$u(x,t) = \sum_{k=1}^{\infty} a_k \cos(k\pi t) \sin(k\pi x); \tag{10.30}$$

see formula (5.15) on page 162. Throughout this problem we assume that $f \in C_{p,o}^2$ with f''' piecewise continuous.

(a) Show that the function $u(x,t)$ defined by (10.30) has the property that $u(\cdot, t) \in C_{p,o}^2$ for any $t \in \mathbb{R}$.

(b) Show that

$$\lim_{t \to 0} ||u(\cdot, t) - f|| = 0.$$

(c) Show that $u(x, \cdot) \in C^2(\mathbb{R})$ for any $x \in [0, 1]$.

(d) Show that $u_{tt} = u_{xx}$ for $x \in [0, 1]$, $t \in \mathbb{R}$.

(e) Show that

$$\lim_{t \to 0} ||u_t(\cdot, t)|| = 0.$$

EXERCISE 10.10 This exercise is a continuation of Exercise 10.9 above. Here we again study the wave equation, but with nonzero data for $u_t(\cdot, 0)$. We consider the initial and boundary value problem

$$\begin{aligned} u_{tt} &= u_{xx} \quad \text{for} \quad x \in (0,1), \quad t > 0, \\ u(0,t) &= u(1,t) = 0, \quad t > 0, \\ u(x,0) &= 0, \quad u_t(x,0) = g(x), \quad x \in (0,1). \end{aligned}$$

Throughout the problem we assume that $g \in C^1_{p,o}$ with g'' piecewise continuous. If

$$g(x) = \sum_{k=1}^{\infty} b_k \sin(k\pi x)$$

is the Fourier sine series of g, then the formal solution $u(x, t)$ is given by

$$u(x,t) = \sum_{k=1}^{\infty} \frac{b_k}{k\pi} \sin(k\pi t) \sin(k\pi x); \tag{10.31}$$

see formula (5.15) on page 162.

(a) Show that $u(x, t)$, defined by (10.31), has the property that $u(\cdot, t) \in C^2_{p,o}$ for any $t \in \mathbb{R}$.

(b) Show that

$$\lim_{t \to 0} ||u(, \cdot, t)|| = 0.$$

(c) Show that $u(x, \cdot) \in C^2(\mathbb{R})$ for any $x \in [0, 1]$.

(d) Show that $u_{tt} = u_{xx}$ for $x \in [0, 1]$, $t \in \mathbb{R}$.

(e) Show that

$$\lim_{t\to 0} ||u_t(\cdot,t) - g|| = 0.$$

EXERCISE 10.11 Consider the problem (10.1) with $f \equiv 1$, i.e. the problem studied in Example 10.1. The formal solution $u(x,t)$ is given by (10.4).

(a) Explain why $u(\cdot,t) \in C^{\infty}_{p,o}$ for any $t > 0$, and explain why

$$\lim_{t\searrow 0} ||u(\cdot,t) - f|| = 0.$$

(b) Explain why

$$||u(\cdot,t) - f||_{\infty} \geq 1 \quad \text{for} \quad t > 0.$$

(c) Show that

$$\lim_{t\searrow 0} u(x,t) = 1 \quad \text{for any } x \in (0,1).$$

You should compare the results of the exercise with the plots of the solution $u(\cdot,t)$, for $t = 0, 0.01$ and 0.1, presented in Fig. 3.4 on page 95.

EXERCISE 10.12 Assume that the initial-boundary value problem (10.1) is approximated by the corresponding explicit finite difference scheme, i.e the scheme (4.2) on page 120. Show that if the stability condition $r = \Delta t/(\Delta x)^2 \leq 1/2$ is satisfied, then the error estimate (10.28) holds.

EXERCISE 10.13 Let $\{v_j^m\}$ denote the finite difference solution for the non-homogeneous problem (10.21) obtained by replacing the scheme (10.22) by the Crank-Nicholson scheme studied in Exercise 4.16 on page 153.

(a) Show that the estimate given in Lemma 10.4 holds for this difference solution.

(b) Assume that the initial function f in (10.1) is in $C^8_{p,o}$ with $f^{(9)}$ piecewise continuous and that this problem is approximated by the Crank-Nicholson scheme. Establish the error estimate

$$||u(\cdot,t) - v^m||_{\Delta,\infty} \leq t_m\left(\frac{(\Delta x)^2}{12}||f^{(4)}||_{\infty} + \frac{(\Delta t)^2}{12}||f^{(8)}||_{\infty}\right).$$

We note that, compared to the estimate (10.28), this error estimate is of second order with respect to both Δx and Δt. However, stronger assumptions on the initial function f are required.

EXERCISE 10.14 For a grid function v let $\|\cdot\|_\Delta$ denote the discrete version of the mean square norm, i.e.

$$\|v\|_\Delta = \left(\Delta x \sum_{j=1}^{n} v_j^2\right)^{1/2},$$

where, as above, $\Delta x = \frac{1}{n+1}$.

(a) Show that $\|v\|_\Delta \leq \|v\|_{\Delta,\infty}$ for any grid function v.

(b) Let $v = \{v_j^m\}$ be a solution of the finite difference scheme (10.22). Show that v satisfies the stability estimate

$$\|v^m\|_\Delta \leq \left(\frac{1}{1+\mu_1\Delta t}\right)^m \|f\|_\Delta + \mu_1^{-1} \max_{1\leq k\leq m} \|g(\cdot,t_k)\|_\Delta,$$

where $\mu_1 = \mu_1(h) = \frac{4}{h^2}\sin^2\left(\frac{\pi h}{2}\right)$.

We note that this result generalizes the result of Exercise 4.26c for an equation where $g \neq 0$. Furthermore, compared to the stability estimate given in Lemma 10.4 we observe that t_m does not appear in front of the term $\max_{1\leq k\leq m} \|g(\cdot,t_k)\|_\Delta$. However, we have replaced $\|\cdot\|_{\Delta,\infty}$ by $\|\cdot\|_\Delta$.

(c) Let u, v, and f be as in Theorem 10.6. Use the results above and the fact that $\mu_1 \geq 4$ (see Exercise 2.27) to establish the error estimate

$$\|u(\cdot,t_m) - v^m\|_\Delta \leq \left(\frac{\Delta t}{8} + \frac{(\Delta x)^2}{48}\right)\|f^{(4)}\|_\infty$$

for any $m \geq 0$.

Note that the right-hand side of this estimate is independent of m. This result therefore represents an improvement of the result given in Theorem 10.6 when t_m is large.

11
Reaction-Diffusion Equations

Reaction-diffusion equations arise as mathematical models in a series of important applications, e.g. in models of superconducting liquids, flame propagation, chemical kinetics, biochemical reactions, predator-prey systems in ecology and so on. Both numerical and mathematical analysis of reaction-diffusion equations are currently very active fields of research. Obviously, we cannot study the subject at an advanced level in the present text, but we can get a general feeling of what these problems are about. Our aim is merely to present some simple models and to explore some of their properties using finite difference schemes and energy estimates. Further examples can be found in the exercises.[1]

11.1 The Logistic Model of Population Growth

We start our discussion of reaction-diffusion equations by considering a simple model arising in mathematical ecology. In order to understand the foundation of this model, we first recapture the *logistic* model of population growth. This model states that the growth of a population facing limited

[1] If you want to read more about reaction-diffusion equations, the book by Smoller [23] is an excellent source. This book is a standard reference in this field. Another excellent, yet less demanding, guide to these problems can be found in the book by Logan [19]. For those interested in models arising in biology, Murray [20] presents a huge collection of interesting models.

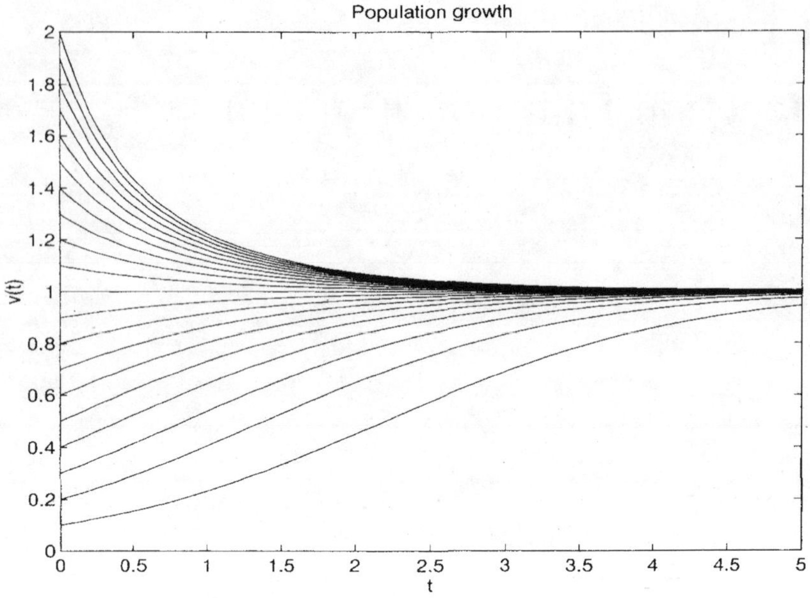

FIGURE 11.1. *The solution of the logistic model of population growth for various initial values* f_0. *We have used* $A = \alpha = 1$ *in these computations.*

resources is governed by the following ordinary differential equation:

$$v'(t) = \alpha v(t)(A - v(t)), \qquad v(0) = f_0. \tag{11.1}$$

Here $v = v(t)$ is the population density, $\alpha > 0$ is the growth rate, and $A > 0$ is the so-called carrying capacity of the environment. The model states that for small populations, we get exponential growth governed by $v'(t) \approx \alpha A v(t)$. But as v increases, the term $-\alpha v^2$ becomes significant, the growth slows down, and the population gradually reaches the carrying capacity of the environment. The problem (11.1) can be solved analytically,[2]

$$v(t) = \frac{A f_0}{f_0 + (A - f_0)e^{-\alpha A t}}, \qquad t \geq 0, \tag{11.2}$$

and we note that $v = A$ is the asymptotic solution as $t \to \infty$ for any initial data $f_0 > 0$. We have plotted this solution[3] for some values of f_0 in Fig. 11.1.

---

[2]The solution formula (11.2) is derived in most courses in ordinary differential equations. If you are not familiar with this formula, you should take a look at Exercise 11.1.

[3]You can read about applications of this model in the book by Braun [5]; see also Murray [20].

11.1.1 A Numerical Method for the Logistic Model

Below we will study the properties of the logistic model when spatial variations in the population are taken into account. This will result in a diffusion term added to the right-hand side of (11.1). In order to prepare ourselves for the study of this model, we shall derive some results for the discrete version of the purely logistic model.

We consider the case of $\alpha = A = 1$, i.e.

$$v'(t) = v(t)(1 - v(t)), \qquad v(0) = f_0, \tag{11.3}$$

and the associated explicit scheme

$$v_{m+1} = v_m + \Delta t v_m(1 - v_m), \qquad v_0 = f_0. \tag{11.4}$$

Here v_m denotes an approximation of v at time $t = t_m = m\Delta t$.

Since the solution of (11.3) is given by

$$v(t) = \frac{f_0}{f_0 + (1 - f_0)e^{-t}}, \qquad t \geq 0,$$

the asymptotic solution is $v = 1$ for any $f_0 > 0$. It is also easy to see that if $0 \leq f_0 \leq 1$, then $0 \leq v(t) \leq 1$ for all $t \geq 0$. Moreover, $v(t)$ is nondecreasing for all $t \geq 0$.

Now, we want to prove similar properties for the discrete solutions generated by (11.4), and we start by considering the invariance property, i.e. that data in $[0, 1]$ imply solutions in $[0, 1]$. We assume that

$$\Delta t < 1, \tag{11.5}$$

and define the polynomial

$$G(v) = v + \Delta t v(1 - v). \tag{11.6}$$

Then

$$G'(v) = 1 + \Delta t(1 - 2v) \geq 1 - \Delta t > 0 \tag{11.7}$$

for all $v \in [0, 1]$.

Consequently, by assuming that $0 \leq v_m \leq 1$ for a given time t_m, we get

$$v_{m+1} = G(v_m) \leq G(1) = 1$$

and

$$v_{m+1} = G(v_m) \geq G(0) = 0.$$

Hence it follows by induction that if f_0 is in the unit interval, then v_m is in the unit interval for all $t_m \geq 0$. It is also easy to see that $f_0 = 0$ implies that $v_m = 0$ for all $t_m \geq 0$, and that $\{v_m\}$ is nondecreasing for initial data in the unit interval.

Next we want to show that, also in the discrete case, $v = 1$ is the asymptotic solution for any $0 < f_0 \leq 1$. Since v_m is in the unit interval for all m, we have

$$v_{m+1} = v_m + \Delta t v_m(1 - v_m) \geq v_m,$$

and thus

$$0 < f_0 = v_0 \leq v_1 \leq v_2 \leq \cdots \leq 1.$$

By using this property, we get

$$\begin{aligned} 1 - v_{m+1} &= 1 - v_m - \Delta t v_m(1 - v_m) \\ &= (1 - v_m)(1 - \Delta t v_m) \\ &\leq (1 - v_m)(1 - \Delta t f_0), \end{aligned}$$

and consequently

$$1 - v_m \leq (1 - v_0)(1 - \Delta t f_0)^m$$

by induction. This implies that

$$1 - (1 - f_0)(1 - \Delta t f_0)^m \leq v_m \leq 1, \quad \text{for} \quad m \geq 1,$$

and then, since $f_0 > 0$, we conclude that v_m converges towards 1 as m goes to infinity. We can summarize our observations concerning v_m as follows:

Lemma 11.1 *Let v_m be the approximate solution of (11.3) generated by the scheme (11.4), and assume that $\Delta t < 1$. Then $\{v_m\}$ has the following properties:*

(a) If $0 \leq f_0 \leq 1$, then $0 \leq v_m \leq v_{m+1} \leq 1$ for all $m \geq 1$.

(b) If $f_0 = 0$, then $v_m = 0$, and if $f_0 = 1$, then $v_m = 1$ for all $m \geq 0$.

(c) If $0 < f_0 \leq 1$, then $v_m \longrightarrow 1$ as $m \longrightarrow \infty$.

These results will be valuable in the discussion of the reaction-diffusion model below.

11.2 Fisher's Equation

In deriving the logistic model (11.1), it is assumed that spatial variation in the density of the population is of little importance for the growth of the population. Thus, one simply assumes that the population is evenly distributed over some area for all time. For real populations, this assumption is often quite dubious. In the next level of sophistication, it is common to take into account the tendency of a population to spread out over the area where it is possible to live. This effect is incorporated by adding a *Fickian diffusion* term to the model. Then we get the following partial differential equation:

$$u_t = du_{xx} + \alpha u(A - u). \tag{11.8}$$

Here d is a diffusion coefficient and $u = u(x,t)$ is the population density. In mathematical ecology, this model of population growth is called *Fisher's equation.* Obviously, the introduction of a diffusion term leads to a partial differential equation which in contrast to the ordinary differential equation (11.1) cannot in general be solved analytically.

We mentioned that the term du_{xx} models the diffusion of the population. Similar terms arise in a lot of applications where we want to capture the tendency of nature to smooth things out. For instance, if you drop a tiny amount of ink into a glass of water, you can watch how the ink spreads throughout the water by means of molecular diffusion. This situation is modeled by the diffusion equation where Fick's law is used to state that there is a flux of ink from areas of high concentration to areas of low concentration. Similarly, if you consider a long uniform rod and start heating it at some fixed location, Fourier's law of heat conduction states that there is a flux of heat from hot areas to cold areas. Similarly again, a Fickian diffusion term in a model of population density states that there is a migration from areas of high population density to areas of low population density.[4]

Usually Fisher's equation (11.8) is studied in conjunction with a Neumann-type boundary condition, i.e.

$$u_x(0,t) = u_x(L,t) = 0, \tag{11.9}$$

where L denotes the length of the domain. The reason for this boundary condition is that we assume the area to be closed, so there is no migration from the domain. We may consider a valley surrounded by mountains, or we can simply think of an island.

Since we are interested in the qualitative behavior of this model rather than the actual quantities, we simplify the situation by putting $d = \alpha = A = L = 1$, and study the following problem:

$$\begin{aligned} u_t &= u_{xx} + u(1-u) \quad \text{for} \quad x \in (0,1), \quad t \in (0,T], \\ u_x(0,t) &= u_x(1,t) = 0, \qquad t \in [0,T], \\ u(x,0) &= f(x), \qquad x \in [0,1], \end{aligned} \tag{11.10}$$

[4] Human beings do not always obey this sound principle.

where $f = f(x)$ denotes the initial distribution of the population. Since $A = 1$, we assume that the initial data satisfy[5]

$$0 \leq f(x) \leq 1 \tag{11.11}$$

for all $x \in [0, 1]$.

11.3 A Finite Difference Scheme for Fisher's Equation

We want to study Fisher's equation using a finite difference scheme. Let u_j^m denote an approximation of $u(x_j, t_m)$; then an explicit finite difference scheme can be written as follows:

$$u_j^{m+1} = ru_{j-1}^m + (1 - 2r)u_j^m + ru_{j+1}^m + \Delta t u_j^m(1 - u_j^m), \; j = 1, \ldots, n, \tag{11.12}$$

where $r = \Delta t/\Delta x^2$ and $m \geq 0$. We initialize the scheme by

$$u_j^0 = f(x_j), \qquad j = 0, \ldots, n+1. \tag{11.13}$$

The boundary conditions of (11.10) at $x = 0$ and $x = 1$ are incorporated by introducing the auxiliary points $x_{-1} = -\Delta x$ and $x_{n+2} = 1 + \Delta x$. Since $u_x(0, t) = u_x(1, t) = 0$, we use the following discrete boundary conditions:

$$\frac{u_1^m - u_{-1}^m}{2\Delta x} = 0 \quad \text{and} \quad \frac{u_{n+2}^m - u_n^m}{2\Delta x} = 0. \tag{11.14}$$

Combining (11.12) and (11.14), we get

$$u_0^{m+1} = (1 - 2r)u_0^m + 2ru_1^m + \Delta t u_0^m(1 - u_0^m), \quad m \geq 0 \tag{11.15}$$

at the left boundary and

$$u_{n+1}^{m+1} = 2ru_n^m + (1 - 2r)u_{n+1}^m + \Delta t u_{n+1}^m(1 - u_{n+1}^m), \quad m \geq 0 \tag{11.16}$$

at the right boundary. The finite difference scheme is now fully specified by the initial condition (11.13), the scheme (11.12), and the boundary conditions (11.15) and (11.16). For ease of reference we summarize the scheme as follows:

[5]Note that it is perfectly reasonable to study this problem with initial population densities exceeding the carrying capacity. Negative initial conditions are, however, beyond any reasonable interpretation.

$$u_j^0 = f(x_j), \qquad j = 0, \dots, n+1 \quad \text{for} \quad m \geq 0$$

$$\begin{aligned} u_0^{m+1} &= (1-2r)u_0^m + 2ru_1^m + \Delta t u_0^m(1-u_0^m), \\ u_j^{m+1} &= ru_{j-1}^m + (1-2r)u_j^m + ru_{j+1}^m + \Delta t u_j^m(1-u_j^m), \quad 1 \leq j \leq n, \\ u_{n+1}^{m+1} &= 2ru_n^m + (1-2r)u_{n+1}^m + \Delta t u_{n+1}^m(1-u_{n+1}^m), \end{aligned} \tag{11.17}$$

In Fig. 11.2 we have plotted an approximate solution of the problem (11.10) using the scheme above with the data

$$\Delta t = 0.001, \quad \Delta x = 0.05, \quad \text{and} \quad f(x) = \cos^2(\pi x).$$

The numerical solution is plotted as a function of x for different values of t, and we observe that the approximate solution seems to remain within the unit interval. Furthermore, the approximate solution seems to converge towards the value $u = 1$ for all x as t increases. The fact that u_j^m remains in the unit interval indicates a kind of a maximum principle. Let us look at one more example of the same flavor in order to investigate this issue a bit further. In Fig. 11.3 we have solved the problem again using the same grid parameters, but we have changed the initial condition to read

$$f(x) = \frac{1}{10}\cos^2(\pi x).$$

Again we note that the numerical solution remains within the unit interval and that it seems to converge towards $u = 1$.

11.4 An Invariant Region

Both the numerical experiments discussed above and also the origin of the model suggest that the solution always will stay within the unit interval; thus the unit interval is referred to as an *invariant region* for this model. We will prove this property provided that the mesh parameters satisfy the requirement

$$\Delta t < \frac{(\Delta x)^2}{2 + (\Delta x)^2}, \tag{11.18}$$

which is slightly more restrictive than the corresponding condition for the heat equation; $r \leq 1/2$ or $\Delta t \leq (\Delta x)^2/2$.

We start by considering a fixed time level t_m and assume that

$$0 \leq u_j^m \leq 1 \quad \text{for} \quad j = 0, \dots, n+1. \tag{11.19}$$

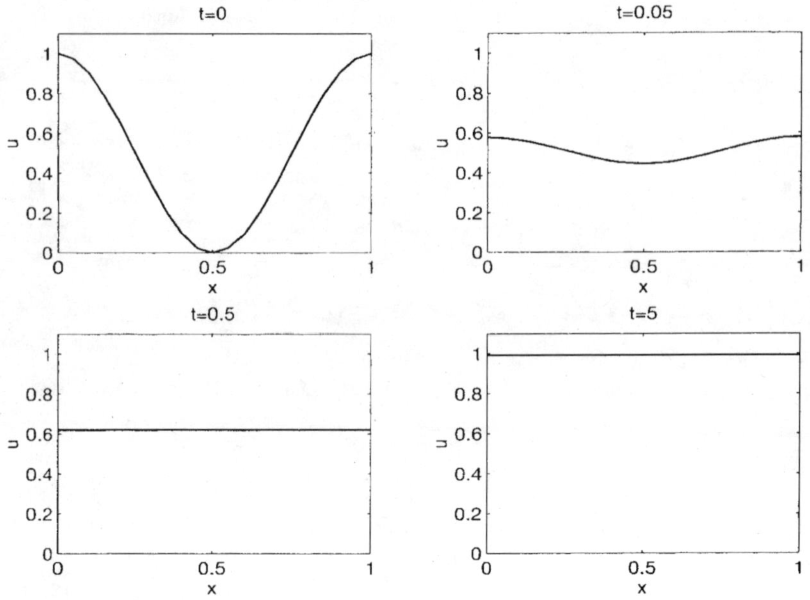

FIGURE 11.2. *An approximate solution of the nonlinear population model using the initial distribution* $f(x) = \cos^2(\pi x)$. *The numerical solution is plotted as a function of* x *for* $t = 0, 0.05, 0.5, 5$.

Furthermore, we define the following auxiliary functions:

$$K(u) = 2r + (1-2r)u + \Delta t u(1-u)$$

and

$$H(u) = (1-2r)u + \Delta t u(1-u).$$

Now it follows, using the scheme defined by (11.17) and the assumption (11.19), that

$$u_j^{m+1} \leq 2r + (1-2r)u_j^m + \Delta t u_j^m(1-u_j^m) = K(u_j^m) \tag{11.20}$$

and that

$$u_j^{m+1} \geq (1-2r)u_j^m + \Delta t u_j^m(1-u_j^m) = H(u_j^m). \tag{11.21}$$

Observe that the stability condition (11.18) implies that

$$1 - 2r - \Delta t > 0,$$

hence

$$K'(u) = H'(u) = (1-2r) + (1-2u)\Delta t \geq 1 - 2r - \Delta t > 0$$

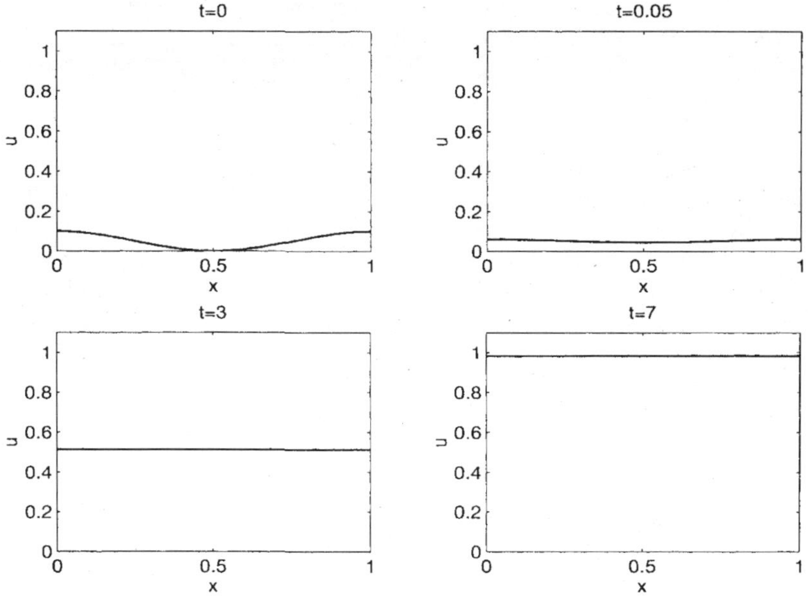

FIGURE 11.3. *An approximate solution of the nonlinear population model using the initial distribution* $f(x) = \frac{1}{10}\cos^2(\pi x)$. *The numerical solution is plotted as a function of* x *for* $t = 0, 0.05, 3, 7$.

for all $u \in [0,1]$. Since K and H are strictly increasing functions, it follows from (11.20) and (11.21) that

$$u_j^{m+1} \leq K(u_j^m) \leq K(1) = 1,$$

and that

$$u_j^{m+1} \geq H(u_j^m) \geq H(0) = 0.$$

Thus we have

$$0 \leq u_j^{m+1} \leq 1$$

for $j = 0, \ldots, n+1$. By induction on the time level, we have the following result.

Theorem 11.1 *Suppose* u_j^m *is generated by the scheme (11.17) and that the mesh parameters satisfy the condition (11.18). Furthermore, we assume that the initial data satisfy*

$$0 \leq f(x) \leq 1 \quad \text{for} \quad x \in [0,1].$$

Then

$$0 \leq u_j^m \leq 1$$

for $j = 0, \ldots, n+1$ *and* $m \geq 0$.

As mentioned above, the unit interval is referred to as an invariant region for the scheme.[6] You should note that a maximum principle and an invariant region are not exactly the same. For the heat equation, which is known to satisfy a maximum principle, the values that the solution can attain are bounded by the data given initially or at the boundaries. Thus, by giving small data, say less than a given $\epsilon \ll 1$ in magnitude, we know that the absolute value of the solution itself is bounded by ϵ. Conversely, for the nonlinear model (11.10), we noticed in the computations presented above that an initial condition bounded by 1/10 gives a numerical solution that converges towards $u = 1$ as time increases. Generally, maximum principles imply the existence of an invariant region, but an invariant region does not necessarily imply a maximum principle.

11.5 The Asymptotic Solution

In the numerical experiments discussed above, we observed that the approximate solutions always stayed within the unit interval and that they approached the state $u = 1$ as time increased. The first observation is fully explained in the light of Theorem 11.1, where it is proved that the unit interval is an invariant region for the discrete solutions. But what about the asymptotics? Is it correct that as t increases, the limiting solution is always $u = 1$? Before we start analyzing this issue, let us challenge this hypothesis.

EXAMPLE 11.1 Motivated by a similar problem above, we choose the initial function

$$f(x_j) = \text{rand}(x_j), \qquad j = 0, \ldots, n+1,$$

where "rand" is a random number in the unit interval. We have used the "rand" function in Matlab. This function generates uniformly distributed random numbers in the unit interval. In Fig. 11.4 we have plotted the numerical solution as a function of x at time $t = 0, 0.05, 1.5, 5$. In the experiment we have used $\Delta t = 0.001$ and $\Delta x = 0.05$, which satisfy the stability condition (11.18). Note that the initial condition is evaluated simply by calling the rand function for each grid point x_j and assigning the result to $f(x_j)$. From the figure, we observe that even for this wild initial distribution the population is smoothed out and converges towards $u = 1$ for all $x \in [0, 1]$ as time increases.

■

[6] The notion of invariant regions plays a fundamental role in the mathematical theory of reaction-diffusion equations. This is carefully discussed in Chapter 14 of Smoller's book [23].

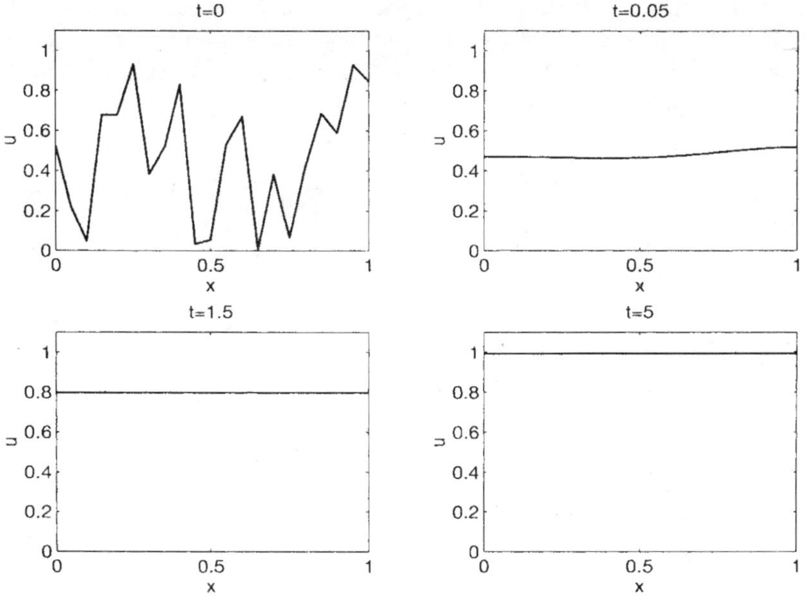

FIGURE 11.4. *The numerical solution of Fisher's equation using random numbers as the initial condition. The numerical solution is plotted as a function of* x *for* $t = 0, 0.05, 1.5, 5$.

Motivated by the numerical experiments, we want to prove that $u = 1$ is the asymptotic solution of the finite difference scheme. We start analyzing this issue by considering some particular cases. First we observe that if the initial density is identically equal to zero, i.e, $f(x) = 0$ for all $x \in [0, 1]$, then it follows from the scheme that $u_j^m = 0$ for all j and m. It also follows from (11.10) that $u = 0$ is a solution of the continuous problem. Thus, in order to prove that $u = 1$ is the asymptotic solution, we have to assume that the initial data is nonzero.

Next we consider the case of nonzero but constant initial data, i.e. $f(x) = f_0 = constant$. Then it follows by induction that $u_j^m = v_m$, where v_m is computed by (11.4). Consequently, the properties of the discrete solution are given by Lemma 11.1.

Finally, we turn our attention to the problem of asymptotic behavior in the case of nonconstant initial data. Our aim is now to prove that the approximate solution of the partial differential equation (11.10) converges towards $u = 1$ as time increases. In order to avoid technical difficulties of limited interest, we assume that the initial density distribution f satisfies the following requirement:

$$0 < f(x) \leq 1 \qquad (11.22)$$

for all $x \in [0,1]$. In Exercise 11.2 we will study what happens if we allow the initial density to be zero or greater than one in parts of the domain.

Let us first recall that by Theorem 11.1, the assumption (11.22) on the initial data implies that

$$0 \leq u_j^m \leq 1$$

for $j = 0, \ldots, n+1$ and $m \geq 0$. In order to analyze the scheme (11.17), we define

$$\bar{u}_m = \min_{j=0,\ldots,n+1} u_j^m \tag{11.23}$$

and observe that, obviously, $0 \leq \bar{u}_m \leq 1$ for all $m \geq 0$. By the assumption (11.22), it also follows that

$$\bar{u}_0 > 0,$$

and by the scheme (11.17), we have

$$u_j^{m+1} \geq 2r\bar{u}_m + (1-2r)u_j^m + \Delta t u_j^m(1-u_j^m)$$

for $j = 0, \ldots, n+1$. By assuming that the mesh parameters Δx and Δt satisfy the stability condition (11.18), it follows that the polynomial

$$P_m(u) = 2r\bar{u}_m + (1-2r)u + \Delta t u(1-u)$$

satisfies

$$P_m'(u) = 1 - 2r + \Delta t(1-2u) \geq 1 - 2r - \Delta t > 0$$

for all $u \in [0,1]$, and then

$$u_j^{m+1} \geq P_m(u_j^m) \geq P_m(\bar{u}_m) = \bar{u}_m + \Delta t \bar{u}_m(1-\bar{u}_m).$$

Since this holds for all $j = 0, \ldots, n+1$, it follows that

$$\bar{u}_{m+1} \geq \bar{u}_m + \Delta t \bar{u}_m(1-\bar{u}_m) \tag{11.24}$$

for $m \geq 0$.

Now we can prove that $\bar{u}_m$ tends to 1 as m tends to infinity by comparing $\bar{u}_m$ with v_m generated by the discrete logistic model, i.e.

$$v_{m+1} = v_m + \Delta t v_m(1-v_m), \qquad v_0 = \bar{u}_0 > 0. \tag{11.25}$$

Assuming that $\bar{u}_m \geq v_m$, we get

$$\bar{u}_{m+1} - v_{m+1} \geq G(\bar{u}_m) - G(v_m) = G'(\tilde{v}_m)(\bar{u}_m - v_m) \geq 0,$$

and thus it follows by induction that

$$\bar{u}_m \geq v_m$$

for any $m \geq 0$. By using part (c) of Lemma 11.1, we conclude that $\bar{u}_m$ tends to 1 as m tends to infinity. We have derived the following result:

Theorem 11.2 *Suppose u_j^m is generated by (11.17) and that the mesh parameters satisfy the condition (11.18). Furthermore, we assume that the initial condition satisfies*

$$0 < f(x) \leq 1 \qquad \text{for} \quad x \in [0,1].$$

Then, for all $j = 0, 1, \ldots, n+1$,

$$u_j^m \longrightarrow 1$$

as $m \longrightarrow \infty$.

The case of $0 < f(x) \leq 1$ is covered by this theorem. Generalizations are studied computationally in Exercise 11.2.

11.6 Energy Arguments

Above we have studied some properties of discrete approximations of Fisher's equation,

$$u_t = u_{xx} + u(1-u) \tag{11.26}$$

with boundary data

$$u_x(0,t) = u_x(1,t) = 0 \tag{11.27}$$

and initial condition

$$u(x,0) = f(x). \tag{11.28}$$

In this section we will derive some results for the continuous model. Throughout this section we will assume that a smooth solution exists[7] and derive properties of such a solution.

Above, we studied discrete approximations of this problem, and some interesting properties were recorded. First we noted that the discrete solutions are bounded in an invariant region. This property was analyzed for the discrete solutions in Section 11.4. Secondly, we noted that $u = 1$ is the asymptotic solution as proved in Section 11.5. A final observation is that the derivatives seem to decay rapidly as time increases. This effect is particulary apparant in the computations graphed in Figure 11.4 on page 347.

In this section, all these three observations will be discussed for the continuous model.

[7] For existence arguments we refer the interested reader to Chapter 14 of Smoller [23]. These arguments are beyond the scope of the present text.

11.6.1 An Invariant Region

As mentioned above, we assume that the problem (11.26)–(11.28) has a solution. More precisely, we assume that there is a unique smooth function[8] u satisfying the requirements (11.26)–(11.28).

We will show that the interval

$$[\varepsilon, 1 + \varepsilon], \qquad 0 < \varepsilon < 1, \tag{11.29}$$

is an invariant region for u. To this end, we assume that

$$0 < \varepsilon \leq f(x) \leq 1 + \varepsilon \tag{11.30}$$

for all $x \in [0, 1]$.

In order to prove that u will remain in the interval $[\varepsilon, 1 + \varepsilon]$, we assume the opposite; specifically, we assume that u exceeds the value $1 + \varepsilon$. Then, by the regularity of u, there must exist a time t_0 such that

$$u(x, t) \leq 1 + \varepsilon$$

for all $x \in [0, 1]$ and $t < t_0$. Furthermore, at $t = t_0$ there must be a location $x = x_0$ such that

(i) $u_t(x_0, t_0) \geq 0$,

(ii) $u_{xx}(x_0, t_0) \leq 0$,

(iii) $u(x_0, t_0) = 1 + \varepsilon$;

(see Fig. 11.5). Using (11.26), (ii) and (iii) we get

$$\begin{aligned} u_t(x_0, t_0) &= u_{xx}(x_0, t_0) + u(x_0, t_0)(1 - u(x_0, t_0)) \\ &\leq (1 + \varepsilon)(1 - (1 + \varepsilon)) \\ &= -\varepsilon(1 + \varepsilon) < 0, \end{aligned}$$

which contradicts (i). Hence, there is no such point (x_0, t_0), and consequently u remains in $[\varepsilon, 1 + \varepsilon]$.

By a similar argument, it follows that u cannot become smaller than ε. We have derived the following result:

Theorem 11.3 *Suppose that u, satifying $u, u_x, u_{xx}, u_t \in C([0, 1] \times [0, \infty))$, solves (11.26)–(11.28). Then, if the initial condition f satisfies (11.30), we have*

$$0 < \varepsilon \leq u(x, t) \leq 1 + \varepsilon$$

for any $x \in [0, 1]$, $t \geq 0$.

---

[8] We assume that $u, u_x, u_{xx}, u_t \in C([0, 1] \times [0, \infty))$

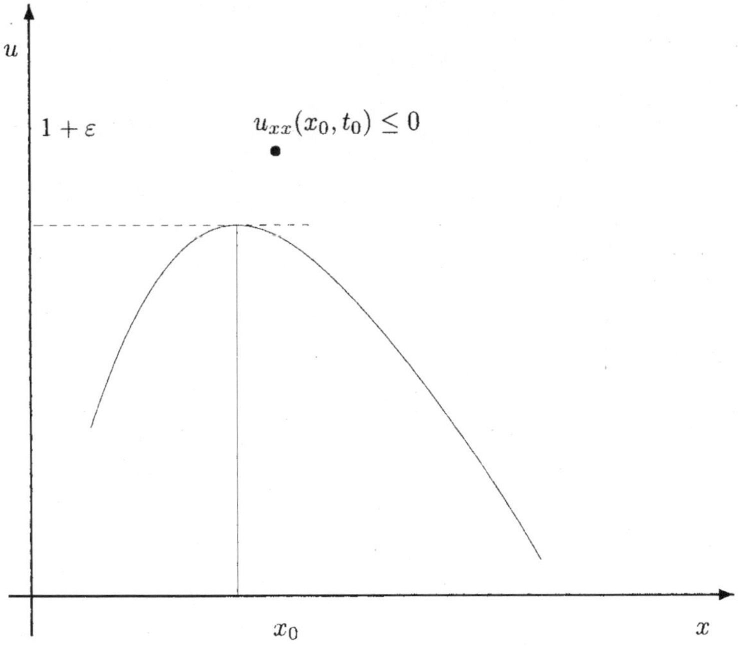

FIGURE 11.5. *The solution u close to a local maximum.*

11.6.2 Convergence Towards Equilibrium

We showed above that the discrete solutions generated by the scheme (11.17) converge towards $u_j^n = 1$ as $t_n \to \infty$. Now we want to show a similar result for the continuous model using an energy estimate.

Let u be the solution of (11.26)–(11.28) for initial data f satisfying (11.30), and define

$$E(t) = \int_0^1 (u(x,t) - 1)^2 dx \tag{11.31}$$

for $t \geq 0$. By using the equation (11.26) and the boundary conditions (11.27), we obtain

$$\begin{aligned} E'(t) &= 2\int_0^1 (u-1)u_t dx \\ &= 2\int_0^1 (u-1)u_{xx} - u(1-u)^2 dx \\ &= -2\int_0^1 (u_x)^2 dx - 2\int_0^1 u(1-u)^2 dx. \end{aligned}$$

Now it follows from Theorem 11.3 that

$$u(x,t) \geq \varepsilon > 0$$

for all $x \in [0,1]$, $t \geq 0$, and consequently

$$E'(t) \leq -2\varepsilon \int_0^1 (1-u(x,t))^2 dx = -2\varepsilon E(t). \tag{11.32}$$

Hence, Gronwall's inequality (see Lemma 8.7) implies that

$$E(t) \leq e^{-2\varepsilon t} E(0), \tag{11.33}$$

and we have the following result:

Theorem 11.4 *Let u be the solution of (11.26)–(11.28) with initial data f satisfying*

$$0 < \varepsilon \leq f(x) \leq 1 + \varepsilon$$

for all $x \in [0,1]$. Then u approaches the asymptotic solution $u = 1$ in the sense that

$$\int_0^1 (u(x,t) - 1)^2 \, dx \leq e^{-2\varepsilon t} \int_0^1 (1 - f(x))^2 \, dx \tag{11.34}$$

for $t \geq 0$.

11.6.3 Decay of Derivatives

Our final discussion of Fisher's equation concerns the decay of the derivatives exposed in Figure 11.4 above. To study this effect, we define

$$F(t) = \int_0^1 (u_x(x,t))^2 \, dx, \qquad t \geq 0, \tag{11.35}$$

where again u solves (11.26)–(11.28). By differentiating the equation

$$u_t = u_{xx} + u(1-u)$$

with respect to x, we get

$$(u_x)_t = (u_x)_{xx} + (u_x - 2uu_x).$$

Hence, if we define

$$v = u_x,$$

it follows that v satisfies

$$v_t = v_{xx} + (1-2u)v \tag{11.36}$$

with boundary conditions

$$v(0,t) = v(1,t) = 0. \tag{11.37}$$

Now,

$$\begin{aligned}
F'(t) &= \frac{d}{dt}\int_0^1 v^2(x,t)dx \\
&= 2\int_0^1 v(x,t)v_t(x,t)dx \\
&= 2\int_0^1 vv_{xx} + (1-2u)v^2 dx \qquad (11.38)\\
&= 2[vv_x]_0^1 - 2\int_0^1 (v_x)^2 dx + 2\int_0^1 (1-2u)v^2 dx \\
&\leq -2\int_0^1 (v_x)^2 dx + 2\int_0^1 v^2 dx,
\end{aligned}$$

where we have used the fact that $\varepsilon \leq u(x,t) \leq 1+\varepsilon$ for all $x \in [0,1]$ and $t \geq 0$. Next we recall Poincaré's inequality stating that if $w(0) = w(1) = 0$, then

$$\pi^2 \int_0^1 (w(x))^2\, dx \leq \int_0^1 (w'(x))^2\, dx \tag{11.39}$$

for any continuously differentiable function $w = w(x)$; see Corollary 10.1 on page 317.

Due to the boundary conditions (11.37), it follows by (11.39) that

$$\int_0^1 (v_x(x,t))^2\, dx \geq \pi^2 \int_0^1 (v(x,t))^2\, dx,$$

and then (11.38) implies that

$$\begin{aligned}
F'(t) &\leq -2\pi^2 \int_0^1 v^2 dx + 2\int_0^1 v^2 dx \\
&= 2(1-\pi^2)F(t). \qquad (11.40)
\end{aligned}$$

From Gronwall's inequality we therefore obtain

$$F(t) \leq e^{2(1-\pi^2)t}F(0),$$

and thus we have the following result:

Theorem 11.5 *Let u be a smooth solution of (11.26)–(11.28) for a continuously differentiable initial function $f = f(x)$ satisfying (11.30). Then the spatial derivative of u decays as follows:*

$$\int_0^1 (u_x(x,t))^2\, dx \leq e^{2(1-\pi^2)t} \int_0^1 (f'(x))^2\, dx.$$

11.7 Blowup of Solutions

So far we have used energy-type arguments to derive various upper bounds for the solutions of reaction-diffusion equations. This strategy will be pursued even further in Project 11.2, where a precise decay estimate is derived. Here, we will take the opposite view and use a kind of energy estimate to show that the solution of a reaction-diffusion equation can blow up in the sense that u goes to infinity for a finite time $t = t^* < \infty$. Such behavior is of course important to characterize.

For a Neumann problem,

$$\begin{aligned} u_t &= u_{xx} + g(u), \\ u_x(0,t) &= u_x(1,t) = 0, \\ u(x,0) &= f(x), \end{aligned} \tag{11.41}$$

it is easy to see that the solution can blow up for certain choices of g. To see this, we note that if the initial condition f is constant, e.g.

$$f(x) = f_0 \tag{11.42}$$

for all $x \in [0,1]$, then

$$u(x,t) = v(t), \qquad x \in [0,1], \quad t > 0,$$

where v is the solution of

$$v'(t) = g(v), \qquad v(0) = f_0.$$

Hence the solution of (11.41) is given by the solution of an ordinary differential equation which is known to blow up in finite time for some functions g. Let, for instance,

$$g(v) = v^3 \quad \text{and} \quad f_0 > 0,$$

then

$$v(t) = \frac{f_0}{\sqrt{1 - 2tf_0^2}},$$

and we note that

$$v(t) \longrightarrow \infty \quad \text{as} \quad t \longrightarrow \frac{1}{2f_0^2}.$$

Hence, the solution of (11.41) blows up in finite time for data satisfying (11.42) when $g(u) = u^3$ and $f_0 > 0$.

Next we consider the Dirichlet problem

$$u_t = u_{xx} + u^3, \tag{11.43}$$
$$u(0,t) = u(1,t) = 0, \tag{11.44}$$
$$u(x,0) = f(x), \tag{11.45}$$

for $x \in [0,1]$, $t \geq 0$.

Obviously, because of (11.44), we cannot use the argument above to show that the solution may blow up; a more sophisticated analysis is needed.

As above we assume that u is a smooth solution and that

$$f(x) \geq 0, \qquad x \in (0,1). \tag{11.46}$$

Then it follows that

$$u(x,t) \geq 0 \tag{11.47}$$

for any (x,t) where the solution exists. This feature is left as an exercise for the reader; see Exercise 11.9. Next we define the quantity

$$\alpha(t) = \int_0^1 u(x,t)\sin(\pi x)dx, \tag{11.48}$$

and we assume that

$$\alpha(0) = \int_0^1 f(x)\sin(\pi x)dx > 2. \tag{11.49}$$

It is our aim to prove that $\alpha(t)$ blows up in finite time. Due to the properties of the sine function, this implies that u also blows up in finite time.

In order to prove that α blows up, we consider

$$\alpha'(t) = \int_0^1 u_t(x,t)\sin(\pi x)dx.$$

By (11.43), we get

$$\alpha'(t) = \int_0^1 u_{xx}\sin(\pi x)dx + \int_0^1 u^3\sin(\pi x)dx,$$

and thus integration by parts implies

$$\alpha'(t) = -\pi^2\alpha(t) + \int_0^1 u^3\sin(\pi x)dx. \tag{11.50}$$

Here we want to relate

$$\int_0^1 u^3\sin(\pi x)dx$$

and $\alpha(t)$. To do this, we need Hölder's inequality,

$$\int_a^b |y(x)z(x)|dx \leq \left(\int_a^b |y(x)|^p dx\right)^{1/p} \left(\int_a^b |z(x)|^q dx\right)^{1/q}, \qquad (11.51)$$

where $\frac{1}{p} + \frac{1}{q} = 1$; see (8.34) on page 266. Choosing $p = 3/2$ and $q = 3$, we get

$$\begin{aligned}
\alpha(t) &= \int_0^1 u \sin(\pi x) dx \\
&= \int_0^1 \left(\sin^{2/3}(\pi x)\right)\left(u \sin^{1/3}(\pi x)\right) dx \qquad (11.52)\\
&\leq \left(\int_0^1 \left(\sin^{2/3}(\pi x)\right)^{3/2} dx\right)^{2/3} \left(\int_0^1 \left(u \sin^{1/3}(\pi x)\right)^3 dx\right)^{1/3} \\
&= \left(\frac{2}{\pi}\right)^{2/3} \left(\int_0^1 u^3 \sin(\pi x) dx\right)^{1/3},
\end{aligned}$$

and consequently

$$\int_0^1 u^3(x,t)\sin(\pi x)dx \geq \frac{\pi^2}{4}\alpha^3(t) \qquad (11.53)$$

for $t \geq 0$.

By using (11.53) in (11.50), we get

$$\alpha'(t) \geq -\pi^2\alpha(t)\big(1 - (\frac{\alpha(t)}{2})^2\big). \qquad (11.54)$$

Hence, if $\alpha(t) > 2$, then $\alpha'(t) > 0$. Since $\alpha(0) > 2$, this implies that $\alpha(t) > 2$ for all $t > 0$ where the solution exists. If the inequality in (11.54) had been an equality, this nonlinear differential equation could be linearized by defining

$$\beta(t) = 1/\alpha^2(t); \qquad (11.55)$$

see Exercise 11.1. By using this definition of β in the inequality (11.54), we obtain

$$\beta'(t) = -2\frac{\alpha'(t)}{\alpha^3(t)} \leq 2\pi^2\big(\alpha(t) - \frac{\alpha^3(t)}{4}\big)/\alpha^3(t),$$

and thus

$$\beta'(t) \leq 2\pi^2\left(\beta(t) - \frac{1}{4}\right).$$

From this differential inequality we immediately obtain an upper bound on $\beta(t)$; see Exercise 8.21. Multiplying this inequality by $e^{-2\pi^2 t}$ and integrating in time, we get

$$\beta(t) \leq \frac{1}{4} + e^{2\pi^2 t}\left(\beta(0) - 1/4\right). \tag{11.56}$$

By (11.49) and (11.55), we have that

$$0 < \beta(0) < 1/4,$$

and then it follows from (11.56) that there is a finite time t_0 such that

$$\beta(t) \leq 0$$

for $t \geq t_0$. Consequently, there is a time $t^* \in [0, t_0]$ such that

$$\alpha(t) \longrightarrow \infty \quad \text{as} \quad t \longrightarrow t^* < \infty.$$

This proves that the solution of (11.43)–(11.46) blows up in finite time if the condition (11.49) is satisfied.

11.8 Exercises

EXERCISE 11.1 Consider the nonlinear ordinary differential equation

$$v'(t) = av(t) + b\big(v(t)\big)^2.$$

Here a and b are given constants.

(a) Assume that $v(t) \neq 0$ and define $w(t) = 1/v(t)$. Show that $w(t)$ satisfies a linear differential equation.

(b) Verify formula (11.2).

(c) Assume that $v(t)$ satisfies a differential equation of the form

$$v'(t) = av(t) + b\big(v(t)\big)^n.$$

Explain how this equation can be linearized by a proper change of variables.

EXERCISE 11.2 The purpose of this exercise is to study the asymptotic behavior of the numerical solution of Fisher's equation generated by the scheme (11.17). In particular, we are interested in initial data not satisfying the requirements of Theorem 11.2.

(a) Implement the scheme (11.17).

(b) Use your computer program to investigate the asymptotic behavior of the initial function defined by $f(x) = 0$ for $x < 3/7$ and for $x > 5/7$ and $f(x) = 1/2$ for $x \in [3/7, 5/7]$.

(c) Repeat (b) for the initial function given by $f(x) = 10(1+\cos(10\pi x))$.

EXERCISE 11.3 Consider the problem

$$\begin{aligned} u_t &= u_{xx} + u(1-u) \quad \text{for} \quad x \in (0,1), \quad t \in (0,T], \\ u(0,t) &= u(1,t) = 1, \qquad t \in [0,T], \\ u(x,0) &= f(x), \qquad x \in [0,1], \end{aligned} \tag{11.57}$$

where $f = f(x)$ denotes the initial data, which we assume to be in the unit interval.

(a) Define an explicit finite difference scheme for this problem.

(b) Show that under appropriate conditions on the mesh parameters, the unit interval is an invariant region for the discrete solutions.

(c) Show that $u = 1$ is the asymptotic solution of the scheme.

(d) Use an energy estimate to show that $u = 1$ is also the asymptotic solution for the continuous model.

EXERCISE 11.4 Consider the problem

$$\begin{aligned} u_t &= u_{xx} - u^2 \quad \text{for} \quad x \in (0,1), \quad t \in (0,T], \\ u(0,t) &= u(1,t) = 0, \qquad t \in [0,T], \\ u(x,0) &= f(x), \qquad x \in [0,1], \end{aligned} \tag{11.58}$$

where $f = f(x)$ denotes the initial data, which we assume to be bounded.

(a) Derive a maximum principle for this problem.

(b) Define an explicit finite difference scheme and derive, under appropriate conditions on the mesh parameters, a discrete version of the maximum principle.

(c) Show that $u = 0$ is the asymptotic solution of the scheme.

EXERCISE 11.5 The following reaction-diffusion equation arises in the modeling of the electrical activity in the human heart:

$$u_t = u_{xx} + u(3u-1)(1-u). \tag{11.59}$$

Here, we consider this model equipped with boundary data

$$u(0,t) = u(1,t) = 0 \tag{11.60}$$

and an initial condition

$$u(x,0) = f(x) \tag{11.61}$$

satisfying

$$0 \leq f(x) \leq 1 \tag{11.62}$$

with $f(0) = f(1) = 0$.

Put $\alpha = \Delta t/(\Delta x)^2$, $p(u) = u(3u-1)(1-u)$, and consider the two schemes

$$v_j^{m+1} = \alpha v_{j-1}^m + (1-2\alpha)v_j^m + \alpha v_{j+1}^m + \Delta t p(v_j^m), \tag{11.63}$$

$$w_j^{m+1} = \alpha w_{j-1}^m + (1-2\alpha)w_j^m + \alpha w_{j+1}^m + \Delta t p(w_j^{m+1}). \tag{11.64}$$

These schemes are referred to as explicit and semi-implicit respectively.

(a) Derive a bound on Δt such that $[0,1]$ is an invariant region for $\{v_j^m\}$ generated by (11.63).

(b) Derive a similar bound for $\{w_j^m\}$.

(c) Discuss the properties of these two schemes with respect to the stability condition and the complexity of the implementation.

EXERCISE 11.6 Prove the following discrete version of Jensen's inequality:

$$g\left(\frac{1}{n}\sum_{i=1}^{n} v_i\right) \leq \frac{1}{n}\sum_{i=1}^{n} g(v_i)$$

for a convex function g. A continuous version of this inequality is derived in Project 11.2 below.

EXERCISE 11.7 The purpose of this exercise is to prove a discrete version of Gronwall's inequality (cf. Lemma 8.7).

Suppose that

$$v_{n+1} \leq v_n + \Delta t \alpha v_n, \qquad n \geq 0, \quad \Delta t > 0, \quad t_n = n\Delta t,$$

for a constant α. Show that

$$v_n \leq e^{\alpha t_n} v_0.$$

EXERCISE 11.8 We observed above that the derivatives of the solution of Fisher's equation decayed rapidly; see Fig. 11.4 and Theorem 11.5. The purpose of this exercise is to show that this is a feature of the solution of many reaction-diffusion equations in the presence of Neumann-type boundary data.

Consider the problem

$$\begin{aligned} u_t &= Du_{xx} + p(u), \\ u_x(0,t) &= u_x(1,t) = 0, \\ u(x,0) &= f(x), \end{aligned} \tag{11.65}$$

where $D > 0$ is a constant and where we assume that

$$\sup_u |p'(u)| \leq M < \infty.$$

(a) Show that if

$$F(t) = \int_0^1 (u_x(x,t))^2 \, dx,$$

we have

$$F'(t) \leq 2\left(M - D\pi^2\right) F(t).$$

(b) Show that if

$$\sup_u |p'(u)| < D\pi^2,$$

then u_x decays to zero as t goes to infinity.

EXERCISE 11.9 Show that the solution of (11.43)–(11.45), under the assumption of (11.46), satisfies (11.47) whenever u exists.

11.9 Projects

Project 11.1 *Population models*

As explained above, Fisher's model can be used to study the evolution of a single species in the presence of limited resources. In this project we will consider some more complicated models. For simplicity, we consider only prototypical models and do not care about the scaling of the variables involved.

Throughout this project u and v denote the density of two populations residing in a common district. A large class of models can be written in the following form:

$$\begin{aligned} u_t &= u_{xx} + uM(u,v), & u(x,0) &= u^0(x), \\ v_t &= v_{xx} + vN(u,v), & v(x,0) &= v^0(x), \end{aligned}$$

where M and N are given functions.

(a) Consider the system above on the unit interval and with the boundary conditions $u_x = v_x = 0$ for $x = 0$ and $x = 1$. Generalize the scheme (11.17) in order to handle this problem.

(b) Implement the finite difference scheme derived above.

(c) Consider an interaction of two predator-prey type species. Let u denote the density of the prey, and let v be the density of the predator. Explain why it is reasonable to assume

$$M_v < 0 \quad \text{and} \quad N_u > 0$$

in this model.

(d) Put $M = 1 - v$ and $N = u - 1$. Derive, under proper conditions on the mesh parameters, an invariant region for the scheme generalized in (a).

(e) Implement the scheme and try to answer the following questions by doing numerical experiments:

- What is the asymptotic solution of the scheme if $u^0(x) = 0$ for all x?
- What is the asymptotic solution of the scheme if $u^0(x) = 1$ and $v^0(x) = \cos^2(5\pi x)$?

(f) Consider next a situation of two competing species; u denotes the density of species S_1 and v denotes the density of species S_2. Explain why the competition of the two species leads to the following requirements:

$$M_v < 0 \quad \text{and} \quad N_u < 0.$$

(g) Put $M = (A_1 - u - v)$ and $N = (A_2 - u - v)$. Here A_1 and A_2 are given positive constants representing the environmental capacities for feeding species S_1 and S_2 respectively. Show that the rectangle defined by $0 \le u \le A_1$ and $0 \le v \le A_2$ is invariant for the scheme generalized in (a).

(h) Explore, using numerical experiments, how the asymptotic behavior of the scheme depends on the values of A_1 and A_2.

(i) Finally, we consider the case of symbiosis. Explain why such an interaction leads to the following requirement:

$$M_v > 0 \quad \text{and} \quad N_u > 0.$$

(j) Put $M = (1 + v - u)$ and $N = (1 + u - v)$ and prove that for this model, the unit square is invariant for the discrete solutions generated by the scheme derived in (a).

Project 11.2 *More on Asymptotics*

The purpose of this project is to show that energy estimates can be applied to get accurate information about the solution of a reaction-diffusion equation. On page 105 we considered the problem

$$u_t = u_{xx} - u^3, \qquad x \in (0,1), \quad t > 0, \tag{11.66}$$

with boundary conditions

$$u(0,t) = u(1,t) = 0, \qquad t \geq 0, \tag{11.67}$$

and initial condition

$$u(x,0) = f(x). \tag{11.68}$$

It was proved there that

$$\int_0^1 u^2(x,t)dx \leq \int_0^1 f^2(x)dx \tag{11.69}$$

for any $t \geq 0$.

In this project our aim is to sharpen this result.

(a) Define an explicit finite difference scheme for the problem (11.66)–(11.68).

(b) Show that under suitable assumptions on the mesh parameters, the interval $[-1, 1]$ is invariant for the numerical solutions.

(c) Define

$$E_\Delta(t_n) = \Delta x \sum_{j=1}^{n} (u_j^n)^2,$$

where $\{u_j^n\}$ is the numerical solution, and plot this quantity as function of t for some grid sizes using

(i) $f(x) = \sin(\pi x)$,

(ii) $f(x) = x^5(1-2x)^6 e^{\sin(3x)}$,

(iii) $f(x_j) = \sin(10\text{rand}(x_j))$.

Here the "rand" function in (iii) is as described in Example 11.1 on page 346. Use these computations to conclude that the estimate (11.69) seems a bit weak. We will now try to sharpen it.

(d) Show that $[-1,1]$ is an invariant region for the continuous solution of (11.66)–(11.68).

(e) Define

$$E(t) = \int_0^1 u^2(x,t)dx,$$

and show that

$$E(t) = -2\int_0^1 (u_x(x,t))^2\, dx - 2\int_0^1 u^4(x,t)dx. \qquad (11.70)$$

Of course, (11.70) directly implies (11.69), but now we want a more accurate estimate. We shall use the inequalities of Poincaré and Jensen to bound the right-hand side of (11.70).

(f) Use Poincaré's inequality to show that

$$E'(t) \le -2\pi^2 E(t) - 2\int_0^1 u^4(x,t)dx, \qquad (11.71)$$

and conclude that

$$E(t) \le e^{-2\pi^2 t} E(0). \qquad (11.72)$$

We note that (11.72) is a much sharper bound than (11.69). But an even better result can be obtained by also taking the second term on the right-hand side of (11.71) into account. In order to do so, we use the inequality of Jensen. This states that if g is a smooth convex function, then

$$g\left(\int_0^1 v(x)dx\right) \le \int_0^1 g(v(x))\, dx. \qquad (11.73)$$

This inequality will be derived below.

(g) Use (11.73) to show that

$$\left(\int_0^1 u^2(x,t)dx\right)^2 \le \int_0^1 u^4(x,t)dx. \qquad (11.74)$$

(h) Use (11.71) and (11.74) to conclude that

$$E'(t) \leq -2\pi^2 E(t) - 2E^2(t). \tag{11.75}$$

(i) Show that

$$E(t) \leq \frac{\pi^2 E(0)}{\pi^2 + E(0)\left(1 - e^{-2\pi^2 t}\right)} e^{-2\pi^2 t}. \tag{11.76}$$

(j) Plot E_Δ defined in (c) together with the bounds defined by (11.72) and (11.76) for the initial conditions (i), (ii), and (iii) also defined in (c). Comment on the sharpness of (11.76) for these initial conditions.

(k) Finally we have to prove the inequality of Jensen. Let g be a smooth convex function. Use a Taylor-series approximation to show that

$$g(t) + (s-t)g'(t) \leq g(s) \tag{11.77}$$

for any $s, t \in \mathbb{R}$. Put

$$s = z(x), \qquad t = \int_0^1 z(y)dy,$$

and integrate (11.77) with respect to x and conclude that

$$g\left(\int_0^1 z(x)dx\right) \leq \int_0^1 g\left(z(x)\right) dx, \tag{11.78}$$

which is Jensen's inequality.

12
Applications of the Fourier Transform

In this chapter, we briefly discuss the Fourier transform and show how this transformation can be used to solve differential equations where the spatial domain is all of $\mathbb{R}$.

In the same way as Fourier series arise in the analysis of linear partial differential equations on an interval, the Fourier transform is an appropriate tool for the corresponding problems when the spatial domain is extended to the whole real line. This can for example be illustrated by the heat equation

$$u_t = u_{xx}. \tag{12.1}$$

We have seen (see Chapter 3) that when the spatial variable x is restricted to an interval, then separation of variables leads to eigenvalue problems of the form

$$-X''(x) = \lambda X(x) \tag{12.2}$$

with proper boundary conditions. For example, the eigenvalue problem (12.2), with Dirichlet boundary conditions, leads directly to Fourier sine series. As we shall see below, the Fourier transform can be used in a similar way to study the pure initial value problem for (12.1), i.e. the initial value problem where the spatial variable x is defined for all of $\mathbb{R}$.

In deriving the properties of the Fourier transform below, we will assume that the functions are sufficiently well behaved to justify our calculations. We will not specify clearly for which class of functions the formulas hold. This would lead to a more technical discussion which is beyond our current scope. Therefore, the present chapter should be seen more as an informal illustration of how the Fourier transform can be applied to partial differential equations, and not as a rigorous discussion of properties of the Fourier transform. The solution formulas for certain differential equations which we derive here are therefore only formal solutions. However, by direct inspection we can of course check the validity of these solutions.

12.1 The Fourier Transform

If f is a function defined on $\mathbb{R}$, then the Fourier transform, $\hat{f}(\omega)$, is a new function defined on $\mathbb{R}$ given by[1]

$$\hat{f}(\omega) = \int_{-\infty}^{\infty} f(x)e^{-i\omega x}\,dx, \tag{12.3}$$

where $i = \sqrt{-1}$.

We note that even if $f(x)$ is real for all x, the new function $\hat{f}$ will in general not be real valued. Also, since the integral in (12.3) is over all of $\mathbb{R}$, the value $\hat{f}(\omega)$ will not exist unless the function $f(x)$ behaves properly for x near $\pm\infty$. However, for well-behaved functions, which tend to zero sufficiently fast at $\pm\infty$, the integral in (12.3) will be well defined.

EXAMPLE 12.1 Let $H(x)$ be the Heaviside function given by

$$H(x) = \begin{cases} 0 & x \leq 0, \\ 1 & x > 0, \end{cases}$$

and let

$$f(x) = H\big(a - |x|\big),$$

where $a > 0$ is a parameter. Alternatively,

$$f(x) = \begin{cases} 1 & \text{for } |x| < a, \\ 0 & \text{otherwise.} \end{cases}$$

The function f, which is usually referred to as a square pulse, is plotted in Fig. 12.1. Since $f(x) \equiv 0$ for $|x| > a$, the Fourier transform $\hat{f}(\omega)$ is given by

$$\hat{f}(\omega) = \int_{-a}^{a} e^{-i\omega x}\,dx = -\frac{1}{i\omega}e^{-i\omega x}\Big|_{x=-a}^{x=a} = \frac{2}{\omega}\sin(a\omega).$$

■

[1] You may find slightly different definitions of the Fourier transform in other texts. In particular, $\hat{f}(\omega)$ will frequently be defined with the scaling factor $\frac{1}{\sqrt{2\pi}}$ in front of the integral.

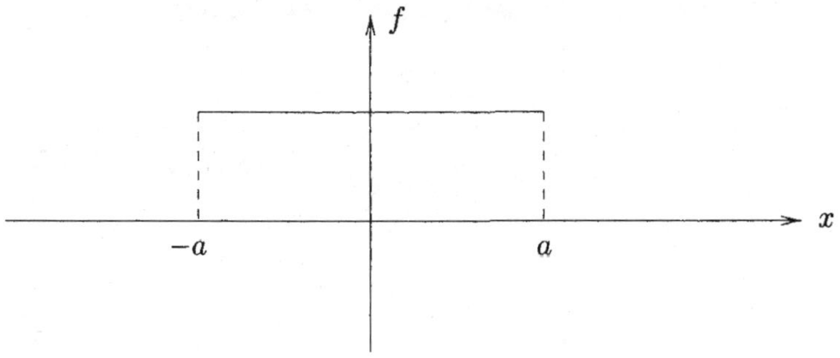

FIGURE 12.1. *A square pulse.*

EXAMPLE 12.2 Let $a > 0$ be a parameter and let

$$f(x) = 2H(x) - H(x+a) - H(x-a).$$

Alternatively (see Fig. 12.2),

$$f(x) = \begin{cases} 0 & \text{for} \quad x \le -a, \\ -1 & \text{for} \quad x \in (-a, 0], \\ 1 & \text{for} \quad x \in (0, a], \\ 0 & \text{for} \quad x > a. \end{cases}$$

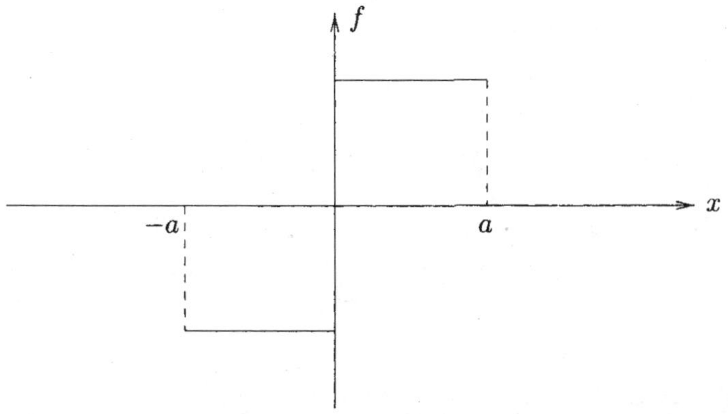

FIGURE 12.2. *A square wave.*

The Fourier transform $\hat{f}(\omega)$ is given by

$$\begin{aligned}\hat{f}(\omega) &= -\int_{-a}^{0} e^{-i\omega x}dx + \int_{0}^{a} e^{-i\omega x}dx \\ &= \frac{1}{i\omega}e^{-i\omega x}\Big|_{x=-a}^{x=0} - \frac{1}{i\omega}e^{-i\omega x}\Big|_{x=0}^{x=a} \\ &= \frac{2}{i\omega}(1-\cos(a\omega)) = -\frac{4i}{\omega}\sin^2(a\omega/2).\end{aligned}$$

■

In both the examples above, the function $f(x)$ is real valued. In Example 12.1 we also obtained a real-valued function $\hat{f}(\omega)$, while $\hat{f}(\omega)$ is purely imaginary in Example 12.2. In fact, by rewriting (12.3) in the form

$$\hat{f}(\omega) = \int_{-\infty}^{\infty} f(x)\cos(\omega x)\,dx - i\int_{-\infty}^{\infty} f(x)\sin(\omega x)\,dx,$$

we easily see that if f is a real-valued even function, then $\hat{f}(\omega)$ is real. On the other hand, if f is a real-valued odd function, then $\hat{f}(\omega)$ is purely imaginary.

Example 12.3 Let f be the real-valued even function

$$f(x) = e^{-b|x|},$$

where $b > 0$ is a parameter. Then $\hat{f}(\omega)$ is given by

$$\begin{aligned}\hat{f}(\omega) &= \int_{-\infty}^{\infty} e^{-b|x|}e^{-i\omega x}dx \\ &= \int_{-\infty}^{0} e^{(b-i\omega)x}dx + \int_{0}^{\infty} e^{-(b+i\omega)x}dx \\ &= \frac{1}{b-i\omega} + \frac{1}{b+i\omega} = \frac{2b}{b^2+\omega^2},\end{aligned}$$

which is real. ■

12.2 Properties of the Fourier Transform

The Fourier transform can be considered as a map which takes functions $f(x)$ into its transform $\hat{f}(\omega)$. To indicate more clearly that the function $\hat{f}$ is derived from f, we sometimes write $\mathcal{F}(f)(\omega)$ instead of $\hat{f}(\omega)$.

From the definition of $\mathcal{F}(f)(\omega)$ it follows that the map $\mathcal{F}$ is linear, i.e.

$$\mathcal{F}(\alpha f + \beta g) = \alpha\mathcal{F}(f) + \beta\mathcal{F}(g), \tag{12.4}$$

where f, g are functions and α, $\beta \in \mathbb{R}$.

EXAMPLE 12.4 Let

$$f(x) = e^{-b|x|} - 4H(a - |x|).$$

Together with the results of Examples 12.1 and 12.3, the property (12.4) leads to

$$\mathcal{F}(f)(\omega) = \hat{f}(\omega) = \frac{2b}{b^2 + \omega^2} - \frac{8}{\omega}\sin(a\omega).$$

■

The property (12.4) is derived directly from the definition (12.3). We have

$$\begin{aligned}\mathcal{F}(\alpha f + \beta g) &= \int_{-\infty}^{\infty} (\alpha f(x) + \beta g(x))e^{-i\omega x}\,dx \\ &= \alpha \int_{-\infty}^{\infty} f(x)e^{-i\omega x}\,dx + \beta \int_{-\infty}^{\infty} g(x)e^{-i\omega x}\,dx \\ &= \alpha\mathcal{F}(f)\omega + \beta\mathcal{F}(g)\omega.\end{aligned}$$

When the Fourier transform is used to solve differential equations, we need a relation between $\mathcal{F}(f)$ and $\mathcal{F}(f')$. From integration by parts we have

$$\begin{aligned}\mathcal{F}(f')(\omega) &= \int_{-\infty}^{\infty} f'(x)e^{-i\omega x}\,dx \\ &= f(x)e^{-i\omega x}\Big|_{x=-\infty}^{\infty} + i\omega \int_{-\infty}^{\infty} f(x)e^{-i\omega x}\,dx.\end{aligned}$$

Hence, if we assume that $|f(x)|$ tends to zero as x tends to $\pm\infty$ such that the boundary terms disappear, we have

$$\mathcal{F}(f')(\omega) = i\omega\mathcal{F}(f)(\omega) = i\omega\hat{f}(\omega). \tag{12.5}$$

This formula expresses that differentiation of f is transformed into a multiplication with the function $i\omega$ by the Fourier transform.

EXAMPLE 12.5 Let us assume that $u = u(x)$ satisfies a differential equation of the form

$$au''(x) + bu'(x) + cu(x) = f(x), \tag{12.6}$$

where f is a given function and $a, b, c \in \mathbb{R}$. Assume we can take the Fourier transform of each side of the identity (12.6). By using the properties (12.4) and (12.5), we then obtain

$$(-a\omega^2 + bi\omega + c)\hat{u}(\omega) = \hat{f}(\omega). \tag{12.7}$$

Hence, the differential equation (12.6) is transformed into the algebraic equation (12.7). Since algebraic equations usually are easier to solve, this example clearly indicates that the Fourier transform is potentially useful in solving differential equations. ∎

The property (12.5) has a counterpart which states that the Fourier transform of the function $xf(x)$ is given by $i\hat{f}'(\omega)$, i.e.

$$\mathcal{F}(xf)(\omega) = i\frac{d}{d\omega}\mathcal{F}(f)(\omega) = i\hat{f}'(\omega). \tag{12.8}$$

At this point we should be a little careful with our notation. In (12.8) the function $\hat{f}'(\omega) = \frac{d}{d\omega}\mathcal{F}(f)(\omega)$ is obtained by first computing $\hat{f}(\omega) = \mathcal{F}(f)(\omega)$ and then differentiating this function with respect to ω. This is not the same as $\mathcal{F}(f')(\omega)$, which is obtained by first differentiating f with respect to x and then computing the Fourier transform.

The property (12.8) follows by differentiating the expression for $\mathcal{F}(f)(\omega)$ with respect to ω. If we assume that we can differentiate under the integral, then we obtain

$$\begin{aligned}\frac{d}{d\omega}\mathcal{F}(f)(\omega) &= \frac{d}{d\omega}\int_{-\infty}^{\infty} f(x)e^{-i\omega x}\,dx \\ &= -i\int_{-\infty}^{\infty} xf(x)e^{-i\omega x}\,dx = -i\mathcal{F}(xf)(\omega).\end{aligned}$$

Property (12.8) follows by multiplying both sides of this equality by i.

Another useful property of the Fourier transform is the following scaling property ($a \neq 0$):

$$\mathcal{F}\big(f(ax)\big)(\omega) = \frac{1}{a}\mathcal{F}(f)\big(\frac{\omega}{a}\big). \tag{12.9}$$

This follows from a change of variables, since

$$\begin{aligned}\mathcal{F}\big(f(ax)\big)(\omega) &= \int_{-\infty}^{\infty} f(ax)e^{-i\omega x}dx \\ &= \int_{-\infty}^{\infty} f(y)e^{-i\omega y/a}\frac{dy}{a} = \frac{1}{a}\mathcal{F}(f)\big(\frac{\omega}{a}\big).\end{aligned}$$

The properties (12.5) and (12.8) are fundamental to the use of the Fourier transform in differential equations. A less obvious application of these properties is given in the following important example.

EXAMPLE 12.6 Consider the function

$$f(x) = e^{-x^2/2}.$$

We want to compute $\hat{f}(\omega)$. A direct evaluation of the integral

$$\hat{f}(\omega) = \int_{-\infty}^{\infty} e^{-x^2/2} e^{i\omega x}\, dx$$

can be done by using the "residue theorem" of complex analysis. Here, we shall instead use an indirect differential equation argument, which is based on the properties (12.5) and (12.8) of the Fourier transform. It is straightforward to check that the function $f(x)$ satisfies the linear initial value problem

$$f'(x) = -xf(x), \quad f(0) = 1. \tag{12.10}$$

Furthermore, $f(x)$ is the unique solution of this problem. In fact, by multiplication of the integral factor $e^{x^2/2}$, the differential equation is reduced to

$$\left(e^{x^2/2} f(x)\right)' = 0.$$

If we take the Fourier transform of both sides of (12.10), we obtain from (12.5) and (12.8) that

$$i\omega \hat{f}(\omega) = -i\hat{f}'(\omega) = -i\frac{d}{d\omega}\hat{f}(\omega)$$

or

$$\hat{f}'(\omega) = -\omega \hat{f}(\omega). \tag{12.11}$$

We note that this equation corresponds exactly to the differential equation in (12.10). Furthermore, from the formula

$$\int_{-\infty}^{\infty} e^{-y^2} dy = \sqrt{\pi} \tag{12.12}$$

(see Exercise 1.11 on page 24), we obtain

$$\hat{f}(0) = \int_{-\infty}^{\infty} e^{-x^2/2} dx = \sqrt{2}\int_{-\infty}^{\infty} e^{-y^2} dy = \sqrt{2\pi}.$$

But the unique solution of (12.11) with $\hat{f}(0) = \sqrt{2\pi}$ is given by

$$\hat{f}(\omega) = \sqrt{2\pi} e^{-\omega^2/2}.$$

Hence, up to a multiplication of the factor $\sqrt{2\pi}$, the functions f and $\hat{f}$ are equal. ■

EXAMPLE 12.7 Let $g(x) = e^{-ax^2}$, where $a > 0$. We would like to compute $\hat{g}(\omega)$. If $f(x) = e^{-x^2/2}$, as in Example 12.6 above, then

$$g(x) = f(x\sqrt{2a}).$$

By property (12.9) we therefore obtain

$$\hat{g}(\omega) = \frac{1}{\sqrt{2a}}\hat{f}\left(\frac{\omega}{\sqrt{2a}}\right) = \sqrt{\frac{\pi}{a}}e^{-\omega^2/(4a)}.$$

■

12.3 The Inversion Formula

As explained in Example 12.5 above, the Fourier transform will replace certain differential equations by corresponding algebraic relations for the transforms. For example, the differential equation

$$-u''(x) + u(x) = f(x)$$

implies the relation

$$(\omega^2 + 1)\hat{u}(\omega) = \hat{f}(\omega)$$

for the corresponding Fourier transforms, and hence

$$\hat{u}(\omega) = \frac{1}{1+\omega^2}\hat{f}(\omega).$$

However, in order to obtain the solution $u(x)$ from this expression, we need to know how we can derive a function from its Fourier transform. In fact, so far in our discussion it is not even clear that a function is uniquely determined by its Fourier transform.

The tool we seem to need is an inverse transform which describes how a function $f(x)$ can be computed from $\hat{f}(\omega)$. The proper *inversion formula* is given by

$$f(x) = \frac{1}{2\pi}\int_{-\infty}^{\infty} \hat{f}(\omega)e^{i\omega x}d\omega. \tag{12.13}$$

We should note the similarity between this inversion formula and the definition (12.3) of the Fourier transform. The formula nearly states that f is the Fourier transform of $\hat{f}$. However, we note the missing minus sign in the term $e^{i\omega x}$ and the extra factor $1/(2\pi)$ in front of the integral. An alternative formulation of (12.13) is therefore

$$f(x) = \frac{1}{2\pi}\mathcal{F}(\hat{f})(-x). \tag{12.14}$$

Before we try to justify the inversion formula, let us show that it is consistent with the result of Example 12.7.

EXAMPLE 12.8 For any $a > 0$ let

$$f_a(x) = e^{-ax^2}.$$

In Example 12.7 we showed that

$$\hat{f}_a \equiv \mathcal{F}(f_a) = \sqrt{\frac{\pi}{a}} f_{1/(4a)} = \sqrt{\frac{\pi}{a}} f_b,$$

where $b = 1/4a$. Hence, since $a = 1/4b$ and $ab = 1/4$, this implies

$$\mathcal{F}(\hat{f}_a) = \sqrt{\frac{\pi}{a}} \mathcal{F}(f_b) = \sqrt{\frac{\pi}{a}}\sqrt{\frac{\pi}{b}} f_{1/(4b)} = 2\pi f_a$$

or

$$f_a(x) = \frac{1}{2\pi}\mathcal{F}(\hat{f}_a)(x).$$

Since $f_a(x) = f_a(-x)$, this is consistent with (12.14). ■

In order to try to justify the inversion formula (12.13), we first recall the complex form of the Fourier series; see Sections 8.1.3 and 8.1.4. If $f(x)$ is a function defined on the interval $(-l, l)$ which can be represented by its Fourier series, then

$$f(x) = \sum_{k=-\infty}^{\infty} c_k e^{ik\pi x/l}, \tag{12.15}$$

where the coefficients are given by

$$c_k = \frac{1}{2l}\int_{-l}^{l} f(y)e^{-ik\pi y/l}\,dy. \tag{12.16}$$

We shall see that the inversion formula (12.13) arises formally as a limit of the Fourier series as $l \to \infty$.

Let

$$\hat{f}_l(\omega) = \int_{-l}^{l} f(y)e^{-i\omega y}dy.$$

Then for sufficiently regular functions f we clearly have

$$\hat{f}_l(\omega) \longrightarrow \hat{f}(\omega) \quad \text{as} \quad l \to \infty. \tag{12.17}$$

Furthermore, the Fourier series (12.15)–(12.16) can be written in the form

$$f(x) = \frac{1}{2l}\sum_{k=-\infty}^{\infty} \hat{f}_l(\omega_k)e^{i\omega_k x}, \tag{12.18}$$

where $\omega_k = k\pi/l$.

Let $\Delta\omega = \pi/l$ denote the distance between these points. The "grid points," $\{\omega_k = k(\Delta\omega)\}_{k=-\infty}^{\infty}$, define a uniform partition of the real line. Therefore, it is more convenient to rewrite (12.18) in the form

$$f(x) = \frac{1}{2\pi}\left[\Delta\omega \sum_{k=-\infty}^{\infty} \hat{f}_l(\omega_k)e^{i\omega_k x}\right]. \tag{12.19}$$

We observe that this expression for $f(x)$ resembles the inverse formula (12.13).

Of course, an expression of the form

$$\Delta\omega \sum_{k=-\infty}^{\infty} g(\omega_k)e^{i\omega_k x}$$

is just a "trapezoidal approximation" of the integral

$$\int_{-\infty}^{\infty} g(\omega)e^{i\omega x}\,d\omega.$$

Note also that if l tends to infinity, then $\Delta\omega$ tends to zero. Hence, together with (12.17) this suggests that

$$\Delta\omega \sum_{k=-\infty}^{\infty} \hat{f}_l(\omega_k)e^{i\omega_k x} \longrightarrow \int_{-\infty}^{\infty} \hat{f}(\omega)e^{i\omega x}\,d\omega \quad \text{as} \quad l \to \infty.$$

By combining this with (12.19), we therefore obtain the inversion formula

$$f(x) = \frac{1}{2\pi}\int_{-\infty}^{\infty} \hat{f}(\omega)e^{i\omega x}\,d\omega.$$

The derivation of the inversion formula outlined above is far from being a strict mathematical proof. We shall not provide a rigorous proof here. However, it is interesting to note that the main tool in a rigorous proof is frequently the fact that the inversion formula holds for the functions e^{-ax^2} studied in Example 12.8 above. The reason for this is roughly that any smooth function can be approximated to any accuracy by weighted integrals of translations of such functions. For a proof of the inversion formula which essentially uses such an argument we refer for example to Rauch [22].

The inversion formula can be used to compute Fourier transforms which may be hard to compute directly.

Example 12.9 Let us recall from Example 12.3 that the function

$$f(x) = e^{-b|x|},$$

where b is positive, has the Fourier transform

$$\hat{f}(\omega) = \frac{2b}{b^2 + \omega^2}.$$

Since $f(-x) = f(x)$, we therefore obtain from the inversion formula that

$$f = \frac{1}{2\pi}\mathcal{F}(\hat{f}).$$

Hence, by reversing x and ω, if

$$g(x) = \frac{1}{\pi}\frac{b}{b^2 + x^2} \quad \text{then} \quad \hat{g}(\omega) = e^{-b|\omega|}.$$

■

12.4 The Convolution

Let us consider the pure initial value problem for the heat equation (12.1), i.e. ,

$$\begin{aligned} u_t &= u_{xx} \quad \text{for} \quad x \in \mathbb{R}, \quad t > 0, \\ u(x,0) &= f(x), \quad x \in \mathbb{R}. \end{aligned} \tag{12.20}$$

For each $t \geq 0$ let $\hat{u}(\omega, t) = \mathcal{F}\big(u(\cdot, t)\big)(\omega)$ be the Fourier transform of $u(\cdot, t)$. Here $u(\cdot, t)$ denotes the function $x \mapsto u(x,t)$ for a fixed value of t. Hence,

$$\hat{u}(\omega, t) = \int_{-\infty}^{\infty} u(x,t)e^{-i\omega x}dx. \tag{12.21}$$

It follows from property (12.5) that

$$\mathcal{F}\big(u_{xx}(\cdot, t)\big)(\omega) = -\omega^2 \hat{u}(\omega, t). \tag{12.22}$$

Furthermore, by differentiation under the integral sign (see Proposition 3.1 on page 107) we obtain

$$\mathcal{F}\big(u_t(\cdot, t)\big)(\omega) = \int_{-\infty}^{\infty} u_t(x,t)e^{-i\omega x}dx = \frac{\partial}{\partial t}\hat{u}(\omega, t). \tag{12.23}$$

However, since we know that $u_t = u_{xx}$ from the differential equation (12.20), we can conclude from (12.22) and (12.23) that

$$\frac{\partial}{\partial t}\hat{u}(\omega, t) = -\omega^2 \hat{u}(\omega, t), \qquad t > 0. \tag{12.24}$$

This last equation can be regarded as an ordinary differential equation with respect to t, where ω is just a parameter. The solution is given by

$$\hat{u}(\omega, t) = e^{-\omega^2 t}\hat{u}(\omega, 0) = e^{-\omega^2 t}\hat{f}(\omega). \tag{12.25}$$

Let now $S(x, t)$ be the function

$$S(x, t) = \frac{1}{\sqrt{4\pi t}} e^{-x^2/4t}, \tag{12.26}$$

and let $\hat{S}(\omega, t) = \mathcal{F}\big(S(\cdot, t)\big)$ be the Fourier transform with respect to x. It follows directly from Example 12.7, with $a = 1/(4t)$, that

$$\hat{S}(\omega, t) = e^{-\omega^2 t}.$$

Hence, the identity (12.25) can be rewritten in the form

$$\hat{u}(\omega, t) = \hat{S}(\omega, t)\hat{f}(\omega), \tag{12.27}$$

which states that the Fourier transform of the solution u is the product of two Fourier transforms. Furthermore, the function S (and $\hat{S}$) is explicitly known, while f is the given initial function. Therefore, the Fourier transform of u is the product of the Fourier transforms of two known functions. From this information we would like to obtain u.

Let us consider a slightly more general situation. Let $f(x)$ and $g(x)$ be given functions. We would like to identify a function h such that $\hat{h} = \hat{f} \cdot \hat{g}$. From the definition of the Fourier transform we have

$$\begin{aligned}
\hat{f}(\omega)\hat{g}(\omega) &= \int_{-\infty}^{\infty} f(y)e^{-i\omega y}\,dy \int_{-\infty}^{\infty} g(z)e^{-i\omega z}\,dz \\
&= \int_{-\infty}^{\infty}\int_{-\infty}^{\infty} f(y)g(z)e^{-i\omega(y+z)}\,dy\,dz \\
&= \int_{-\infty}^{\infty}\int_{-\infty}^{\infty} f(x-z)g(z)e^{-i\omega x}\,dx\,dz,
\end{aligned}$$

where the last identity is obtained from the substitution $x = y+z$. However, by changing the order of integration we obtain

$$\begin{aligned}
\hat{f}(\omega)\hat{g}(\omega) &= \int_{-\infty}^{\infty}\int_{-\infty}^{\infty} f(x-z)g(z)\,dz\,e^{-i\omega x}\,dx \\
&= \int_{-\infty}^{\infty} h(x)e^{-i\omega x}\,dx = \hat{h}(\omega),
\end{aligned} \tag{12.28}$$

where

$$h(x) = \int_{-\infty}^{\infty} f(x-z)g(z)\,dz.$$

The function h is usually referred to as the *convolution* of the functions f and g, and is usually denoted by $f * g$. Hence, the function $(f * g)(x)$ is given by

$$(f * g) = \int_{-\infty}^{\infty} f(x-y)g(y)\,dy = \int_{-\infty}^{\infty} f(y)g(x-y)\,dy, \qquad (12.29)$$

where the last identity follows by a change of variables. From (12.28) we obtain that the Fourier transform of $f * g$ is the product of $\hat{f}$ and $\hat{g}$, i.e.

$$\mathcal{F}(f * g)(\omega) = \mathcal{F}(f)(\omega)\mathcal{F}(g)(\omega) = \hat{f}(\omega)\hat{g}(\omega). \qquad (12.30)$$

Let us now return to the pure initial value problem for the heat equation (12.20). As a consequence of (12.27) and (12.30), we obtain the solution formula

$$u(x,t) = \big(S(\cdot,t) * f\big)(x) = \int_{-\infty}^{\infty} S(x-y,t)f(y)\,dy, \qquad (12.31)$$

where the function $S(x,t)$ is defined by (12.26). Hence, we have obtained a formal solution of the pure initial value problem (12.20).

We should remark here that we have encountered the function $S(x,t)$ and the formula (12.31) already in Chapter 1. In Exercise 1.17 we established the solution formula (12.31) when the initial function f is a step function. Below we will check the validity of this solution for more general initial functions.

12.5 Partial Differential Equations

In the discussion above we have derived most of the important properties of the Fourier transform which are used in differential equations. In this final section of this chapter we will illustrate the use of the Fourier transform by considering two examples. First we will complete the discussion of the pure initial value problem for the heat equation, and afterwards we will study Laplace's equation in a half-plane.

12.5.1 The Heat Equation

The formal solution $u(x,t)$ of the pure initial value problem for the heat equation (12.20) is given by (12.31) above, i.e.

$$u(x,t) = \int_{-\infty}^{\infty} S(x-y,t)f(y)dy = \frac{1}{\sqrt{4\pi t}}\int_{-\infty}^{\infty} e^{-\frac{(x-y)^2}{4t}} f(y)dy. \qquad (12.32)$$

The function $S(x,t)$, given by

$$S(x,t) = \frac{1}{\sqrt{4\pi t}}e^{-x^2/4t}, \qquad (12.33)$$

is usually referred to as the *fundamental solution* of the heat equation. We observe that when the initial function f is known, $u(\cdot, t)$ can be derived from a convolution of f and the fundamental solution $S(\cdot, t)$.

Before we check the validity of the solution (12.32), let us observe some properties of the function $S(x, t)$.

For any $t > 0$ we have

$$S(x,t) > 0 \quad \text{and} \quad \int_{-\infty}^{\infty} S(x,t)\, dx = 1. \tag{12.34}$$

The first of these claims is obvious, while the integral property follows since

$$\begin{aligned} \int_{-\infty}^{\infty} S(x,t) dx &= \frac{1}{\sqrt{\pi}} \int_{-\infty}^{\infty} e^{-x^2/4t} \frac{dx}{\sqrt{4t}} \\ &= \frac{1}{\sqrt{\pi}} \int_{-\infty}^{\infty} e^{-z^2} dz = \frac{1}{\sqrt{\pi}} \sqrt{\pi} = 1. \end{aligned}$$

Here we have used the identity (12.12).

Because of the two properties (12.34), the formula (12.32) has the interpretation that $u(x, t)$ is a proper weighted average of the initial function f.

Another interesting property of the function S is that

$$\lim_{t \to 0} S(x,t) = 0 \quad \text{for} \quad x \neq 0, \tag{12.35}$$

while

$$\lim_{t \to 0} S(0,t) = \infty.$$

Hence, as t tends to zero, the "mass" of the function will be concentrated close to zero. In Fig. 12.3 the function $S(x, t)$ is plotted for three different values of t.

A final property we shall note is that the function $S(x, t)$ satisfies the heat equation, i.e.

$$S_t(x,t) = S_{xx}(x,t) \quad \text{for} \quad t > 0. \tag{12.36}$$

This property should be of no surprise, since its Fourier transform

$$\hat{S}(\omega, t) = e^{-\omega^2 t}$$

satisfies the equation

$$\hat{S}_t = -\omega^2 \hat{S},$$

and by the property (12.5) this is consistent with (12.36). A direct verification of (12.36) is also straightforward and is left to the reader as an exercise (see Exercise 12.5).

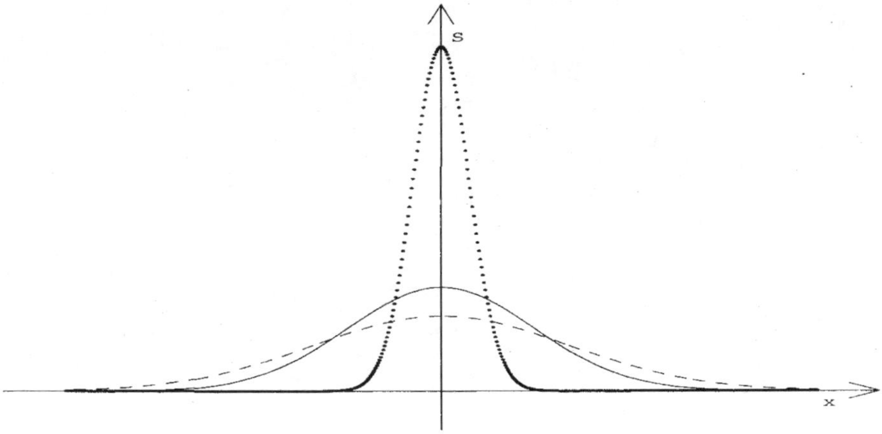

FIGURE 12.3. *The function* $S(x,t)$ *for* $t = 0.1$ $(\cdots)$, $t = 1.1$ $(-)$, *and* $t = 2.1(--)$.

In order to verify that the formal solution (12.32) is a solution of the pure initial value problem (12.20), we have to show that this solution satisfies the differential equation and the initial condition. Observe that the integral in (12.32) is with respect to y. Hence, the variables x and t act as parameters with respect to this integral, and for proper functions f we should have that

$$\begin{aligned} u_t(x,t) &= \int_{-\infty}^{\infty} S_t(x-y,t)f(y)\,dy, \\ u_{xx}(x,t) &= \int_{-\infty}^{\infty} S_{xx}(x-y,t)f(y)\,dy. \end{aligned} \tag{12.37}$$

In fact, the proper tool for verifying these formulas is a generalization of Proposition 3.1 on page 107 to integrals over all of $\mathbb{R}$ (instead of a bounded interval). Such a generalization is fairly straightforward and will not be discussed further here. However, if the formulas (12.37) hold, then it follows immediately from (12.36) that u given by (12.32) satisfies the heat equation $u_t = u_{xx}$. We can therefore conclude that the formal solution (12.32) satisfies the heat equation in a strict mathematical sense as long as the initial function f allows differentiation under the integral sign in the variables x and t.

We also have to check that the function $u(x,t)$ satisfies the initial condition. It is of course straightforward to see that as long as the Fourier transforms $\hat{u}(\cdot,t)$ and $\hat{f}$ exist, then

$$\lim_{t \searrow 0} \hat{u}(\omega,t) = \lim_{t \searrow 0} e^{-\omega^2 t}\hat{f}(\omega) = \hat{f}(\omega).$$

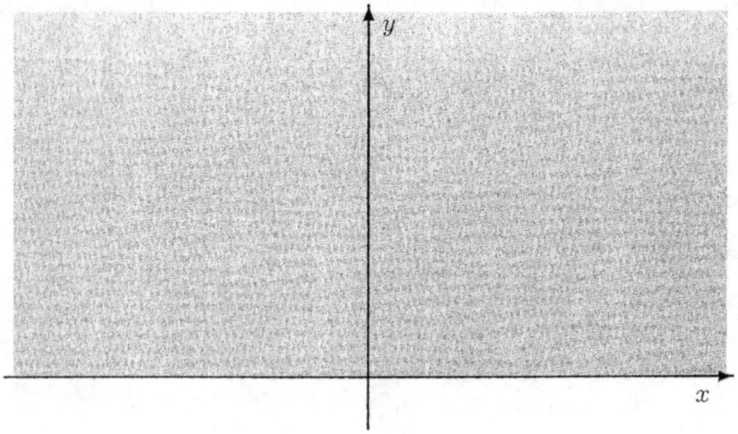

FIGURE 12.4. *The upper half-plane.*

Hence, the Fourier transform of $u(\cdot, t)$ converges pointwise to the Fourier transform of the initial function f. However, a more reasonable requirement seems to be that

$$\lim_{t \searrow 0} u(x,t) = f(x) \quad \text{for} \quad x \in \mathbb{R}, \tag{12.38}$$

i.e. we require that u converges pointwise to f. In Exercise 12.10 an outline of a proof for (12.38) is given under proper assumptions on the initial function f.

12.5.2 Laplace's Equation in a Half-Plane

In this section we will use the Fourier transform to obtain a formal solution of Laplace's equation

$$\Delta u = u_{xx} + u_{yy} = 0 \quad \text{for} \quad x \in \mathbb{R},\ y > 0. \tag{12.39}$$

Hence, the solution will be a harmonic function in the upper half-plane; see Fig. 12.4. On the x-axis we require Dirichlet boundary conditions of the form

$$u(x,0) = f(x). \tag{12.40}$$

Furthermore, u should tend to zero as y tends to infinity in the sense

$$\int_{-\infty}^{\infty} |u(x,y)|\, dx \longrightarrow 0 \quad \text{as} \quad y \to \infty. \tag{12.41}$$

In order to find a formal solution of the problem (12.39)–(12.41), we let

$$\hat{u}(\omega, y) = \int_{-\infty}^{\infty} u(x,y)e^{-i\omega x}\,dx.$$

Hence, $\hat{u}$ is the Fourier transform of u with respect to x. The differential equation (12.39) will be transformed into

$$-\omega^2\hat{u}(\omega, y) + \hat{u}_{yy}(\omega, y) = 0. \tag{12.42}$$

For each fixed value of ω this is an ordinary differential equation with respect to y, with general solution

$$\hat{u}(\omega, y) = c_1(\omega)e^{-\omega y} + c_2(\omega)e^{\omega y}. \tag{12.43}$$

We note that c_1 and c_2 are allowed to depend on ω.

The "boundary condition" (12.41) implies that

$$|\hat{u}(\omega, y)| \le \int_{-\infty}^{\infty} |u(x,y)|\,dx \longrightarrow 0 \quad \text{as} \quad y \to \infty.$$

Therefore, we must choose

$$c_1(\omega) = 0 \quad \text{for} \quad \omega < 0$$

and

$$c_2(\omega) = 0 \quad \text{for} \quad \omega > 0.$$

Furthermore, since the boundary condition (12.40) implies that $\hat{u}(\omega, 0) = \hat{f}(\omega)$, this leads to the representation

$$\hat{u}(\omega, y) = e^{-|\omega| y}\hat{f}(\omega). \tag{12.44}$$

Let $P(x, y)$ be given by

$$P(x,y) = \frac{1}{\pi}\frac{y}{x^2 + y^2}.$$

From Example 12.9 we recall that

$$\hat{P}(\omega, y) = \int_{-\infty}^{\infty} P(x,y)e^{-i\omega x}\,dx = e^{-|\omega| y}.$$

Hence, the formula (12.44) can be written as

$$\hat{u}(\omega, y) = \hat{P}(\omega, y)\hat{f}(\omega),$$

and by property (12.30) this implies that

$$u(x,y) = (P(\cdot,y) * f)(x) = \int_{-\infty}^{\infty} P(x-z,y)f(z)\,dz. \tag{12.45}$$

The function $P(x,y)$ is called the *Poisson kernel.* This function has properties which resemble the properties of the fundamental solution $S(x,t)$ for the heat equation. For example, it is straightforward to show that

$$P(x,y) \geq 0 \quad \text{and} \quad \int_{-\infty}^{\infty} P(x,y)\,dx = 1. \tag{12.46}$$

Therefore the formula (12.45) has the interpretation that $u(x,y)$ is a proper weighted average of the boundary function f. The reader is asked to verify a number of properties of the Poisson kernel P and of the solution formula (12.45) in Exercise 12.11.

12.6 Exercises

EXERCISE 12.1 Find the Fourier transform of the following functions ($a > 0$):

(a)

$$f(x) = \begin{cases} \cos(x) & |x| < \frac{\pi}{2}, \\ 0 & \text{otherwise.} \end{cases}$$

(b)

$$f(x) = \begin{cases} x & |x| < a, \\ 0 & \text{otherwise.} \end{cases}$$

(c)

$$f(x) = \begin{cases} a - |x| & |x| < a, \\ 0 & \text{otherwise.} \end{cases}$$

EXERCISE 12.2 Compute the function $g(x) = (f * f)(x)$ when ($a > 0$).

(a)

$$f(x) = \begin{cases} 1 & |x| < a, \\ 0 & \text{otherwise.} \end{cases}$$

(b)

$$f(x) = e^{-|x|}.$$

EXERCISE 12.3 Assume that $\hat{f}(\omega) = e^{-\omega^2}/(1+\omega^2)$. Determine $f(x)$.

EXERCISE 12.4 Let $f(x)$ be a given function and define $g(x)$ by

$$g(x) = f(x-a),$$

where a is constant. Show that $\hat{g}(\omega) = e^{-i\omega a}\hat{f}(\omega)$.

EXERCISE 12.5 Let $S(x,t)$ be the fundamental solution of the heat equation given by (12.26). Show by a direct computation that

$$S_t = S_{xx} \quad \text{for} \quad t > 0.$$

EXERCISE 12.6 Use formula (12.31) to find the solution of the pure initial value problem (12.20) when

(a)

$$f(x) = H(x) = \begin{cases} 0 & x \le 0, \\ 1 & x > 0. \end{cases}$$

(b)

$$f(x) = e^{-ax^2}, \quad \text{where} \quad a > 0.$$

Compare your solution in (a) with the discussion in Section 1.4.4.

EXERCISE 12.7 Let a be a constant. Use the Fourier transform to find a formal solution of the problem

$$\begin{aligned} u_t &= u_{xx} + au_x \quad \text{for} \quad x \in \mathbb{R},\ t > 0 \\ u(x,0) &= f(x). \end{aligned}$$

EXERCISE 12.8 Consider the Laplace problem (12.39)–(12.41). Assume that the Dirichlet condition (12.40) is replaced by the Neumann condition

$$u_y(x,0) = f(x), \quad x \in \mathbb{R}.$$

Use the Fourier transform to find a formal solution in this case.

EXERCISE 12.9 Consider the Laplace problem:

$$\begin{aligned} (\Delta u) &= 0 \quad \text{for} \quad x \in \mathbb{R},\ 0 < y < 1, \\ u(x,0) &= 0, \quad x \in \mathbb{R}, \\ u(x,1) &= f(x), \quad x \in \mathbb{R}. \end{aligned}$$

Use the Fourier transform to find a formal solution of this problem.

EXERCISE 12.10 The purpose of this exercise is to analyze the pointwise limit (12.38). We assume that $f(x)$ is a continuous and bounded function, i.e.

$$|f(x)| \leq M \quad \text{for} \quad x \in \mathbb{R},$$

where M is a positive constant.

(a) Show that $u(x,t) - f(x)$ has the representation

$$u(x,t) - f(x) = \int_{-\infty}^{\infty} (f(x-y) - f(x))S(y,t)dy.$$

(b) Show that

$$\lim_{t \searrow 0} u(x,t) = f(x).$$

(Hint: $|u(x,t) - f(x)| \leq \int_{-\infty}^{\infty} |f(x-y) - f(x)|S(y,t)dy$. Break the integral up into two pieces, $|y| \leq \delta$ and $|y| \geq \delta$.)

EXERCISE 12.11

(a) Show that the Poisson kernel $P(x,y)$ satisfies the properties (12.46).

(b) Show by a direct computation that

$$\Delta P = 0 \quad \text{for} \quad (x,y) \neq (0,0).$$

(c) Discuss the validity of the formal solution (12.45) of the boundary value problem (12.39)–(12.41).

References

[1] H. Anton, Elementary Linear Algebra, Wiley, 1987.

[2] W. Aspray, John von Neumann and the Origins of Modern Computing, MIT Press, 1990.

[3] W. E. Boyce, R. C. DiPrima, Elementary Differential Equations and Boundary Value Problems, Wiley, 1986.

[4] S. C. Brenner, L. R. Scott, The Mathematical Theory of Finite Element Methods, Springer-Verlag, New York 1994.

[5] M. Braun, Differential Equations and Their Applications, Springer-Verlag 1992.

[6] D. Colton, Partial Differential Equations, Random House, 1988.

[7] S.D. Conte, C. de Boor, Elementary Numerical Analysis, an Algorithmic Approach, McGraw-Hill, 1972.

[8] G. Dahlquist, Å. Björck, Numerical Methods, Englewood Cliffs, Prentice-Hall, 1974.

[9] P. J. Davis, R. Hersh, The Mathematical Experience, Birkhauser, 1980.

[10] S. K. Godunov, V. S. Ryabekii, Difference Schemes, North-Holland, 1987.

[11] G. H. Golub, C. F. van Loan, Matrix Computations, North Oxford Academic Publishing, 1983.

[12] D. Gottlieb, S. A. Orszag, Numerical Analysis of Spectral Methods: Theory and Applications, Siam, Regional Conference Series in Applied Mathematics, 1977.

[13] W. Hackbusch: Iterative Solution of Large Sparse Systems of Equations, Springer Verlag 1994.

[14] E. Isaacson, H. B. Keller, Analysis of Numerical Methods, Wiley, 1966.

[15] C. Johnson, Numerical Solution of Partial Differential Equations by the Finite Element Method. Cambridge University Press, Cambridge, 1987.

[16] H. B. Keller, Numerical Methods for Two-Point Boundary-Value Problems, Blaisdell Publ. Comp. 1968.

[17] H. O. Kreiss, J. Lorenz, Initial-Boundary Value Problems and the Navier-Stokes Equations, Academic Press, 1989.

[18] J. D. Logan, Applied Mathematics, A Contemporary Approach, Wiley-Interscience, 1987.

[19] J. D. Logan, An Introduction to Nonlinear Partial Differential Equations, Wiley-Interscience, 1994.

[20] J. D. Murray, Mathematical Biology, Springer-Verlag, Biomathematics Texts, second ed. 1993.

[21] M. H. Protter, H. F. Weinberger, Maximum Principles in Differential Equations, Springer-Verlag 1984.

[22] J. Rauch, Partial Differential Equations, Springer Verlag 1991.

[23] J. Smoller, Shock Waves and Reaction-Diffusion Equations, 2nd ed, Springer-Verlag 1994.

[24] J. Stoer, R. Bulirsch, Introduction to Numerical Analysis, Springer-Verlag, 1980.

[25] W. A. Strauss, Partial Differential Equations, Wiley, 1992.

[26] J. C. Strikwerda, Finite Difference Schemes and Partial Differential Equations, Wadsworth & Brooks/Cole, 1989.

[27] V. Thomee, Finite Difference Methods for Linear Parabolic Equations, Handbook of numerical analysis, vol. I, editors: P. G. Ciarlet nad J. L. Lions. North-Holland 1990.

[28] H. F. Weinberger, A first course in partial differential equations, Wiley,1965.

[29] G. B. Whitham, Linear and Nonlinear Waves, Wiley-Interscience, 1973.

[30] E. Zauderer, Partial Differential Equations of Applied Mathematics, 2nd ed, Wiley-Interscience, 1989.

[31] O. C. Zienkiewicz, The Finite Element Method in Engineering Science, McGraw-Hill, New York 1977.

Index

Texts in Applied Mathematics

(continued from page ii)

29. *Tveito/Winther:* Introduction to Partial Differential Equations: A Computational Approach.
30. *Gasquet/Witomski:* Fourier Analysis and Applications: Filtering, Numerical Computation, Wavelets.
31. *Brémaud:* Markov Chains: Gibbs Fields, Monte Carlo Simulation, and Queues.
32. *Durran:* Numerical Methods for Wave Equations in Geophysical Fluid Dynamics.
33. *Thomas:* Numerical Partial Differential Equations: Conservation Laws and Elliptic Equations.
34. *Chicone:* Ordinary Differential Equations with Applications.
35. *Kevorkian:* Partial Differential Equations: Analytical Solution Techniques, Second Edition.
36. *Dullerud/Paganini:* A Course in Robust Control Theory: A Convex Approach.
37. *Quarteroni/Sacco/Saleri:* Numerical Mathematics.
38. *Gallier:* Geometric Methods and Applications: For Computer Science and Engineering.
39. *Han/Atkinson:* Theoretical Numerical Analysis: A Functional Analysis Framework.
40. *Brauer/Castillo-Chávez:* Mathematical Models in Population Biology and Epidemiology.
41. *Davies:* Integral Transforms and Their Applications, Third Edition.
42. *Deuflhard/Bornemann:* Scientific Computing with Ordinary Differential Equations.
43. *Deuflhard/Hohmann:* Introduction to Scientific Computing.
44. *Knabner/Angermann:* Numerical Methods for Elliptic and Parabolic Partial Differential Equations.
45. *Larsson/Thomée:* Partial Differential Equations with Numerical Methods.
46. *Pedregal:* Introduction to Optimization.
47. *Ockendon/Ockendon:* Waves and Compressible Flow.
48. *Hinrichsen/Pritchard:* Mathematical Systems Theory I.
49. *Bullo/Lewis:* Geometric Control of Mechanical Systems.